Nietzsche und der französische Existenzialismus

Nietzsche-Lektüren

Band 7

Nietzsche und der französische Existenzialismus

Herausgegeben von
Alfred Betschart, Andreas Urs Sommer, Paul Stephan

DE GRUYTER

ISBN 978-3-11-076010-1
e-ISBN (PDF) 978-3-11-076016-3
e-ISBN (EPUB) 978-3-11-076026-2
ISSN 2568-2229

Library of Congress Control Number: 2021950134

Bibliografische Information der Deutschen Nationalbibliothek
Die Deutsche Nationalbibliothek verzeichnet diese Publikation in der Deutschen Nationalbibliografie; detaillierte bibliografische Daten sind im Internet über http://dnb.dnb.de abrufbar.

Druck und Bindung: CPI books GmbH, Leck

MIX
Papier aus verantwortungsvollen Quellen
FSC® C083411

www.degruyter.com

Inhalt

Siglenverzeichnis — IX

Alfred Betschart, Paul Stephan und Andreas Urs Sommer
Vorwort — 1

Sartre und Nietzsche im Überblick

Nik Farrell Fox
Nietzsche and Sartre: Twin Philosophers of Paradox — 7

Elisa Reato
Sartre lecteur de Nietzsche — 35

Camus und Nietzsche im Überblick

Erminio Maglione
Albert Camus et la philosophie de Nietzsche — 53

Der Nihilismus und seine Überwindung

Alfred Betschart
Nietzsche, Sartre und der Vorwurf des Nihilismus — 69

Eike Brock
Camus' Nietzsche. *Amor fati* und Revolte — 89

Philipp Schwab
Überwindungen des Nihilismus? Kierkegaard – Nietzsche – Camus — 105

Emanuel Seitz
Der Wille zur Macht ist die letzte Karte: Leben und Skepsis bei Cioran und Nietzsche — 123

Freigeistigkeit und Individualismus

Paul Stephan
Zwischen Verschmelzungswunsch und Selbstbehauptung. Das Problem der Alterität als Grundproblem der Existenzphilosophie von Kierkegaard bis Camus – und seine Gegenwärtigkeit — 141

Andrew Inkpin
Mit oder ohne die Anderen? Beauvoir contra Nietzsche über Freiheit — 161

Hans-Martin Schönherr-Mann
Das über sich Hinausgehen. Aber wohin? Nietzsches Wiederkünftige und der Übermensch als sich Transzendierende im Sinn von Sartre und de Beauvoir? — 175

Freiheit und Determinismus

Jaanus Sooväli
From Responsibility to Great Responsibility of Innocence: Nietzsche, Sartre, and Existentialism — 191

Anthony Kosar
„Alles ist *in actu*“: Existentielles ‚Freiheitsbewusstsein‘ zwischen absoluter Verantwortung und völliger Unverantwortlichkeit — 203

Philosophie und Ästhetik

Josephina Bierl
Die ewige Wiederkunft und ihre Rezeption im *Mythe de Sisyphe* bei Albert Camus — 219

Jens Bonnemann
Die Geburt der Kunst aus dem Ekel am Dasein. Die philosophische Relevanz ästhetischer Erfahrungen bei Nietzsche und Sartre — 233

Selbstwerdung, Ökologie, Leiblichkeit

Marco Brusotti
"Choice" *or* "Gift"? "Choice" *and* "Gift"? Nietzsche, Jaspers and Ricœur in a Text by Georges Canguilhem on Becoming Oneself — 249

Damon Boria
Reading Nietzsche and Sartre Toward an Eco-Existentialism — 271

Christian Niemeyer
Sartres Flaubert im Vergleich zu jenem Nietzsches. Oder: Über einen Existenzialisten *avant la lettre* unter Bezug auf Nietzsches Madame Bovary und seinem auf sie bezüglichen „Experimentalismus" — 283

Personenregister — 299

Siglenverzeichnis

Für Nietzsches Werke auf Deutsch:

AC	Friedrich Nietzsche: „Der Antichrist. Fluch auf das Christenthum“. In: KSA 6, S. 165–254.
CV 3	Friedrich Nietzsche: „Der griechische Staat (1872) – Fünf Vorreden zu fünf ungeschriebenen Büchern – Nachgelassene Schriften“. In: KSA 1, S. 764–777.
DD	Friedrich Nietzsche: „Dionysos-Dithyramben“. In: KSA 6, S. 375–410.
DS	Friedrich Nietzsche: „David Strauss“. In: KSA 1, S. 157–242.
DW	Friedrich Nietzsche: „Die dionysische Weltanschauung“. In: KSA 1, S. 551–577.
EH	Friedrich Nietzsche: „Ecce Homo. Wie man wird, was man ist“. In: KSA 6, S. 255–374.
eKGWB	Friedrich Nietzsche: Digitale Kritische Gesamtausgabe. Werke und Briefe. Digital Critical Edition of the Complete Works and Letters. Based on the Critical Text by G. Colli and M. Montinari, Berlin, New York, de Gruyter 1967-, edited by Paolo D'Iorio, 2009-, http://www.nietzschesource.org/#eKGWB.
FW	Friedrich Nietzsche: „Die fröhliche Wissenschaft“. In: KSA 3, S. 343–651.
GD	Friedrich Nietzsche: „Götzen-Dämmerung oder Wie man mit dem Hammer philosophirt“. In: KSA 6, S. 55–161.
GM	Friedrich Nietzsche: „Zur Genealogie der Moral“. In: KSA 5, S. 245–412.
GT	Friedrich Nietzsche: „Die Geburt der Tragödie“. In: KSA 1, S. 9–156.
JGB	Friedrich Nietzsche: „Jenseits von Gut und Böse“. In: KSA 5, S. 9–243.
KGB	Friedrich Nietzsche: Briefwechsel. Kritische Gesamtausgabe. Giorgio Colli/Mazzino Montinari (Hrsg.). Berlin, New York 1975 ff.
KGW	Friedrich Nietzsche: Werke. Kritische Gesamtausgabe. Giorgio Colli/Mazzino Montinari (Hrsg.). Berlin, New York 1967 ff.
KSA	Friedrich Nietzsche: Sämtliche Werke. Kritische Studienausgabe in 15 Einzelbänden. Giorgio Colli/Mazzino Montinari (Hrsg.). 3. Auflage. München, Berlin, New York 1999.
KSB	Friedrich Nietzsche: Sämtliche Briefe. Kritische Studienausgabe in 8 Bänden. Giorgio Colli/Mazzino Montinari (Hrsg.). 2. Auflage. München, Berlin, New York 2003.
M	Friedrich Nietzsche: „Morgenröthe“. In: KSA 3, S. 9–331.
MA	Friedrich Nietzsche: „Menschliches, Allzumenschliches“. In: KSA 2, S. 9–366.
NL	Friedrich Nietzsche: Nachlass. Zitiert nach KSA oder KGW.
NW	Friedrich Nietzsche: „Nietzsche contra Wagner“. In: KSA 6, S. 413–445.
SE	Friedrich Nietzsche: „Schopenhauer als Erzieher“. In: KSA 1, S. 335–427.
SGT	Friedrich Nietzsche: „Sokrates und die Tragoedie“. In: KSA 1, S. 601–640.
UB	Friedrich Nietzsche: „Unzeitgemäße Betrachtungen“. In: KSA 1, S. 157–510.
WA	Friedrich Nietzsche: „Der Fall Wagner. Ein Musikanten-Problem“. In: KSA 6, S. 9–53.
WB	Friedrich Nietzsche: „Richard Wagner in Bayreuth“. In: KSA 1, S. 429–510.
WL	Friedrich Nietzsche: „Ueber Wahrheit und Lüge im aussermoralischen Sinne“. In: KSA 1, S. 873–890.
WS	Friedrich Nietzsche: „Zweite Abtheilung: Der Wanderer und sein Schatten“. In: KSA 2, S. 535–704.
Za	Friedrich Nietzsche: „Also sprach Zarathustra“. In: KSA 4, S. 9–408.

https://doi.org/10.1515/9783110760163-001

Für Nietzsches Werke auf Englisch:

A	Friedrich Nietzsche: The Antichrist. Cambridge 2005.
BGE-1	Friedrich Nietzsche: Beyond Good and Evil. New York 1966.
BGE-2	Friedrich Nietzsche: Beyond Good and Evil. New York 1990.
BGE-3	Friedrich Nietzsche: Beyond Good and Evil. Cambridge Texts in the History of Philosophy. Cambridge 2002.
D	Friedrich Nietzsche: Daybreak. Cambridge 1997.
EH-1	Friedrich Nietzsche: Ecce Homo. Cambridge 2005.
EH-2	Friedrich Nietzsche: „Ecce Homo". In: Friedrich Nietzsche: On the Genealogy of Morals and Ecce Homo. New York 1969, pp. 199–338.
EH-3	Friedrich Nietzsche: Ecce Homo: How To Become What You Are. Oxford et. al. 2007.
GM-1	Friedrich Nietzsche: On the Genealogy of Morality. Cambridge Texts in the History of Political Thought. Cambridge 2006.
GM-2	Friedrich Nietzsche: „On the Genealogy of Morals". In: Friedrich Nietzsche: On the Genealogy of Morals and Ecce Homo. New York 1969, pp. 1–198.
GS-1	Friedrich Nietzsche: The Gay Science. New York 1974.
GS-2	Friedrich Nietzsche: The Gay Science. Cambridge Texts in the History of Philosophy. Cambridge 2001.
HH-1	Friedrich Nietzsche: Human, All Too Human. Cambridge 1986.
HH-2	Friedrich Nietzsche: Human, All Too Human. Cambridge 1996.
LN	Friedrich Nietzsche: Writings from the Late Notebooks. Cambridge 2003.
PTAG	Friedrich Nietzsche: Philosophy in the Tragic Age of the Greeks. Washington 1998.
TI	Friedrich Nietzsche: „Twilight of the Idols". In: The Anti-Christ, Ecce Homo, Twilight of the Idols. Cambridge Texts in the History of Philosophy. Cambridge 2005, pp. 153–250.
UM	Friedrich Nietzsche: Untimely Meditations. Cambridge 1997.
WP	Friedrich Nietzsche: The Will to Power. New York 1967.
Za-1	Friedrich Nietzsche: Thus Spoke Zarathustra. Middlesex, New York 1969.
Za-2	Friedrich Nietzsche: Thus Spoke Zarathustra: A Book for All and None. Adrian Del Caro/Robert Pippin (eds.). Cambridge, New York et al. 2006.

Für Beauvoirs Werke[1]:

DS	Le deuxième sexe / Das andere Geschlecht / The Second Sex (1949).
EBS	Entretiens de S. de Beauvoir avec J.-P. Sartre / Gespräche mit Jean-Paul Sartre. / Conversations with Jean-Paul Sartre (1974).
FA	La force de l'âge / In den besten Jahren / The Prime of Life (1960).
FBS	Faut-il brûler Sade ? / Soll man de Sade verbrennen? / Must We Burn Sade? (1951–52).
MJFR	Mémoires d'une jeune fille rangée / Memoiren einer Tochter aus gutem Hause / Memoirs of a Dutiful Daughter (1958).

1 Titel auf Französisch, Deutsch und Englisch; in Klammer Datum der Erstpublikation (resp. Datum der Niederschrift, falls Publikation verspätet erfolgte). Alternative Titel aus früheren Übersetzungen sind ebenfalls in Klammern angeführt.

PMA Pour une morale de l'ambiguïté / Für eine Moral der Doppelsinnigkeit / The Ethics of Ambiguity (1947).

Für Camus' Werke:

Cal Caligula / Caligula / Caligula (1944).
Chu La chute / Der Fall / The Fall (1956).
DerInt Dernière interview / – / A Final Interview (1959).
E L'étranger / Der Fremde / The Stranger (1942).
EE L'envers et l'endroit / Licht und Schatten / Betwixt and Between (The Wrong Side and the Right Side) (1937).
Eté L'été / Heimkehr nach Tipasa / Summer (1954).
Gide Rencontres avec André Gide (1951).
HR L'homme révolté / Der Mensch in der Revolte / The Rebel (1951).
Îles Préface aux *Îles*/ Vorwort *Die Inseln* (1959).
MS Le mythe de Sisyphe / Der Mythos des Sisyphos / The Myth of Sisyphus (1942).
P La peste / Die Pest / The Plague (1947).
RemRév Remarque sur la révolte (1945).
RevPro51 L'Europe de la fidélité. In: La Révolution Prolétarienne (1951).

Für Sartres Werke:

AllR Aller et retour / Hin und zurück / Departure and Return (1944).
Arc66 Jean-Paul Sartre répond (1966).
ASD Autoportrait à soixante-dix ans / Selbstporträt mit siebzig Jahren / Self-Portrait at Seventy (1975).
Carn Carnets de la drôle de guerre / Tagebücher / The War Diaries (1939–1940).
CLut Cahier Lutèce (1954).
CM Cahiers pour une morale / Entwürfe für eine Moralphilosophie / Notebooks for an Ethics (1947/1948).
CMid Carnet Midy (1924).
CP Les Communistes et la paix / Die Kommunisten und der Frieden / The Communists and Peace (I + II 1952, III 1954).
CRD Critique de la raison dialectique, t. 1 / Kritik der dialektischen Vernunft, Bd. 1 / Critique of Dialectical Reason, vol 1. (1960).
DBD Le diable et le bon Dieu / Der Teufel und der liebe Gott / The Devil and the Good Lord (Lucifer and the Lord) (1951).
Déf Une défaite (1927).
DetL Détermination et liberté / Determination und Freiheit / Determinism and Freedom (1964).
EBS Entretiens de S. de Beauvoir avec J.-P. Sartre / Gespräche mit Jean-Paul Sartre / Conversations with Jean-Paul Sartre (1974).
EH L'existentialisme est un humanisme / Der Existentialismus ist ein Humanismus / Existentialism Is a Humanism (Existentialism / Existentialism and Humanism) (1946).
Emp Empédocle (1926–1927).
EN L'être et le néant / Das Sein und das Nichts / Being and Nothingness (1943).
ErA Er l'Arménien (1927–1928).

Ero	Érostrate / Herostrat / Erostratus (1936).
HC	Huis clos / Geschlossene Gesellschaft (Bei geschlossenen Türen) / No Exit (In Camera / The Vicious Circle) (1943).
I-aire	L'imaginaire / Das Imaginäre / The Imaginary (Psychology of the Imagination) (1940).
I-tion	L'imagination / Die Imagination (Über die Einbildungskraft) / Imagination (1936).
IF	L'Idiot de la famille. Gustave Flaubert de 1821 à 1857 / Der Idiot der Familie / The Family Idiot (I+II 1971 / III 1972).
ItinT	Sartre par Sartre / Sartre über Sartre / Itinerary of a Thought (1969).
L-Cast	Lettres au Castor et à quelques autres / Briefe an Simone de Beauvoir / The Letters of Jean-Paul Sartre to Simone de Beauvoir (1926–1963).
L-EA	Une lettre sur les étudiants d'aujourd'hui (1929).
L-GM	Lettre à Gabriel Marcel de Jean-Paul Sartre (1943).
LibCar	La liberté cartésienne / Die cartesianische Freiheit / Cartesian Freedom (1945).
MAme	La mort dans l'âme / Der Pfahl im Fleische / Sick at Heart (1948–1949).
Mall	Mallarmé / Mallarmés Engagement / Mallarmé, or the Poet of Nothingness (1947–1949).
MatR	Matérialisme et révolution / Materialismus und Revolution / Materialism and Revolution (1946).
MorH	Morale et histoire (1964–1965).
Mon64	Jean-Paul Sartre s'explique sur ‚Les mots' / Was bedeutet Literatur, in einer Welt die hungert? / Jean-Paul Sartre Speaks (A Long, Bitter, Sweet Madness) (1964).
Mots	Les mots / Die Wörter / The Words (1963).
Mou	Les mouches / Die Fliegen / The Flies (1943).
MS	Les mains sales / Die schmutzigen Hände / Dirty Hands (Red Gloves / Crime Passionel) (1948).
N	La nausée / Der Ekel / Nausea (1938).
Nizan	Paul Nizan / Über Paul Nizan / Paul Nizan (1960).
NouvM	Un nouveau mystique / Ein neuer Mystiker (1943).
Obl79	L'écriture et la publication (1979).
PropE	A propos de l'existentialisme: mise au point Sartre / Zum Existentialismus. Eine Klarstellung / What is Existentialism?; A More Precise Characterization of Existentialism (1944).
Schilpp	The Interview (for *The philosophy of Jean-Paul Sartre*; 1975).
PréTM	Présentation des Temps modernes / Vorstellung von Les Temps Modernes / Introduction to Les Temps Modernes (The Case for Responsible Literature) (1945).
QL	Qu'est-ce que la littérature ? / Was ist Literatur? / What is Literature? (1947).
QM	Questions de méthode (Question de méthode) / Fragen der Methode (Marxismus und Existentialismus) / Search for a Method (Problem of Method) (1957).
RacE	Les racines de l'éthique (1964).
RépSil	La république du silence / Die Republik des Schweigens / The Republic of Silence (1944).
RQJui	Réflexions sur la question juive / Betrachtungen zur Judenfrage / Anti-Semite and Jew (Portrait of the Anti-Semite) (1946).
SemSca	La semence et le scaphandre (1923).
SéqV	Le séquestré de Venise (1957).
SFilm	Sartre. Un film / Sartre. Ein Film / Sartre by Himself (1977).
SG	Saint Genet (1952).

TE	La transcendance de l'ego / Die Transzendenz des Ego / The Transcendence of the Ego (1934).
VérE	Vérité et existence / Wahrheit und Existenz / Truth and Existence (1948).

Alfred Betschart, Paul Stephan und Andreas Urs Sommer

Vorwort

Nietzsche elektrisiert noch immer. Und der französische Existenzialismus elektrisiert nicht minder. Eindrücklich vor Augen geführt wurde uns dies, als wir einen *Call for Papers* für eine Tagung zum Thema *Nietzsche und der französische Existenzialismus* lancierten: Noch nie gab es bei den ohnehin immer rege nachgefragten Naumburger Nietzsche-Veranstaltungen eine derartige Fülle von Vortragsangeboten, von denen leider nur eine kleine Anzahl berücksichtigt werden konnte.

So fand im Nietzsche-Dokumentationszentrum in Naumburg vom 31. Juli bis 2. August 2020 physisch sowie gleichzeitig online eine trotz Corona-Einschränkungen überaus erfolgreiche Tagung mit Teilnehmerinnen und Teilnehmern aus fünf Kontinenten statt. Behandelt wurden die verschiedensten Gesichtspunkte zum Verhältnis zwischen Nietzsche und den Vertretern des französischen Existenzialismus, vor allem Jean-Paul Sartre, Albert Camus und Simone de Beauvoir, aber auch anderen Intellektuellen aus deren Umfeld wie Georges Canguilhem und Emil Cioran. Kernprobleme waren die unterschiedlichen Freiheits- und Wahrheitskonzepte von Nietzsche und den französischen Existenzialisten und Existenzialistinnen, der jeweilige Zusammenhang zwischen Politik und Philosophie, (Anti-)Nihilismus und Individualismus als Bindeglied zwischen Nietzsche und dem französischen Existenzialismus sowie die Gründe für ein mögliches Veralten oder eine mögliche Aktualität des Existenzialismus und Nietzsches.

Der vorliegende Band dokumentiert eine Auswahl der an der Tagung gehaltenen Vorträge in überarbeiteter Form. Obwohl das Thema „Nietzsche und der französische Existenzialismus" so naheliegt, ist der vorliegende Band doch der erste Versuch in Buchform, das Thema von verschiedenen Seiten und mit verschiedenen Herangehensweisen umfassend zu erschließen.

Die Themen des Bandes sind alles andere als ‚abgegrast' und so kam es im Laufe der Tagung zu einigen Kontroversen, die sich auch in dem Band widerspiegeln. Insbesondere umstritten war und ist, inwiefern Nietzsche als Vordenker des französischen Existenzialismus anzusehen ist oder nicht. Während manche Nietzsche als Teil einer ‚existenzphilosophischen' Traditionslinie betrachten, versuchen andere, ihn eher davon zu distanzieren oder gar als Antipoden dieser Richtung darzustellen – und Ähnliches gilt aus der Perspektive der Existenzialismus-Forschung. Der Band vereinigt Aufsätze, die in beide Richtungen argumentieren, und wichtige Anhaltspunkte für künftige Debatten in dieser für die Geistesgeschichte der Moderne nicht unerheblichen Frage geben.

https://doi.org/10.1515/9783110760163-002

Auch die Frage, welchen Stellenwert Nietzsche in den unterschiedlichen intellektuellen Kulturen Frankreichs und Deutschlands gewonnen oder nicht gewonnen hat, lässt sich am Umgang des Existenzialismus mit Nietzsche erörtern: Ist es so, wie die Forschungen Herbert Schnädelbachs nahelegen könnten, dass das in Frankreich vorherrschende szientistische Verständnis von Philosophie die Aufnahme Nietzsches eher begünstigt hat als in der ‚reinen' deutschen Philosophie, deren Ideal vom Neukantianismus über die Phänomenologie bis zu Heidegger vorherrschte? Oder hat umgekehrt erst die intensive philosophische Rezeption Nietzsches in Deutschland bei Autoren wie Alfred Baeumler, Karl Löwith und Karl Jaspers die französische philosophische Adaption Nietzsches möglich gemacht? Und wie hat diese Adaption wiederum auf die philosophische Theoriebildung in Deutschland zurückgestrahlt?

Der Band demonstriert zudem, dass sich die Frage nach einer Kontinuität zwischen Nietzsche und dem französischen Existenzialismus kaum abgelöst von aktuellen Diskussionen behandeln lässt. Löwith wählte, um den Gegensatz zwischen Kierkegaard und Nietzsche herauszustellen, das Begriffspaar ‚Leben und Existenz', entsprechend lautete der Haupttitel unserer Tagung *Zwischen Leben und Existenz*. Löwith meinte damit: Während Kierkegaard, der Denker der ‚Existenz', den Menschen vor allem in christlicher Tradition als geistiges Wesen in seinem Verhältnis zu Gott bestimmt und seine leibliche Seite abgelehnt habe, kehre Nietzsche, der Denker des ‚Lebens', jene Seite gerade hervor und betrachte den Menschen vor allem als biologisches Wesen. Trotz dieser Differenz betrachtet Löwith die beiden Denker als verbunden in ihrer Wahl des radikal Vereinzelten als denkerischen Ausgangspunkt und ihrer Diagnose eines zu überwindenden Nihilismus, der die europäische Gegenwartskultur beherrsche. Die Aufgabe des Denkens sei es nun, ausgehend von den Ansätzen beider eine neue Anthropologie zu entwickeln; zu überwinden sei der Nihilismus freilich nur, wenn man jenen radikalindividualistischen Ausgangspunkt aufgebe. Damit beschreibt Löwith, was die Existenzialisten des 20. Jahrhunderts umtreiben sollte: Eine von den Vereinzelungsdenkern inspirierte, allerdings nun atheistisch gewendete Weltsicht mit Einsichten der Humanwissenschaften zu verbinden, um auch der biologisch-triebhaften und der sozialen Seite des Menschen gerecht zu werden.

In diesem Koordinatensystem lassen sich alle Beiträge dieses Bandes verorten: zwischen Leben und Existenz, zwischen Individualität und Kollektivität, zwischen Bejahung und Verneinung des Nihilismus, zwischen Atheismus und Gottessuche. Wie man sich auch immer darin verorten mag, scheint es sich doch um einen Problemkreis zu handeln, der an Aktualität nichts eingebüßt hat und zu dem man sich denkerisch wie praktisch in irgendeiner Form zu verhalten hat.

Ermöglicht wurde die Tagung durch Mittel der Deutschen Forschungsgemeinschaft sowie durch die organisatorische und z.T. auch finanzielle Unterstützung der Friedrich-Nietzsche-Stiftung, der Nietzsche Gesellschaft e. V., der Sartre Gesellschaft e. V., des Nietzsche-Forschungszentrums der Albert-Ludwigs-Universität Freiburg im Breisgau und der Deutschen Gesellschaft für französischsprachige Philosophie, bei denen sich die Herausgeber dieses Bandes und Organisatoren der Veranstaltung bedanken möchten. Ihr Dank gilt insbesondere auch den Mitarbeitenden des Nietzsche-Dokumentationszentrums unter Leitung von Dr. Ralf Eichberg und Susanne Rettenwander sowie in Freiburg Frau Louisa Estadieu, ohne deren tatkräftige Unterstützung die Tagung nicht hätte abgehalten werden können. Frau Estadieu wirkte zudem an der Endredaktion des Bandes mit. Alle Beteiligten haben sich vom Thema und vom gemeinsamen Gespräch elektrisieren lassen.

Chur, Leipzig und Freiburg
im Sommer 2021

Sartre und Nietzsche im Überblick

Nik Farrell Fox

Nietzsche and Sartre: Twin Philosophers of Paradox

Abstract: Nietzsche and Sartre: Twin Philosophers of Paradox. The *Nietzsche-Sartre connection* is a topic that is often alluded to but rarely discussed in a systematic way. In this paper I set out to break this relative silence and pursue the following objectives: firstly, to establish Nietzsche and Sartre as *parallel thinkers* with a common vision across the range of their philosophy, including the key areas of the self, ontology, and ethics. Secondly, to present Nietzsche and Sartre as twin philosophers of a *deconstructive form of paradoxical thinking* or *sublimating logic* that overturns metaphysical humanism and anticipates many key ideas in contemporary posthumanist thinking.

An Imaginary Contamination

> We were friends and have grown distant from one another... There is probably a tremendous invisible curve and star orbit within which our so different paths and destinations may be *included* as tiny stretches of the way – let us raise ourselves to this thought! (Nietzsche, GS-1, §279)

It is fair to say that Sartre was ambivalent about what he thought of Nietzsche, wavering between adulation, indifference and some suspicion. It is well documented that the prankster Sartre used to throw urine-filled condoms at his foes at the École Normale Supérieure (ENS), shouting 'Thus pissed Zarathustra', as was his intense aversion to moustaches, of which Nietzsche's was, of course, a more than immodest specimen.[1] In 1924 Sartre describes Nietzsche in his notebooks as more a poet than a philosopher, whose form of thought is better than his actual thought himself: "Nietzsche. He is a poet who had the misfortune of being taken for a philosopher... he will always have success with those who

1 According to Burnier, one of Sartre's self-professed most successful ventures was to deter a number of young men from growing moustaches (in Boulé 2005, p. 89). Of course, Roquentin leaves us in no doubt how he feels about moustaches: "The fine gentleman exists Legion of Honour, exists moustache, that's all... he sees the two pointed ends of a moustache on both sides of the nose; I do not think therefore I am a moustache" (Sartre 1965a: N, p. 147).

https://doi.org/10.1515/9783110760163-003

prefer the form of ideas to their exchange" (Sartre 1990: CMid, p. 471).[2] Sartre's direct comments about Nietzsche in his texts and interviews are mixed, expressing both negativity and positivity. Nietzsche is the first philosopher mentioned in *Being and Nothingness* for his critique of "the illusion of backworlds" (Sartre 2018: EN, p. 2) but he criticizes Nietzsche's idea of will-to-power (which he understands in biological terms as a brute will to dominate others and as a "psychobiological residue" [Sartre 2018: EN, p. 741]) and he dismisses the *Übermensch* as the culmination of evolution in which only the strongest survive. In *Nausea*, Roquentin's description of the general frailty and feebleness of existence is often read as an early critique of Nietzsche's Romantic vitalism: "There were those idiots who came to tell you about will-power and struggle for life. Hadn't they ever seen a beast or a tree?" (Sartre 1965a: N, p. 133)[3] In *Saint Genet* he rejects the idea of eternal return, which he understands in a literal fashion and regards as a form of nihilism. In conversation with John Gerassi in the 1970s, he recalls that in his youth he took a dislike to Nietzsche's elitist jargon and viewed him as not particularly important – "I hated him. I think his crap about the elite, his übermensch [sic], radicalized us a lot" (Gerassi 2009, p. 53), while in interview in 1975 he was more conciliatory, commenting how at the ENS he "interested me like many others" but ultimately "never stood for anything in particular in my eyes" (Sartre 1981: Schilpp, p. 9).

Despite these reservations, Sartre would openly identify with Nietzschean aspirations.[4] Although, as Noudelmann remarks, "[e]verything would seem to oppose the aristocratic thinker Nietzsche to the anti-elitist philosopher Sartre",

2 It is likely Sartre was acquainted with Nietzsche's writings through Charles Andler's six-volume biography – Contat and Rybalka believed he had read at least to volume two, as well as the biography by Daniel Halévy. He may also have read a selection of aphorisms published by Jean Bolle in 1934 and by Geneviève Bianquis in the late 1930s. In his *Ecrits de Jeunesse* he claims to have read Nietzsche but does not specify which texts he has read, but Cohen-Solal believes he did read *Ecce Homo* around the time of writing *Une Défaite*. See Daigle (2009, p. 57).

3 Conversely, Plank (1998, p. 375) is one of the few to recognize the strongly Nietzschean character of Sartre's phenomenological descriptions in *Nausea* as shown in Roquentin's encounter with the chestnut tree, for instance, when he loses his individuality to become submerged in Nature, giving rise to the 'terror' Nietzsche describes in *The Birth of Tragedy* when we look directly into the undifferentiated Dionysian realm.

4 While at the Lycée Louis-le-Grand, for instance, Sartre and Nizan would occasionally refer to themselves as 'supermen'. One should also not take at face value Sartre's oft-quoted youthful remark about Nietzsche in the *Carnets Midi* – "He is a poet who had the bad fortune to be taken for a philosopher" (Sartre 1990: CMid, p. 471). In Aron's *Mémoires* he reports that Sartre wrote an essay for one of Brunschvicg's seminars arguing against this view that Brunschvicg himself espoused (See Flynn 2014, p. 28).

Sartre's closeness to Nietzsche nonetheless "involves foremost an imaginary contamination based on admiration and parody" (Noudelmann 2012, p. 46). Sartre's dismissals of Nietzsche tell only a partial story for Nietzsche could hardly be said to have been unimportant to him. As Noudelmann argues, Nietzsche is a thinker "whom we can indeed find smuggled into Sartre's work" (Noudelmann 2012, p. 45) at every corner to the point of imaginary identification and contamination. Daigle concurs on this point, describing Sartre as the "Unaware Nietzschean" and Nietzsche who, despite Sartre's misunderstandings of his philosophy, as being "very present in Sartre's intellectual universe" (Daigle 2009, p. 57), sharing together the project of an affirmative creative ethics as a response to a crisis of nihilism in the wake of the Death of God. This is strongly evidenced by the fact that Nietzsche didn't really leave Sartre's philosophical brain from beginning to end. His first published essay, *The History of Truth*, written as a young student and published in a French Lycée review in 1925, was a typical Nietzschean study, while towards the end of his life, he was still working on a long text of Nietzsche's ethics that he began around the time of his *Notebooks (*1947/8) as part of his ethical research (which Simone de Beauvoir described as "a very fine study" [Beauvoir 1984: EBS, p. 180]). From the period 1929–1937, as Flynn describes, Sartre is very much "under the spell of a Nietzschean exuberance" (Flynn 2014, p. 149). His literary compositions, *A Defeat* (1927), *Er the Armenian* (1928) and *The Legend of Truth* (1931) all reveal "a proto-existentialist, quasi-Nietzschean character" with a valorisation of freedom that extends through his entire work (Flynn 2014, p. 27). His first (unfinished) novel, *A Defeat*, where he addresses the explicitly Nietzschean themes of power, autonomy, freedom, nature and embodiment, was a story in which he even sympathetically imagined himself to be Nietzsche. His essay *Legend of Truth* followed Nietzsche in showing the inadequacy of science and ideology as collective representations of belief as compared with the individual judgement of exceptional artists, philosophers and writers. The central theme of these works, culminating in *Nausea*, was the contingency of the world, the notion central to Nietzsche's philosophy,[5] that there is no guiding hand of necessity that rules existence – "every existing thing is born without reason, prolongs itself out of weakness and dies by chance" (Sartre 1965a: N, p. 133). The world, including the sphere of morality and political life, does not have to be as it is as configurations are always pos-

5 C.f. Zarathustra's paean to "Lord Contingency" – "'Lord Contingency' – that is the oldest nobility of the world, which I restored to all things when I redeemed them from their bondage under Purpose" (Nietzsche, Za-1, 3.4).

sible. Contingency, for both Sartre and Nietzsche, is the "perfect, free gift" that can lead to nihilism but also to freedom and creativity.

Philosophers of Paradox

The paradoxes of an author. The so-called paradoxes of an author, which a reader objects to, are often not at all in the author's book but rather in the reader's head.
(Nietzsche, HH-1, §185)

On the surface, Nietzsche and Sartre appear as philosophers of paradox and contradiction. Not only does Nietzsche frequently contradict himself, he even seems to delight in doing so: "Does the good historian not, at bottom, constantly *contradict?*" (Nietzsche, D, P §2). More than this, he believed that a virtue can be made of contradiction as it is essentially a sign of spiritual abundance. Conversely, the inability to tolerate contradiction – to contain or negotiate opposites in the right way – is seen as an aspect of cultural illness (Nietzsche, GS-1, §297). *Zarathustra*, Nietzsche's literary gift of a joyful, agonistic and bestowing friendship to his readers, is a book offered in contradictory fashion as one "for all and none". In EH he declares "I contradict as has never been contradicted before" (Nietzsche, EH-1, 4 §1) and presents his earthly prophet Zarathustra as a paragon of contradiction: "This most affirmative of all spirits contradicts with every word he speaks; all oppositions are combined into a new unity in him" (Nietzsche, EH-1, 3 §6). As Karl Jaspers complained, this can pose a problem for the unsuspecting reader: "One finds it insufferable that Nietzsche says first this, then that, and then something entirely different" (Jaspers 1965, p. xi). For any proposition in Nietzsche one can usually find its opposite somewhere else in his work: "*Self-contradiction* is the fundamental ingredient in Nietzsche's thought" (Jaspers 1965, p. 10). Similarly, in Sartre's philosophy, as Ally comments, "[c]ontradictions abound, and paradox lurks" (Ally 2020, p. 373). Contradiction and self-contestation are, as Boulé (2005) notes, essential elements in any proper understanding of Sartre. Indeed, his autobiography is littered with accounts of his self-transfigurations enacted through the regular shifting of perspective: "I never stopped creating myself" (Sartre 1967: Mots, p. 23); "Above all, I had to renounce myself" (Sartre 1967: Mots, p. 102); "I have often written them against myself" (Sartre 1967: Mots, p. 103); "I came to think systematically against myself to the point of weighing the evidence for an idea by how much I disliked it" (Sartre 1967: Mots, p. 156).

The difficulties in understanding Nietzsche and Sartre extend to their recommendations on how to read their work that, like their thought in general, veer ambiguously between the poles of humanism and anti-humanism. In *Beyond Good and Evil*, for instance, Nietzsche tells us "there is absolutely nothing impersonal about the philosopher" (Nietzsche, BGE-1, §6), a theme he reiterates often in his writings, and yet in HH he acknowledges that "[e]very writer is surprised anew when a book, as soon as it is separated from him, begins to take on a life of its own" (Nietzsche, HH-1, §208). With similar ambiguity, Sartre asserts in an interview in 1979 "You have to take into account the life of people who write: it is projected in the writing in one way or another" (Sartre 1979: Obl79, p. 26) and yet, in *What is Literature?* we are told that "the text towers above its author" (Sartre 1965b: QL, p. 34). One might say in this light that their texts are, for the unwary, "booby-trapped", and therefore destined to easily confuse or to draw the reader to hasty conclusions and it is this, in part, that has contributed to much of the distortion and noise that surrounds them and their interrelation.[6] This is exacerbated by the fact that, too often, scholars take an entirely synchronic view of these philosophers, fixating on individual passages or tomes and failing to view their work as a whole, as containing subtle shifts within a discernible line of development and unity – a kind of "living entelechy" (Hutter 2006, p. 4) or "enrichment within continuity" (Perna 2007, p. 47) in which later works expand earlier ones and earlier ones hold in themselves the seeds of their future unfolding. It is only with a totalizing view of the entirety of their works that we can fully appreciate the rich ambiguities and resolutions in their thinking. As Jaspers remarked, to understand Nietzsche fully we need to read the entirety of his works in a comparative way, experiencing "both the systematic possibilities" of his thinking as well as the likely possibility of "their collapse" (Jaspers 1965, p. 3).

6 Betschart (2020) uses communication theory in explaining the dissemination of Sartre's thought internationally in terms of 'senders', 'receivers', 'messages', 'noise', 'jammers' and 'distortion'.

New Nietzsche and New Sartre

What the father kept silent, that comes in the son to be spoken
and often I found the son to be the father's unveiled secret.
(Nietzsche, Za-1, 2 §7)

The author cannot be the contemporary of his contemporaries
unless he is, on the whole, behind them and ahead of them.
(Sartre 1972: IF, p. 424)

In death, as Sartre claimed, we become "prey to the living" (Sartre 2018: EN, p. 706), and there has certainly been no shortage of appropriations of the deceased Nietzschean and Sartrean minds. After the "Nazification" of Nietzsche in the first half of the twentieth century by German philosophers, such as Alfred Baeumler and Richard Oehler, who, taking the lead from Elisabeth Förster-Nietzsche, Nietzsche's sister, emphasized his system of will to power as a political ontology of force and domination, the "New Nietzsche" was born in reaction to this in France after the Second World War. Rescuing Nietzsche from his "blood and thunder" personification and from the reading of Heidegger (1987) who cast him as the "last metaphysician", the French poststructuralists portrayed the "New Nietzsche" as a playful deconstructionist who overturned the faith in oppositional and metaphysical thinking, problematizing the notions of truth, objectivity, language, the subject and interpretation. Derrida (*Spurs: Nietzsche's Styles*), Deleuze (*Nietzsche and Philosophy*), Foucault (*Nietzsche, Genealogy, History*) and Kofman (*Nietzsche and Metaphor*) used Nietzsche's genealogical method to deconstruct relations of power in modern society, as well as to expose the anthropological and logocentric bias of Western philosophical thinking as derived from Plato, Descartes, Kant and Hegel.

Like Nietzsche's, Sartre's image has evolved significantly in the years since his death. Moving beyond the confines of the "Old Sartre" who was presented primarily as a Cartesian dualist thinker, the "New Sartre" has emerged in full, precipitated initially by the *volte-face* of poststructuralists, like Derrida and Foucault, who changed their attitude to Sartre from one of parricidal attack in the 1960s into one of homage in the 1980s. Despite his previous criticisms of Sartre, in the lead article for the commemorative fiftieth anniversary issue of *Les Temps Modernes*, Derrida expresses the "boundless gratitude" and acknowledges the "immense debt" he and others owe to Sartre. He confesses that in previous years he "wouldn't have dared" admit his affection for Sartre and *Les Temps Modernes* but that he is now moved to "do justice" to them and recognize the value of Sartre's philosophical œuvre (Derrida 1996, pp. 44, 40). He also confesses to an amnesia for forgetting that Sartre had put into question the rhetoric of "fraternity" (Derrida 1996, p. 12), and claims to be the heir of "Sartre's opposition to himself and his antinomies". It is Sartre's self-contradiction that most touches him, Derrida confesses, and his 'disaccord' or disagreement with self that he most admires (Derrida 1996, pp. 23, 32). Furthermore, he acknowledges that "Sartre's refusal of antinomies" prefigured his own deconstruction of oppositional thinking (Derrida 1996, p. 14). In the past twenty years or so, several scholars have explored the many ways in which Sartre's thinking has inspired, presaged, anticipated, suggested or confirmed poststructuralist and postmodernist thinking (just as Nietzsche's has) and given birth to the "New Sartre", significantly recasting him in philosophical terms as resolutely anti-Cartesian and as the precursor and originator of many of the themes that the poststructuralists later themselves adopted [7], such as, as Howells lists them:

> the decentred subject, the rejection of a metaphysics of presence, the critique of bourgeois humanism and individualism, the concept of the reader as producer of the text's multiple meanings, the recognition of language and thought structures as masters rather than mastered in most acts of discourse and thinking, [and] a materialist philosophy of history as detotalized and fragmented. (Howells 1992, p. 2)

Exploding Dualisms

> This *dual* series of experiences, this access to apparently separate worlds, is repeated in my nature in every respect: I am a *Doppelgänger*, I have a "second" face in addition to the first. *And* perhaps also a third. (Nietzsche, EH-1, 1 §3)

> [A] complex movement of contradictions which are posited and surpassed. (Sartre 1968a: QM, pp. 98–99)

Despite some distortions and vulgarizations that posit the French philosopher as a Cartesian, Sartre and Nietzsche are united as resolutely *anti-dualistic* thinkers. With its "faith in opposite values" (Nietzsche, BGE-1, §2), metaphysical philosophy, in Nietzsche's view, strictly divides opposites, demanding separability of them, and claiming that phenomena emerge from things in themselves, "assuming for the more highly valued thing a miraculous source in the very kernel and being of the 'thing in itself'" (Nietzsche, HH-1, §1). Unlike metaphysical philosophy, Nietzsche's historical philosophy follows Heraclitean principles in denying

7 For a full account of Sartre's relationship to poststructuralism, see Farrell Fox (2003), and for a general survey of the recent literature on this, see Farrell Fox (2022).

"the duality of totally diverse worlds" (Nietzsche, PTAG, §5) and discovering that there are no opposites but only differences of degree: "There are no opposites: only from those of logic do we derive the concept of the opposite – and from them falsely transfer it to things" (Nietzsche, WP, §552). In *The Birth of Tragedy* Nietzsche argues for the indissolubility of Apollo and Dionysus, analytically separating them while showing them to be inextricable "the expression of two interwoven artistic impulses" (Nietzsche, BT, §12). They are necessarily interdependent in the sense that "Apollo could not live without Dionysus" (Nietzsche, BT, §4). Denouncing the "hemiplegia" of the "good human" who separates off one side of a dualism and insists on it alone, teaching "that it is a higher thing to be efficient on only one side" (Nietzsche, WP, §351), he proclaims that the greatest are those like Zarathustra who combine opposites (Nietzsche, EH-1, 3 §6.6).

Aiming "to eliminate a number of troublesome dualisms", Sartre argues strongly in his early phenomenology for the "monism of the phenomenon" (Sartre 2018: EN, p. 1) and presents the *pour-soi / en-soi* as "a duality that *is* a unity" (Sartre 2018: EN, p. 125). He makes it clear that body and consciousness are two aspects of the ambiguity of lived experience that cannot be separated until after the event: "We know that there is not, on the one hand, a for-itself, and, on the other, a world, as two closed wholes, whose means of communication it is necessary afterwards to seek. Rather, the for-itself is by itself a relation to the world" (Sartre 2018: EN, p. 413). In the *Critique* and beyond, Sartre reinforces this dialectical method of strategically employing and bending traditional dualisms (e.g. subject / object, consciousness / body, freedom / necessity, conscious / unconscious, reason / emotion). Ally, for instance, highlights the way in which Sartre dissolves binarisms in the *Critique*, employing a deconstructionist method in a strategic form of "ternary" post-metaphysical thinking – "We might call this the *immanent imminence of the ternary*, both immanent and imminent. The third is *always coming* and *always nearby* and *always already there*. When we think with Sartre, if we look for a Third, we will find one" (Ally 2020, p. 365). Following the threefold stages of "contradictions, surpassing and totalization" (Sartre 1968a: QM, p. 34), Sartre's dialectical method provides a way beyond metaphysical dualism and protects against analytical closure by multiplying opposing pairs into a displaced or "sublimated third" in the same way Nietzsche sought to.[8] In Ally's view, it is Sartre's "index of complementarity" between op-

8 Within his surpassing of opposites, for Sartre attempts to establish how, *real-imaginary* becomes *possible*, *pour soi-en soi* becomes *for others*, *praxis-practico inertia* becomes *praxis-process*, *totality-totalization* becomes *totalization of envelopment*, and *need-scarcity* is sublimated into *abundance* (see Ally 2020, p. 365). For Nietzsche, in *Beyond Good and Evil*, §4 he inverts the relation of the truth / falsity opposition by saying that untruth might be more conducive

posing pairs and his "sensitivity to parts and wholes" that puts him on the cusp of a new direction in thinking as we pass from a Holocene to an Anthropocene phase in human and natural history (Ally 2020, pp. 365, 372).

In casting Nietzsche and Sartre as metaphysical humanists, Heidegger was correct in the respect of recognizing the "inverted Platonism" of their philosophical thinking. However, he missed their crucial third step and was blind to their shared sublimating logic which clearly anticipated Derrida's deconstructive method of surpassing dualisms. Within this method, none of these pairs are entirely suppressed, though the traditionally dominant term (e. g. self, reason, consciousness) is subjugated in a movement of inversion and incorporated by the suppressed term (body, affect, unconscious) into a new nuanced unity (displacement). Employing dualistic categories in a strategic or deconstructive way, Nietzsche and Sartre use them posthumanistically as liquid, shifting and intra-changing – "an explicit duality expressing an implicit unity" (Ferrando 2020, p. 60).

The Decentred Subject

> What separates me most deeply from the metaphysicians is: I don't concede that the 'I' is what thinks. Instead, I take the *I itself to be a construction of thinking.* (Nietzsche, LN, p. 20)

> [T]his superfluous *I* is actually a hindrance. If it existed, it would violently separate consciousness from itself; it would divide it, slicing through each consciousness like an opaque blade. The transcendental I is the death of consciousness. (Sartre 1957: TE, p. 7)

A cardinal theme that unites the thinking of Nietzsche and Sartre is the "decentred subject". This can be compartmentalized into three analytical strands (i) their critique of the *bewitched ego*, (ii) their intersectional delineation of the *three ecologies of self* (material, social, psychological) and (iii) their affirmative projects of creating "new versions and refinements of the soul hypothesis" (Nietzsche, BGE-1, §12) in the form of *self-cultivation* (Nietzsche) and *authenticity* (Sartre) based on invention and play.

---

to human wellbeing (it protects us against the stark realities of existence) than truth, not by simply asserting the authority of untruth over truth, but by displacing the grounds for their opposition by reading them in terms of their value for life. Similarly, his move beyond good and evil does not simply invert the good / evil hierarchy but displaces these in the favour of a revaluation of all values (see Woodward 2011, p. 92).

(i) Both Nietzsche and Sartre critique the ego (as conceived by Descartes, Kant, Husserl and "ego-psychologists") as a form of mystification or bewitchment, leading to "impure" forms of reflection and debilitating tropes of bad faith. They posit the "I" as derivative and as a temporal synthetic construct of consciousness rather its primary origin – "The 'I' (which is *not* one with the unitary management of our being!) is indeed only a conceptual synthesis" (Nietzsche, WP, §371). Sartre describes how "it is in exclusively magical terms that we have to describe the relations between the *me* and consciousness" (Sartre 1957: TE, p. 26). Magic arises from "impure reflection" (Sartre 1957: TE, p. 23) that reverses the real phenomenological order of things, making primary things secondary and secondary reflected-upon things primary. My actions are seen to flow from me whereas in fact my me is the reflected synthesis of my actions – unlike pure reflection that "stays with the given... [and] disarms unreflected consciousness by giving it back its instantaneous character" (Sartre 1957: TE, p. 24), impure reflection effects a magical reversal whereby it carries out "an infinitization of the field", creating a transcendent object that serves as an imagined origin and cause "in the past and the future" (Sartre 1957: TE, p. 23). Thus, for Sartre, reversal, hypostasis and pseudo-activity signify the magical way of relating to our own ego in which a more elemental, pre-reflective consciousness instils its own creation with a pseudo-power that allows consciousness to bewitch, escape and even suppress aspects of itself where all our states, emotions and actions through a relation of "creation" are "attached directly (or indirectly, through quality) to the Ego as to its origin" (Sartre 1957: TE, p. 32).[9] Against the Cartesian notion of interiority, Sartre and Nietzsche emphasize the "public" nature of the ego and the fact that there are objective psycho-physiological aspects to ourselves of which we are most often unaware. The fact that the nature of reflective consciousness makes itself an object for such consciousness means that the "I is *an other*" (Sartre 1957: TE, p. 46). In transcending metaphysical idealism, both philosophers theorize the self as an emergent, synthetic construct of consciousness necessarily incarnated and situated in the world of others.

(ii) In general terms, both philosophers argue for a view of the *decentred self* as composed of the dialectical interplay of three ecologies:[10] the *material* (body and environment), the *social* (others and language) and the *psychological* (proj-

9 See O' Shiel (2019) for a clear and illustrative account of the ways in which this bewitchment can manifest itself.

10 Guattari writes in *Les Trois Ecologies* of a "psycho-social-somatic unity" composed of (i) *a material environment in the process of being reinvented* (ii) *a constantly mutating socius* (iii) *a nascent subjectivity* (Guattari 2000, p. 69).

ects of artistic self-cultivation).[11] In "inverting Platonism", both philosophers assign central importance to the body and the influence of environment. In his rejection of "soul atomism" – "the belief which regards the soul as something indestructible, eternal, indivisible, as a monad, as an *atomon*" (Nietzsche, BGE-1, §112) – Nietzsche extols the "great wisdom" of the body (Nietzsche, Za-1, 1 §4) and debunks the rationality of the soul. He posits an intimate connection of transversal exchange between self and world, using the guiding metaphors of *incorporation* and *assimilation* to explain this. The body is a vehicle of assimilation that digests all kinds of material that are processed and modified into emotions and psychological states. Our character is susceptible and modifiable through the effects of inheritance, diet, exercise and environment: "These small things – nutrition, location, climate, recreation, the whole casuistry of selfishness – are inconceivably more important than everything one has taken to be important so far. Precisely here one must begin to *relearn*" (Nietzsche, EH-1, 2 §10). He connects environmental and dietary factors firmly to certain psychological states in the form of a general metabolism. Conceiving the self as a "psycho-somatic continuum", Nietzsche emphasizes the strong effects of physiology – "life style, nutrition, or digestion, perhaps a deficit or excess of inorganic salts in their blood and brain; in brief, in their *physis* – on judgements, feelings, moods and tastes" (Nietzsche, GS-1, §39).

Throughout his philosophical trajectory Sartre gives strong emphasis to the body as a key constituent of the self conceived as a "psycho-somatic unity". Consciousness, he insists, "*exists its body*" (Sartre 2018: EN, p. 441) and is "in its entirety body" (Sartre 2018: EN, p. 413) since the body is its "pure centre of reference" (Sartre 2018: EN, p. 438). Presenting consciousness and body as a living unity, he argues against the idea that "the psyche is *united* to a body" since "the body is its substance and its constant condition of possibility" (Sartre 2018: EN, p. 452). Invoking the difference between *Leib* and *Körper* (as Nietzsche did), he insists that "[t]he body is *lived*, and not *known*" (Sartre 2018: EN, p. 435). "We do not employ this instrument", he adds, "for we *are* it" (Sartre 2018: EN, p. 434). When there is a world, we are necessarily saying that there is a body (Sartre 2018: EN, pp. 413, 436). The *Critique* heightens Sartre's focus on the body as a vehicle of praxis and as a functioning biological organism – a person is "wholly matter" (Sartre 1976: CRD, p. 180), an organism symbiotically related to its environment through need, and engaged in a reciprocal dialectical process of "transsubstantiation" (Sartre 1976: CRD, p. 178).

---

11 These three factors wax and wane in relative importance at different points in their writings. See Farrell Fox (2022) chapter 3.

Alongside the body, Nietzsche and Sartre assign great importance to the social Other and language in constituting subjectivity. Throughout his writings, Nietzsche emphasizes the "heaviness" of our social conditioning and the way in which social and historical forces produce, penetrate, assemble but also limit the individual. His early work presents the individual as a cipher and singularization of the collective (a "Universal Singular" in Sartre's words) bound by an Objective Spirit which ushers in a horizon of myths that "locks a whole cultural movement into unity" (Nietzsche, BT, p. 136). The past has layered structures within us, deposited through language, constituted meaning and memory: "the human carries around the memory of all previous generations" (Nietzsche, PTAG, §35) unaware that she is voicing "[t]he spiritual activity of millennia laid down in language" (Nietzsche, PTAG, §31). Culture is a supersession of Nature, evidenced by Greek culture that offered a "new and improved *physis*" (Nietzsche, UM, p. 123). Although he puts more emphasis on biological / Lamarckian factors in certain passages (Nietzsche, BGE-1, §264; BGE, §200; D, §30; D, §540), he often views genetic inheritance as modifiable through socialization. Linking self-consciousness with sociality (Nietzsche, D, §26), he demonstrates how, paradoxically, self-consciousness arose through the process of socializing or herding individuals (Nietzsche, GS-1, §354). In his middle writings he weighs the formidable forces of social acculturation in the formation of our "second skin" (Nietzsche, D, §455) – a socially shaped self that enables one to function in the world moulded by familial influence, education, interpersonal relationships and one's social positioning. In Nietzsche, D, §34 he views familial influence more on the level of *modeling* and *behavioural transmission* than genetic inheritance. Moral feelings and dispositions are passed onto children through observation of their parents and siblings (Nietzsche, GS-1, §95; Z, 2 §7). Beyond the family, the force of circumstance and the necessity to survive in the social world can lead to the desuetude and loss of one's primary talents and traits: "Circumstances do not only conceal and reveal... they magnify and diminish" (Nietzsche, D, §326). Learned traits and social roles become habituated traits that come to appear as naturally given. As in Sartre's bad faith, Nietzsche observes how people over-identify with their social role such that "they become victims of their own good performance: they themselves have forgotten how much accidents, moods, and caprice disposed of them when the question of their 'vocation' was decided... the role has actually *become* character; and art, nature" (Nietzsche, GS-1, §356).

It was the syncategorematic or social nature of Sartrean consciousness in which we find embedded "the structure of the Other" (Deleuze 2004, p. 383) that attracted Deleuze to Sartre's early phenomenology. Sartre argues in *Being and Nothingness* that Kant and Husserl misunderstood the determining presence of the Other. In their idealist ontology, the constituting negation is "an external negation", separating me from the Other "by a real or ideal space" in which the Other is only an "indifferent exteriority" who cannot affect me in my being (Sartre 2018: EN, p. 421). Sartre insists that interpersonal relations are, by contrast, an "internal negation", that is, a negation which posits the original distinction between the Other and myself as being such that it determines me by means of the Other and determines the Other by means of me. When fixed by the gaze of the Other, this decentres my world and causes me to experience an "internal haemorrhage" in which all the constituents of my world flow toward the Other (Sartre 2018: EN, p. 358). Forms of self-experience thus depend internally on the constitutive importance of the Other. In his later work Sartre deepens the social encoding of the self, showing the ways in which language and culture (as constituent features of the practico-inert) "are not inside the individual like stamps registered by his nervous system. It is the individual who is inside culture and inside language" (Sartre 1968a: QM, p. 113). By the time of *The Family Idiot*, Sartre conceives the subject, formed predominantly by the opaque forces of history and family destiny, as

> a function of the society in which he lives, of the mode of production, of the technical knowledge available at the time, of the structure of the family, of antecedent circumstances, of the historic *future* which reveals itself as his destiny, but also of the singularity of his own previous history and of his biological characteristics, inherited or acquired... (Sartre 1972: IF, p. 442)

(iii) Despite decentring the self in terms of biological, social and linguistic forces, neither Nietzsche nor Sartre abandons the self as a centre of value for a reflexive ethical transfiguration. We can, Sartre maintains, always make something out of what is made of us, through a form of "purifying reflection" (*The Transcendence of the Ego; Being and Nothingness*) that allows us to escape the chains of bad faith. Equally, for Nietzsche, it is possible to achieve "the intellectual conscience", a "rare and difficult art" (Nietzsche, GS-1, §290), but one that allows us some awareness of the complexities of our unconscious motivations and to impose some kind of artistic unity on our affective drives and resist the limiting forms of our social forces: "*Our opinion of ourself*, however, which we have arrived at by this erroneous path, the so-called ego is thenceforth a fellow worker in the construction of our character and our destiny" (Nietzsche, D, §115). The *via regia* to achieving this is through art and *play* which they propose as antidotes to the "spirit of seriousness" and patterns of bad faith that are obstructive to practices of freedom and authentic self-fashioning, and it is this *ludic model of self* that forms the core of their affirmative ethical thinking.

A Creative Ethics of Play

> I know of no other way of associating with great tasks than *play*: as a sign of greatness, this is an essential presupposition. (Nietzsche, EH-1, 2 §10)

> The essential moment therefore is that of creation, that is, the moment of the imaginary, of invention. (Sartre 1992: CM, p. 464)

The principal target of Nietzsche's *revaluation of values* was customary morality which, he argued, is stifling to freedom and self-determination. Customary morality defends conformity to the "sanctity" and "inscrutability" of tradition (Nietzsche, D, §19), an "inexplicable, indeterminate power" behind our thinking that lies "beyond the personal" (Nietzsche, D, §9). It requires a regime of habit and obedience that turns humanity into a self-undermining "perpetual sacrifice" (Nietzsche, D, §18). To combat this, we must become "free spirits", he urges, who "think differently from... the dominant views of the age" (Nietzsche, HH-1, §225) through "novel experiments in living" and diverse "practices of self-cultivation" that resist social conditioning (Nietzsche, D, §164).

From his earliest to his final writings, Nietzsche placed *play* at the centre of his affirmative thinking, associating it directly with creation, a cosmological game in which the events of the universe occur like with a child who builds and scatters pieces of stone (Nietzsche, BT, p. 24). He consistently links the child with innocence and creativity (Nietzsche, HH-1, 2 §270; BGE, §57) and urges us to emulate this – in the "genuine man, a child is hidden: it wants to play" (Nietzsche, Za-1, 1 §18). The laughter of the child is an antidote to Christian seriousness – to "live and laugh gaily" is the mark of one who has learned to live without metaphysical consolations (Nietzsche, GS-1, §324). In *Of the Three Metamorphoses*, Zarathustra judges that the child's play makes possible a revaluation of values and a passage to a higher level of moral thinking – where the camel could only follow rules given to it (passive nihilism) and the lion was only capable of saying no (reactive nihilism), the child playfully creates its own values (active nihilism): "The child is innocence and forgetting, a new beginning, a game, a self-propelled wheel, a first movement, a sacred Yes. For the play of creation, my brothers, a sacred Yes is needed: the spirit now wills *its own will*, and he who had been lost to the world now conquers *his own world*" (Nietzsche, Za-1, 1 §1). Zarathustra himself, as teacher, is a "prelude to better players" and it is the *Übermensch* who, as the superior player, masters the play of forces to give meaning to her self and to the earth (Nietzsche, Za-1, 3, §12.20).

In the entry for March 9, 1940 in his *War Diaries*, Sartre echoes Schiller's remark that "man is fully a man only when he plays", asking "[h]ow can we fail to see that play, by its very nature, excludes the very idea of seriousness?". Anticipating his analysis of bad faith in *Being and Nothingness*, he contrasts play with the seriousness of the engineer and posits it as an integral element in self-identity: "'game'... is the happy metamorphosis of the contingent into the gratuitous... why the assumption of oneself is itself a game" (Sartre 1984: Carn, pp. 313–314). Near the end of *Being and Nothingness* Sartre suggests there is a way out of bad faith and the contradictory desire to become God through play, an activity in which we are the "first origin" and in which we set our own rules – "play liberates subjectivity" (Sartre 2018: EN, p. 753). He associates play with a kind of purifying praxis that works against seriousness: "play, in opposing itself to the spirit of seriousness, seems to be the least possessive attitude" (Sartre 2018: EN, p. 752). It is able "to make manifest... the absolute freedom that is the person's very being" (Sartre 2018: EN, p. 754), leading to psychological instants that "provide the clearest and most moving image of our freedom" (Sartre 2018: EN, p. 622), moments of a "twofold nihilation" (Sartre 2018: EN, p. 572) where past and present, self and world change together in new directions. The quintessential expression of play is through art: "the artist is, on the one hand, the man who chooses really to create imaginary objects, but he is also and above all (if we place ourselves in the ontological point of view) the man who chooses to create imaginarily the real world; he is the one for whom perception is already unreal creation" (Sartre 1992: CM, p. 554). Sartre suggests that it is through art and play that we are able to overcome the separation of subject (*pour-soi*) and object (*en-soi*), the aesthetic object (e.g. the novel) representing an end product and grand fusion that is "fully myself and fully beyond myself" (Sartre 1965b: QL, p. 53). Unlike the psychological freedom which torments what it lacks (i.e. being), the freedom of art and play uplifts: "[t]he recognition of freedom by itself is joy" (Sartre 1965b: QL, p. 52).

Smooth Ontology

> [M]agic formula.... PLURALISM = MONISM.
> (Deleuze/Guattari 1987, p. 20)

A third major area of Nietzschean and Sartrean parallel thinking is ontology. Despite their reputation as individualists, Nietzsche and Sartre are thoroughly relational thinkers reflected in their ontological schemes of the will to power (Nietzsche) and emergent interactionism (Sartre) that follow a Heraclitean logic[12] based on change, becoming, the interplay of forces, combination, emergence and dissipation.[13] This *relationality* is developed most fully by Nietzsche in the *Nachlass* where he defines a thing as "the sum of its effects, united by a concept, an image" (Nietzsche, WP, §551) – if one removes other things, an object has no properties. Therefore "there is no things without other things" and no "things-in-themselves" (Nietzsche, WP, §557). Since independently existing things are inventions "owing to the requirements of logic" (Nietzsche, WP, §558), if everything we projected onto the world in order to make it intelligible were eliminated, "no things remain but only dynamic quanta, in a relation of tension to all other dynamic quanta: their essence lies in their relation to all other quanta, in their 'effect' upon the same" (Nietzsche, WP, §635). The world is thus "essentially a world of relationships" (Nietzsche, WP, §568). As Schrift argues, Nietzsche sought to keep the will to power multiple so that it might appear in multiple forms, "at once producer and product, a monism and a pluralism" (Schrift 1996, p. 341). In his attempt to think difference differently, Nietzsche's recasting was not reductive, nor should it be seen as privileging exclusively one analytic framework. Instead, "the monistic framework of will to power supports Nietzsche's pluralist response to the privileging of oppositional thinking" (Schrift 1996, p. 345). In Nietzsche's philosophy of "originary multiplicity", things or unities are derivative from a complex and rich multiplicity of forces:

> All unity is only as *organization and interplay* unity: not otherwise than how a human community is a unity: so, *opposite* of atomistic anarchy; therewith a *complex of rule*, which *signifies* One, but *is* not one. (Nietzsche, NL 1885, KSA 12, 2[87])

For the "subject as multiplicity" (Nietzsche, WP, §490), Nietzsche tells us,

> physiology gives only an imitation of an astounding intercourse among this multiplicity and subordination and integration of parts into a whole. But it would be false to conclude from a state the necessity of an absolute monarch (the unity of the subject). (Nietzsche, NL 1884, KSA 11, 27[8], tr. by NFF)

His concept of unity is equidistant from an originary unity on one side, and from atomistic chaos on the other. Self-organization is an emergent play among constituents, all seeking out the resistance of antagonists in order to overpower

---

12 See Meyer (2018, p. 371) and Flynn (2014, p. 200).

13 Plank (1998, p. 433) conceives this as an ontology of *Dissipative Systems* in which there is no need for hidden variables or metaphysical postulations, simply a process of combination, recombination, alignment and dissonance between objects, particles and energy centres.

(subordinate and integrate, or functionalize) them.[14] His idea of the *subject as emergent* represents a striving to think post-metaphysically. By conceiving of multiplicities and unities in tandem as evolving and ephemeral but not as mutually exclusive, his position aligns closely with Sartre and with philosophical posthumanism.[15]

In a footnote in *What is Literature?* Sartre signals his interest in the development of dialectical history, an area that his earlier phenomenology had left untouched and one he would excavate fully in the *Critique:* "Some day I am going to try to describe that strange reality History, which is neither objective, nor ever quite subjective, in which the dialectic is contested, penetrated, and corroded by a kind of antidialectic, but which is still a dialectic" (Sartre 1965b: QL, pp. 333–334). He continues this dialectical search in *Notebooks for an Ethics* where he proposes an "existential dialectic" or a "dialectic with holes in it" (Sartre 1992: CM, p. 449), a historical process where individual agency and moral responsibility can produce "an irrational leap into another dimension of being, a new dialectic and a new leap" (Sartre 1992: CM, p. 458).[16] As Badiou describes it, the *Critique* was Sartre's attempt to provide a dialectical comprehension of history, composed "as a symphony in two movements", a regressive movement, theory of practical ensembles and a progressive movement, a rational reconstruction of history as a "totalization without a totalizer" (Badiou 2009, p. 43). Out of this symphony crystallizes Sartre's version of the dialectic which, according to Badiou, "is not the synthetic neutralization of two pre-existing but contradictory terms, but the discovery of the articulation that deploys the dimension along which they suddenly emerge as 'sides'". It designates "the active unity of an operation as a preliminary to any determination of a duality". The dialectic is hence "never a dualized sequence of concepts... [but] is the polarization of a space that is articulated" (Badiou 2009, pp. 167–168).

For Sartre, "reciprocal ternary relations are the basis of *all* relations between men whatever form they take" (Sartre 1976: CRD, p. 111). Ternary reciprocities ground and mediate all abstract dualities and form a bond of sociality – "a ternary relation, as the mediation of man amongst men, is the basis on which reciprocity becomes aware of itself as a reciprocal connection" (Sartre 1976: CRD,

---

14 See Siemens (2015, pp. 97–98) for a cogent account of Nietzsche's "agonism" of opposing forces.
15 As Ferrando states, in philosophical posthumanism relata do not precede relations, but relations do not precede relata either: "relata and relations generate out of co-constitutive, embodied, agential processes, situated in specific spatio-temporal environments", therefore "neither is primary" (2020, p. 165).
16 See Flynn (2014, p. 316).

p. 109). In explaining totality and multiplicity by this ternary logic, Sartre demonstrates a sensitivity to parts and wholes that is missing from atomistic or holistic accounts. A totality is never fixed or static but is always in motion and always in process. Totalization hence relates to becoming as an activity of an intentional, integrative, and open-ended development – it "delineates a practical field [and] attempts the most rigorous synthesis of the most differentiated multiplicity" (Sartre 1968a: QM, p. 46). As individuals or collectivities we are never completed totalities but "detotalized totalities" that are both constituted and dissolved by the transcendence towards future possibilities (Sartre 1992: CM, p. 543). Hegel's dialectic assumes the viewpoint of the whole or Absolute with no respect to particular determinations or individual instances, however "no totalitarian and unifying synthesis of 'Others' is possible" (Sartre 1992: CM, p. 339). The group is dispersed by contingency into a plurality of beings who nonetheless mutually determine each other, a "detotalized totality" (Sartre 1992: CM, p. 17). Thus, Sartre's ontology reveals an intrinsic *relationality* or interconnectedness – "[e]verything is always revealed as united to everything" (Sartre 1976: CRD, p. 360) in a kind of feedback system of mutually constitutive parts, but also accounts for singularity and individuality as emergent phenomena.

Unlike methodological individualism, Sartre's social theory of "emergent interactionism" (Farrell Fox 2020) or "Dialectical Nominalism" (Flynn 2014) recognizes the synthetic enrichment of social phenomena and their irreducibility to their parts. Viewing subjectivity and intersubjectivity as *emergent* explains how a higher order is founded on a lower one and in a sense contains it, but at the same time takes it over and integrates it into new structures which cannot be explained by those that are taken over. This is not just a simple mixing of elements but more a chemical reaction that yields a new substance with different qualities from its constituents. For Sartre, consciousness emerges on the intersubjective level as an ongoing totalization between subjectivities that becomes a "totality" but is always open to detotalization (Sartre 1976: CRD, p. 404). Community is not a simple collection of discrete individuals but is informed by a unity of existence that penetrates and transcends their private worlds. When two or more selves enter into relationship there is a reciprocity that is an existential modification of each, forming an inter-individual reality which transforms and circumscribes individual praxis.

This accords with Nietzsche's (and Deleuze's) ontologico-ethical position, articulating an ethico-political subjectivity which is elaborated neither in terms of a monadic individual nor of a collective entity, but of an essentially self-creating, internally related multiplicity of singularities that interrelate according to intrinsic, spontaneous movements. When they move towards common ends, the singularities give rise to ontological dislocations; that is, the formation of more complex singularities in "constellations of reciprocities" (Sartre 1976: CRD, p. 367). All of reality is a congregation of individual singularities or "energy centres" freely configuring and reconfiguring themselves according to different, spontaneous modalities – "a play of energies and waves of energy at the same time one and many, increasing here and at the same time decreasing there" (Nietzsche, WP, §1067). Sartre's ontology is thus kaleidoscopic – singularities in constant becoming that coalesce and merge together in the production and creation of a composite whole – a "pluralistic monism". Free praxis becomes constituent power in this process and it is within this continuous movement of becoming that the multitude or group discovers itself as subject. This gives us a rationality that goes beyond the dualisms of modernity, outlining a new form of thinking that represents itself in a logic of the singularities in process, in fusion, and in continual surpassing.

In broad ontological terms Nietzsche and Sartre propose an ontology of pluralistic monism that in many ways anticipates and informs philosophical posthumanism and Deleuze and Guattari's "magic formula" outlined in *A Thousand Plateaus*, of "PLURALISM = MONISM" (Deleuze/Guattari 1987, p. 20).[17] It also illustrates a postmodern conception of space as "smooth space" (Deleuze/Guattari 1987, p. 481) in which consciousness and body, self and Other, humans and the earth, intersect and mutually constitute one another through a process of assimilation / incorporation (Nietzsche, EH-1, 2 §10) or "trans-substantiation" (Sartre 1976: CRD, p. 178).

Posthumanist Humanists

What is man? A volatile illusion flittering over matter in movement. (Sartre 1988: Mall, p. 135)

If we search the writings of Nietzsche and Sartre, the negative comments on humanism and "humanity" most probably outweigh the positive ones. Sartre does at various points propose variants of humanism that align with his ethical thinking – "humanism" as "the human condition of every individual" (Sartre 1984: Carn, p. 25), a "humanism of work" (Sartre 1968b: CP, pp. 25, 37) and a "flourish-

17 Schrift describes Nietzsche's ontology as a "pluralist or polyvalent monism" (Schrift 1995, p. 257) while, in Eshleman's view, Sartre subscribes to "a version of substance monism (materialism) conjoined with a modal pluralism" (Eshleman 2011, p. 33). Sartre describes his ontology in the *Critique of Dialectical Reason* as a "monism of materiality" (Sartre 1976: CRD, p. 29) and as a "realistic materialism" (Sartre 1976: CRD, p. 181).

ing humanity" (Sartre 1992: CM, p. 407). In general though, like Nietzsche's critique of "the herd" and the "last man", Sartre's antipathy towards humanism stems principally from its tendency to homogenize human plurality and ignore difference, defining humanity in abstract Kantian terms. Their humanism, best described as a *posthumanist humanism*,[18] was of a highly provisional kind and was predicated on sublimating humanism as they knew it to a higher level that released its creative possibilities from the shackles of metaphysical thinking and its "Otherizing" tendencies.

A key area in which their posthumanist perspective is at play is the theorization of nature where they evince a strong ecological viewpoint, prompting us to rethink our conceptions of nature through a form of thinking that can be described as "terrasophy" (Manschot 2021, p. 75) or "existential ecology" (Ally 2017, p. 6), anticipating several strands of posthumanist theory, such as eco-phenomenology, vibrant materialism and object-oriented ontology.[19] For Nietzsche, our relation to the environment is one of incorporation and assimilation (Nietzsche, EH-1, 2 §10). As a philosopher of "atmospherics" (Manschot 2021, p. 151) or "sensism" (Plank 1998, p. 102), and fascinated by the metabolic processes that take place between people and the environment, Nietzsche provides a phenomenology of nature by delineating space as an affective medium of interchange and transfer. All things are dependent on environment for their flourishing and the expansion of their creative will to power: "Every living thing needs to be surrounded by an atmosphere, a mysterious circle of mist: if one robs it of this veil, if one condemns a religion, an art, a genius to orbit as a star without an atmosphere: then one should not wonder about its becoming withered, hard, and barren" (Nietzsche, UM 2, §7). Nietzsche's encomium to nature in *Zarathustra* is woven around Zarathustra's injunction to "Remain faithful to the earth" (Nietzsche, Za-1 1, §22.3) and is Nietzsche's attempt to develop a new philosophy in which the earth would be central, rather than the instrumental designs of humankind. As Zarathustra marvels, nature not only provides food for nourishment but does so in an aesthetic way, with beautiful and beckoning colours, tastes and aromas. However, human despoliation of the earth has caused it to suffer through exploitation and pollution: "The earth has a skin; and this skin has diseases. One of these diseases for example is called 'human being'" (Nietzsche, Za-1 2, §18). It is the *last human*, in particular, who threatens environmental health and the shrinking of the planet: "For the earth has now become small, and upon it hops the last human, who makes everything small" (Nietzsche, Za-1, §5). How-

---

18 This is exactly how Butterfield (2012) describes Sartre's social theory.

19 See Ferrando (2020, pp. 35, 60, 163) and Farrell Fox (2022, chapter 4).

ever much the earth is ravaged and polluted, however, it retains tremendous possibilities for growth and renewal – as Zarathustra says to the world-crazed fire hound: "the heart of the earth is of gold" (Nietzsche, Za-1 2, §19). In his more optimistic moments he imagines the earth as a vast health resort and gigantic tree of life but voices apprehension for the *Menschen-Erde* (human earth) that can manifest itself as a hell or a garden.[20]

Shapiro (2016), Richardson (2020) and Manschot (2021) view Nietzsche as a "philosopher of the earth" aligning with Deleuze, Guattari and Japanese Nietzscheans, such as Nishitani, who conceive the will to power as an abundant, creative, aesthetic and magical force in nature, perceiving in Nietzsche's Dionysian "fidelity to the earth", a celebratory form of pantheistic or panpsychic thinking.[21] In Sartre's case, the ecological movement in his thinking is less explicit than Nietzsche's, and is sometimes obscured by a residual or lurking anthropocentrism, but is nonetheless strongly evident from a careful or "retrievalist" reading of his philosophy.[22] Ally's book *Ecology and Existence* is by far the most systematic account of the ecological credentials of Sartre's thinking in which he retrieves Sartre from the charges commonly arraigned against him of "reflexive anthropocentrism", "heuristic exceptionalism", "categorical exclusivism" and "naïve instrumentalism" (Ally 2017, p. 26). These charges ignore certain passages in Sartre's work (particularly the *Critique*), misunderstand his dialectical method or, generally speaking, "overlook the subtlety and nuance of his thought and obscures the interpretive latitude that is the hallmark of his manner of thinking things through" (Ally 2017, p. 385). This is evident in readings of Sartre's *Nausea*, for instance, that often make note of his depictions of the plethoric unpleasantness of nature but don't recognize the phantasmagoric or vibrant conception of nature in his rich phenomenological descriptions and the metabolic way in which Roquentin is sunken in the world in a process of vicarious causation:

---

20 See Shapiro (2016, p. 13).

21 Ferrando aligns philosophical posthumanism with panpsychism – "Panpsychism is the recognition that this doubleness of privacy and relationality is not just a human predicament but the condition of all entities in the universe" (Ferrando 2020, p. 107).

22 Heidegger (1987) famously associated Nietzsche's will to power with a technological urge to "enframe" and dominate nature, beckoning humanity to become "masters of the earth". Kirsner assigns a similar instrumentalist logic to Sartre – "the logic of domination which views the world as there to be subdued and controlled and uses an instrumental, managerial form of rationality that regards oneself, others and the environment as objects to be quantified and manipulated" (Kirsner 1985, p. 225). Manschot (2021) and Ally (2017), however, present a strong rebuttal of this logic of instrumentality as supposedly being intrinsic to their thinking.

> The chestnut tree pressed itself against my eyes... The soft sound of the water... flowed into my ears and made a nest there. (Sartre 1965a: N, p. 183)

> My eyes never met anything but repletion. There were swarms of existences at the end of the branches, existences which constantly renewed themselves and were never born. The existing wind came and settled on the tree like a big fly; and the tree shivered... a thing-shiver flowed into the tree, took possession of it, shook it, and suddenly abandoned it. (Sartre 1965a: N, p. 190)

> I was inside; the black stump *did not pass*, it stayed there, in my eyes, just as a lump of food sticks in a windpipe. (Sartre 1965a: N, pp. 188–189)

Sartre develops this *affective dialectic* into philosophical form, describing "the real space of the world" as "hodological" space (Sartre 2018: EN, p. 415). Hodological space denotes the dynamic relation of transference between agent and environment. Through purposive activity the individual delineates a world in which objects function as "valences" that can be either positive and attractive or negative and unappealing. As we move towards or away from the valenced objects, we trace path or "vectors" which are a function of the strength or direction of the agent. These valences fluctuate and shift during the normal process of experience as the agent's goals change, and so vectors are constantly modified relative to changing valences. In this way, the subject generates a constantly fluctuating "field of force" in the course of interacting with the environment.[23] As Sartre explains: "around the Other, an entire space is grouped, and this space is made *with my space*; it is a regrouping, at which I am Present, and which escapes me, of all the objects that populate my universe" (Sartre 2018: EN, p. 350). Space is thus not striated and separated with strict divisions and boundaries, but an open smooth space in which forces converge and coalesce, causing ontological dislocations in one another through a state of transfer: "Originally, space is a qualitative and magnetic field, because it is traversed, because one pursues there, because one flees there, and one is always on the plane... of the unveiling of the new" (Sartre 1992: CM, p. 361). Unlike Euclidean space that has fixed coordinates, hodological space is generated relative to each subject and is constantly fluid, not merely an anatomical entity possessing clearly delineable boundaries, but coextensive with one's range of possibilities and field of force:

> [My body] is everywhere... [it] is always extended through the tool that it utilizes... it is at the end of the stick on which I am leaning... it is at the end of the telescope which shows me the stars; it is... in the entire house, because it is my adaptation to these tools. (Sartre 2018: EN, p. 436)

23 See Mirvish (1996, pp. 65, 69).

Like Nietzsche, Sartre theorizes nature posthumanistically in the active terms of *autopoiesis* and *panpsychism.*[24] Both philosophers, for instance, take a strongly non-Cartesian view of animals and thus elide differences used by forms of humanism to distinguish humans from the rest of nature through the exclusive possession of rationality or consciousness: "I do think that animals exist as consciousness, and as such, nihilate... I am not at all Cartesian on this point, I've always believed in the intelligence and passions of animals" (Sartre 2014: L-GM, p. 63). Indeed, "animal psychism" / panpsychism is a neglected aspect of Sartre's philosophy and one he returned to in interviews in the 1970s:

> Nature is not exclusively in-itself. A plant that is growing is no longer altogether in-itself. It is more complex. It is alive. (Sartre 1981: Schilpp, p. 40)
>
> I think animals have consciousness. In fact I have always thought so. (Sartre 1981: Schilpp, p. 28)

Nietzsche, of course, is well known for waking us from our anthropological sleep. More than any other philosopher, animals play a central role in his philosophical thinking, and no more so than in *Zarathustra*. Significantly, after conversing with the "higher men" in his cave, Zarathustra chooses to escape to talk to his animals: "Tell me, my animals: these higher men all together – do they perhaps not *smell* good. Oh clean fragrances around me! Only now do I know and feel how I love you, my animals" (Nietzsche, Za-1 4, §14.1). As the *Übermensch* appears only on the distant horizon, Zarathustra can only feel at home with his animals who provide him with philosophical solace. Alongside Darwin, Nietzsche was a significant voice of the nineteenth century in eliding the Christian or Enlightenment distinction between humans (as possessors of souls or rationality) and animals (as creatures of pre-programmed instinct and desire). The will to power is conceived as a natural motivation or creative force that applies to *all* organisms. "To understand what life is", Nietzsche writes, "the formula must apply to trees and plants as well as to animals" (Nietzsche, WP, §704) and "for each of these there is a small corner from which it measures, is aware, sees and doesn't see" (Nietzsche, LN, p. 129). Indeed, he insists, "[*t*]*he organic process constantly presupposes interpretations*" (Nietzsche, WP, §643). All organisms have "a biological intentionality or perspectivity", a minimal consciousness that forms "the foundation of the affects" (Nietzsche, BGE-1, §258).

24 See Ally (2017) and Plank (1998). Ally describes how Sartre's concept of praxis in the *Critique* (CDR) captures the idea of "autopoiesis in a nutshell" (Ally 2017, p. 444) and how, on the opening page of *Principles of Biological Autonomy*, Varela quotes Sartre's footnote (Sartre 1976: CRD, p. 34n.16) in its entirety.

He even extends this beyond animals and plants to include all organic and inorganic things: "the will to power... guides the inorganic world as well. Or rather, that there *is* no inorganic world" (Nietzsche, LN, p. 15). In Nietzsche's view, it is erroneous to think that we are different to other animals – what we think of as the protocols of "social morality" can be found everywhere in a simple form "even down to the deepest depths of the animal world" (Nietzsche, D, §26).

In broad terms, through their critique of Cartesian rationalism and mechanistic science, Nietzsche and Sartre in many ways anticipated a "new" (posthumanist) science that incorporated the idea of the will to power as a creative force (Nietzsche) or the principles of dialectical reason (Sartre).[25] Their smooth *relational ontology of forces* can be seen both to be consistent with, and to prefigure important developments in diverse fields of contemporary posthumanist scientific thinking, such as quantum physics, dialectical biology, epigenetics, living-systems theory, earth system science and niche-construction theory.[26] This demonstrates once more their ternary or sublimating philosophizing – instead of thinking of nature and human as a dualism of determinism and freedom, they are *both* posited covalently within an evolving and shifting freedom / determinism dialectic. Within this conception, Nature becomes art and creativity enmeshed within the "contingency of necessity". In Nietzschean terms, this may be seen as a sublimation of "two chambers of the brain... one to experience science and the other nonscience" (Nietzsche, HH-1, §251) or, as he suggests in *The Birth of Tragedy*, a new synthesis in the form of the "musical Socrates" (Nietzsche, BT, §15). He envisages a time when science is not abandoned but is cross-fertilized with art into a *gay science* in which "artistic energies and the practical wisdom of life will join the scientific thinking to form a higher organic system" (Nietzsche, GS-1, §113).

Twin Thinkers

Reading the "New Nietzsche" and "New Sartre" together as resolutely anti-metaphysical twin thinkers of a sublimating deconstructive logic allows us to travel with added insight along all of their winding ethical and ontological paths, and to fully appreciate the fecundities of their paradoxes and "sublimations" which presaged many axial themes in contemporary posthumanist thinking.

---

25 Nietzsche talks about a "new science" of forces (Nietzsche, WP, §619) and Sartre of the prospects of a "new" dialectical science (Sartre 1976: CRD, p. 34).
26 See Ally (2017, pp. 160 – 180).

Whether consciously or not, Sartre enriched and fertilized many Nietzschean insights, continuing and developing his decentring of the self in the intersectional terms of its "three ecologies", his relational ontology of forces, his affective phenomenology of "atmospheric" space, as well as his ethics of invention and play. Against common dissociative accounts that, through a process of vulgarization (usually of Sartre), stress the difference of their thinking, it is clear to see throughout the full trajectory and denouement of their writings the many multifarious threads that interweave, reinforce, and bind together their common mode of philosophizing. Their twin thinking can be seen, I suggest, as a form of quintessential Deleuzian "adsorbsion" that closely mirrors their ontology – a gathering of elements in a way that both forms a powerful assemblage or coalition and yet preserves something of the unique agential impetus of each.

Bibliography

Ally, Matthew C. (2017): *Ecology and Existence: Bringing Sartre to the Water's Edge*. Lanham.

Ally, Matthew C. (2020): "The Logics of the *Critique*". In: Matthew C. Eshleman & Constance L. Mui (eds.): *The Sartrean Mind*, London, pp. 362–375.

Badiou, Alain (2009): *Pocket Pantheon: Figures of Postwar Philosophy*. London.

Beauvoir, Simone de (1984): *Adieux: A Farewell to Sartre*. New York.

Betschart, Alfred (2020): "An Overview of the International Reception of Existentialism: The Existentialist Tsunami". In: Alfred Betschart/Juliane Werner: *Sartre and the International Impact of Existentialism*. London, pp. 1–41.

Boulé, Jean-Pierre (2005): *Sartre, Self-Formation and Masculinities*. London.

Butterfield, Elizabeth (2012): *Sartre and Posthumanist Humanism*. Oxford.

Daigle, Christine (2009): "Sartre and Nietzsche: Brothers in Arms". In: Benedict O'Donohoe/Roy Elveton (eds.): *Sartre's Second Century*. Newcastle, pp. 56–72.

Deleuze, Gilles (2004): *The Logic of Sense*. London.

Deleuze, Gilles/Guattari, Félix (1987): *A Thousand Plateaus*. Minneapolis.

Derrida, Jacques (1996): "Il courait mort: Salut, salut, Notes pour un courrier aux *Temps Modernes*": *Les Temps Modernes*, 587, pp. 7–54.

Eshleman, Matthew C. (2011): "What is it like to be free?". In: Jonathan Webber (ed.): *Reading Sartre: On Phenomenology and Existentialism*. London, pp. 31–47.

Farrell Fox, Nik (2003): *The New Sartre: Explorations in Postmodernism*. London.

Farrell Fox, Nik (2020): "Posthuman horizons: Contemporary responses to Sartre's philosophy". In: Matthew C. Eshleman/Constance L. Mui (eds.): *The Sartrean Mind*, pp. 487–500.

Farrell Fox, Nik (2022): *The Parallel Philosophies of Sartre and Nietzsche: Ethics, Ontology and the Self*. London.

Ferrando, Francesca (2020): *Philosophical Posthumanism*. London.

Flynn, Thomas (2014): *Sartre: A Philosophical Biography*. Cambridge.

Fretz, Leo (1980): "An Interview with Jean-Paul Sartre". In: Hugh J. Silverman/Frederick Elliston (eds.): *Jean-Paul Sartre: Contemporary Approaches to His Philosophy.* Pittsburgh.
Gerassi, John (2009): *Talking with Sartre: Conversations and Debates.* New Haven.
Guattari, Félix (2000): *The Three Ecologies.* London.
Heidegger, Martin (1987): *Nietzsche, 4 volumes.* San Francisco, 1979–1987.
Howells, Christina (1992): "Introduction". In: Christina Howells (ed.): *The Cambridge Companion to Sartre.* Cambridge, pp. 1–10.
Hutter, Horst (2006): *Shaping the Future: Nietzsche's New Regime of the Soul and Its Ascetic Practices.* Oxford.
Jaspers, Karl (1965): *Nietzsche: An Introduction to the Understanding of his Philosophical Activity.* Chicago.
Kirsner, David (1985): "Sartre and the Collective Neurosis of Our Time". In: Frederic Jameson (ed.): *Sartre After Sartre.* Yale French Studies, no. 68, pp. 206–225.
Manschot, Henk (2021): *Nietzsche and the Earth: Biography, Ecology, Politics.* London.
Meyer, Matthew (2018): "Nietzsche's Ontic Structural Realism?". In: Paul Katsafanas (ed.): *The Nietzschean Mind.* London, pp. 365–380.
Mirvish, Adrian (1996): "Sartre and the Problem of Other Embodied Minds". In: *Sartre Studies International* 2:2, pp. 65–84.
Noudelmann, François (2012): *The Philosopher's Touch: Sartre, Nietzsche and Barthes at the Piano.* New York.
O' Shiel, Daniel (2019): *Sartre and Magic: Being, Emotion and Philosophy.* London.
Perna, Maria (2007): "Spinozean Multitude Radical Italian Thought vis-à-vis Sartrean Existential Marxism". In: *Sartre Studies International* 13:1, pp. 35–61.
Plank, William (1998): *The Quantum Nietzsche: The Will to Power and the Nature of Dissipative Systems.* Lincoln.
Richardson, John (2020): *Nietzsche's Values.* Oxford.
Sartre, Jean-Paul (1957): *The Transcendence of the Ego.* New York.
Sartre, Jean-Paul (1965a): *Nausea.* Harmondsworth.
Sartre, Jean-Paul (1965b): *What is Literature?* New York.
Sartre, Jean-Paul (1967): *Words.* London.
Sartre, Jean-Paul (1968a): *Search for a Method.* New York.
Sartre, Jean-Paul (1968b): *The Communists and Peace.* New York.
Sartre, Jean-Paul (1972): *L'Idiot de la famille: Gustave Flaubert de 1821 à 1857, tome III.* Paris.
Sartre, Jean-Paul (1976): *Critique of Dialectical Reason: Theory of Practical Ensembles*, Vol.1. London.
Sartre, Jean-Paul (1979): "L'écriture et la publication" (interview with Michel Sicard). In: *Obliques*, 18–19, pp. 9–29.
Sartre, Jean-Paul (1981): "An Interview with Jean-Paul Sartre". In: Paul Arthur Schilpp (ed.): *The Philosophy of Jean-Paul Sartre.* La Salle, pp. 2–51.
Sartre, Jean-Paul (1984): *The War Diaries of Jean-Paul Sartre, November 1939 – March 1940.* New York.
Sartre, Jean-Paul (1988): *Mallarmé, or the Poet of Nothingness.* University Park.
Sartre, Jean-Paul (1990): "Carnet Midy". In: Michel Contat/Michel Rybalka (eds.): *Écrits de jeunesse.* Paris, pp. 443–497.
Sartre, Jean-Paul (1992): *Notebooks for an Ethics.* Chicago.

Sartre, Jean-Paul (2014): "Lettre à Gabriel Marcel de Jean-Paul Sartre". In: *BNF Revue*, no. 248. Paris, pp. 62–63.
Sartre, Jean-Paul (2018): *Being and Nothingness: An Essay on Phenomenological Ontology*. London.
Schrift, Alan D. (1995): "Putting Nietzsche to Work: Deleuze". In: Peter R. Sedgwick (ed.): *Nietzsche: A Critical Reader*. Oxford, pp. 250–275.
Schrift, Alan D. (1996): "Nietzsche's French Legacy". In: Bernd Magnus/Kathleen Higgins (eds.): *The Cambridge Companion to Nietzsche*. Cambridge, pp. 323–355.
Shapiro, Gary (2016): *Nietzsche's Earth: Great Events, Great Politics*. London.
Siemens, Herman (2018): "Nietzsche's Agon". In: Paul Katsafanas (ed.): *The Nietzschean Mind*. London, pp. 314–333.
Woodward, Ashley (2011): *Understanding Nietzscheanism*. Durham.

Elisa Reato

Sartre lecteur de Nietzsche

Abstract: Sartre as a Reader of Nietzsche. Sartre's famous formula, «nothing in my hands, nothing in my pockets,» reveals the notion of contingency. As every action is creation of the world and of man, the philosopher's central concern is an ontological as well as an ethical and a political one. Based on Sartrean argumentation, this article provides a chronological overview of Sartre as a reader of Nietzsche and offers an extensive sample of several ways that he differs from his predecessor. We focus on the ambiguity of man's condition in order to highlight Sartre's conception of commitment.

Proposer comme titre de cet article «Sartre lecteur de Nietzsche », c'est déjà suggérer, quelle que soit la thèse qu'on peut soutenir par ailleurs, qu'il existe au moins un rapport possible entre ces deux philosophes et que ce rapport vaut sans doute la peine d'être examiné. Mais parler de ce rapport ne va pas de soi pour deux raisons : la première est que l'essentiel de ce que Jean-Paul Sartre a écrit sur Friedrich Nietzsche nous manque. «Une grande, longue et très belle étude» (Beauvoir 2008 : EBS, p. 259), *dixit* Simone de Beauvoir, a été perdue[1]. La seconde raison est que Sartre se révèle très vite un héritier subversif. À cet égard, ses propos statuant sur les «âneries sur l'élite» (Gerassi 2011, p. 107) semblent explicites. Tel est le point de divergence le plus saillant entre les deux penseurs et les conséquences politiques de cette friction philosophique sont immédiates. On ne peut dissimuler, hélas, le mépris de Nietzsche pour la démocratie et le socialisme, alors que Sartre, dans l'après-guerre, travaille à faire advenir une «collectivité sans clivages» (Sartre 1948b : QL, p. 194)[2]. Or les choses ne sont pas aussi simples dans la mesure où la célèbre fin des *Mots* (1963) indiquerait, tout en professant une morale de l'homme moyen, la difficulté à se défaire de l'élitisme nietzschéen. Bref, tous les ponts qui rejoindraient l'œuvre de Nietzsche et l'œuvre de Sartre ne sont pas coupés et toute réévaluation d'un tel dialogue sur les cimes n'est pas intempestive.

---

1 Il s'agit des notes philosophiques de 1947–1948, écrites dans le prolongement des *Cahiers pour une morale*, qui auraient dû aboutir à une biographie.

2 Nietzsche dit clairement « racaille socialiste » : « L'injustice [...] se trouve dans la prétention à des droits ‹ *égaux* › ... » (Nietzsche 1993e, *L'Antéchrist*, p. 1097). Sur la démocratisation de l'Europe préparant à l'esclavage, cf. Nietzsche 1993i, *Par-delà le bien et le mal*, p. 689–690.

https://doi.org/10.1515/9783110760163-004

Nietzsche est un des rares philosophes dont le nom évoque un style de pensée et de vie ; il n'a jamais existé d'école nietzschéenne à proprement parler, mais l'œuvre de Nietzsche est une référence polémique pour les artistes, les littéraires, les politiques tout autant que pour les philosophes, un appel à renverser la tradition morale d'origine platonicienne et la civilisation occidentale chrétienne. Il en va de même pour l'œuvre de Sartre. Dans l'œuvre nietzschéenne, l'autobiographie affleure sans cesse, la philosophie est vécue, ressentie avant d'être pensée ou écrite et ne donne pas l'occasion de traiter séparément de l'homme et de l'œuvre[3]. Quant à Sartre, son projet philosophique, universalisant, se greffe sur le projet initial qui consiste à rendre compte de l'existence singulière, d'un vécu, dans l'écriture romanesque ; ainsi a-t-il pu affirmer que «vie et philo ne font plus qu'un» (Sartre 2008a : L-Cast, p. 39). Sans s'arrêter aux débats dans lesquels Nietzsche n'a pu éviter de s'engager et auxquels Sartre aurait pu souscrire (anti-antisémitisme, anti-nationalisme, anti-darwinisme), rappelons-nous quelques ruminations communes aux deux penseurs : le primat de la philosophie sur la science, le caractère non fixé de l'homme, la question du style, etc. De Nietzsche, Sartre recueille deux enjeux fondamentaux, l'anti-christianisme[4] et l'anti-humanisme[5], plus précisément pour les deux penseurs la mort de Dieu est aussi la mort de l'Homme, autrement dit s'il y a une exigence dans les deux œuvres, elle est celle pour l'homme de faire sens, d'inventer les valeurs là où plus rien ne lui est dicté. Or, il faut préciser que tout au long de la pensée sartrienne les allusions à Nietzsche ont un aspect critique. En effet, la divergence est radicale sur les terrains de l'ontologie, de la morale et de la politique. Ainsi nous voudrions suivre une approche chronologique du corpus sartrien, car cela nous permettra à la fois de retracer la continuité des références à la pensée nietzschéenne et d'en montrer la déprise, ou comment Sartre se libère tour à tour des cadavres de Dieu, de l'Homme et... de Nietzsche.

---

3 Cf. Nietzsche : « Mes écrits *ne* parlent *que* de mes victoires : [...] *ego ipsissimum* » (Nietzsche 1993d, *Humain, trop humain*, p. 699).

4 On rappellera le sous-titre de *L'Antéchrist*, « Imprécation contre le christianisme », et que l'œuvre de Sartre fut mise à l'Index en 1948.

5 Cf. les propos de Nietzsche sur l'homme, « *sublime avorton* » (Nietzsche 1993i, *Par-delà le bien et le mal*, p. 612), « plus singe qu'un singe » (Nietzsche 1993a, *Ainsi parlait Zarathoustra*, p. 291). Cf. aussi la critique de l'humanisme dans *La Nausée* (Sartre 2009c : N, p. 138–139).

Retour sur une proximité d'apparence

Dans les années 1920, Sartre fut un grand lecteur de Nietzsche[6], néanmoins il ne le lit pas en historien de la philosophie mais bien en philosophe[7]. En témoigne par excellence le *Carnet Midy* (1924), où Sartre révèle sa volonté de penser dans le discontinu et son goût pour les formules, mais à propos de Nietzsche il note : «C'est un poète qui a eu la mauvaise fortune d'être pris pour un philosophe. Ses livres traduisent les rêves d'un conquérant, d'un homme né pour être conquérant mais malade et faible et qui se console par l'imagination» (Sartre 1990a : CMid, p. 471). Notons toutefois que ces critiques ne sont pas à prendre au mot : à l'École normale supérieure, il conclut un pacte de surhumanité avec Paul Nizan et pourtant c'est bien la figure du surhomme que les deux amis récusent[8]. Cette fascination pour la figure du surhomme qui enseigne l'autodépassement permet à Sartre de se refonder contre un caractère «lâche et douillet» (Sartre 2014 : L-Cast, p. 10). C'est ainsi qu'il adopte un «ton Frédéric» et une «morale nietzschéenne de la joie» (Sartre 2010a : Carn, p. 442, 370), nommée ailleurs du «Grand Désir» (Sartre 1980a : Nizan, p. 168)[9]. Ce ton tranchant ne va pas sans la «confiance en [s]on étoile» (Sartre 2010a : Carn, p. 356) et fait allusion au protagoniste d'un roman inachevé, *Une défaite* (1927), qui s'inspire de l'Idylle de Tribschen. Sartre s'incarne en Frédéric, qui rêve de gloire et chemine «avec l'étrange illusion d'être au sein d'un peuple inconnu» (Sartre 1990d : Déf, p. 209). Et pourtant ce «héros, dur et nietzschéen» (Sartre 2010a : Carn, p. 278)[10] aperçoit une «secrète ‹ faiblesse ›» (Beauvoir 2018a : MJFR, p. 322) de l'homme et des choses, «la tristesse et l'ennui» (Sartre 1990e : L-EA, p. 434). Il trouve son maître en un écrivain et compositeur célèbre et s'éprend de sa femme (Cosima),

6 Au début des années 1920, Sartre lit *Also sprach Zarathustra* (cf. Sartre 2010c : CLut, p. 919) et, en 1926, *Die Geburt der Tragödie* et *Menschliches, Allzumenschliches* (cf. Dassonneville 2018, p. 279). Michel Contat et Michel Rybalka signalent aussi les biographies de Nietzsche écrites par Daniel Halévy et Charles Andler (cf. Sartre 1990d : Déf, p. 192–193).

7 Pour Sartre, étudier un philosophe c'est « l'épuiser », penser « à ses dépens » (Sartre 2010a : Carn, p. 467), c'est-à-dire réfléchir dans ses perspectives pour mettre au jour ses idées originelles. C'est ainsi qu'il décrit son rapport à Husserl.

8 Cf. Sartre 1980a : Nizan, p. 143. Cf. aussi : « La Morale fut longtemps celle du Surhomme » (Sartre 1990c : SemSca, p. 142). Sur les canulars contre les élites cf. Gerassi 2011, p. 108.

9 En 1926, selon Sartre, il faut étouffer la pitié et la mélancolie afin d'« acquérir la plus grande puissance possible », être « *absolument libre* » et connaître « la *vraie joie* » (Sartre 2014 : L-Cast, p. 30). Cf. Nietzsche 1993a, *Ainsi parlait Zarathoustra*, p. 459–461.

10 « Ô mes frères, je place au-dessus de vous cette table nouvelle : DEVENEZ DURS ! » (Nietzsche 1993a, *Ainsi parlait Zarathoustra*, p. 453). Quand il imposait ses idées, Sartre été surnommé le « lamentable Frédéric » (Beauvoir 2018b : FA, p. 384).

mais est déçu par son ami faible et son amour malheureux ; il s'impose alors la règle de la Joie, convaincu que ses échecs seront des victoires grâce à l'écriture d'une œuvre nouvelle. On le voit, loin d'une tentation de se choisir un maître, l'identification du jeune Sartre au jeune Nietzsche passe par le défi à la figure du grand écrivain sous la forme d'un serment de devenir à son tour un Grand Homme. On sait que Nietzsche grava sur un rocher la devise grecque « deviens ce que tu es », pour commémorer son illumination du Retour Éternel, et que le jeune Sartre avait « la tête farcie » de « cette liberté du grand homme – libre pour son destin» (Sartre 2010a : Carn, p. 357)[11]. Ce qui le séduit dans sa « période d'incubation dans la surhumanité » (Sartre 2010a : Carn, p. 365), c'est la figure de l'« homme seul» (Sartre 1976 : ASD, p. 176) qui, en pensant en dehors de toute autorité et contre lui-même, dépasse sa contingence originelle vers l'*ens causa sui* que constitue la vie de grand artiste. Non seulement Sartre attaque le sérieux du grand écrivain, mais il s'arrache à la fascination pour le surhomme par une profession de foi dans l'œuvre d'art. Bien que le roman soit rythmé par des thèmes nietzschéens (la grande santé, le refus de l'humanisme, etc.), à l'impotent solitaire descendu de ses montagnes la solution qui se profile, exposée aussi dans *Er l'Arménien* (1928), est le salut par la création. À partir de ces critiques, jaillit une conception de la liberté comme création de soi hors la loi. La sentence de Prométhée, « Quand les Dieux seront vaincus, il n'y aura plus de mal sur terre» (Sartre 1990b : ErA, p. 322), semble faire l'écho de l'aphorisme du *Gai Savoir*, « Bien et mal sont les préjugés de Dieu » (Nietzsche 1993 h, *Le Gai Savoir*, p. 162), mais dans l'idéal de vie bonne et belle proposé par Er, l'accent est mis sur le choix, c'est-à-dire sur la liberté et l'entière responsabilité de l'homme dans ses actes – « Rien n'est beau sauf ce que font les hommes » (Sartre 1990b : ErA, p. 330). Dieu sera le grand absent du *Diable et le Bon Dieu* (1951), où Gœtz « seul et [s]on propre ciel» (Sartre 2010c : CLut, p. 919) décidera du Mal et inventera le Bien. On retrouvera dans *L'Existentialisme est un humanisme* (1946) cette idée de la responsabilité de la liberté lorsque Sartre compare le choix moral à la construction d'une œuvre d'art. Peut-on soutenir qu'il ne fait que reprendre la conception de la liberté de Nietzsche ? Une telle objection manque l'originalité de la conception sartrienne de la liberté qui entend redéfinir celle-ci indépendamment de l'opposition du volontaire et de l'involontaire.

11 Sur le « serment » de Nietzsche cf. Sartre 1990d : Déf, p. 212. À cette époque, Sartre s'intéresse aux récits de la vie des grands hommes pour y trouver une prophétie de la sienne, c'est son « illusion biographique » (cf. Sartre 2010a : Carn, p. 363 et 2014 : L-Cast, p. 14). C'est la période d'optimisme, de 1921 à 1929, où il est « mille Socrates », c'est-à-dire potentiellement tous les grands hommes possibles.

Ni foudre ni joie, seulement un vertige devant la liberté

Dans les années 1930, à l'aide de la phénoménologie, Sartre fait *tabula rasa* des idéologies et de l'ordre existant. La théorie de l'homme seul, qui pense le cours des choses sans ancrages et sans le secours d'autrui, débouche sur la critique de l'humanisme dans *La Nausée* (1938), où Sartre substitue à l'ontologie nietzschéenne de la surabondance, incluant tout autant la joie dionysiaque que les terreurs titanesques, une ontologie des mollesses de la contingence[12]. «Quels sommets n'attendrais-je point si ma *propre vie* faisait la matière de la mélodie» (Sartre 2009c : N, p. 48), se demande Roquentin, qui se déprend des illusions qui s'interposent entre un individu et sa liberté, mais est solitaire, mélancolique et ne fait pas preuve de volonté. Lisons un passage du philosophe plus musical que les musiciens : «Mon corps tout entier, que *demande-t-il* en fin de compte à la musique ? [...] Ma mélancolie veut se reposer dans les cachettes et dans les abîmes de la *perfection*» (Nietzsche 1993 h, *Le Gai Savoir*, p. 238). Impossible de ne pas penser à l'air de jazz qui fait disparaître la nausée : «Elle n'existe pas, puisqu'elle n'a rien de trop [...]. Elle *est*» (Sartre 2009c : N, p. 206). On le voit, pour Sartre, la musique échappe à l'existence, alors que, pour Nietzsche, elle renvoie à la pulsation de vivre. La découverte de la contingence est opposée à la volonté de puissance et à la lutte pour la vie prônées par des «imbéciles» (Sartre 2009c : N, p. 158) et se trouve déjà dans *Empédocle* (1926)[13]. «[La contingence] ne s'exprime pas par des mots. Elle est vie. La nature est une faible vie. Et c'est pourquoi elle n'a pas de sens. Laisse là ceux qui disent que ‹ la vie est volonté de puissance ›» (Sartre 2016a : Emp, p. 30). Sans doute, Sartre conteste l'esprit de sérieux «avec autant de vigueur que Nietzsche, et pour des raisons analogues ; elles allégeaient le monde en le projetant dans l'imaginaire et [lui] permettaient de le tenir à distance» (Beauvoir 2018b : FA, p. 366). Mais, dans *Par-delà le bien et le mal*, Nietzsche écrit que «nous aurons besoin un jour d'une nouvelle espèce de philosophes et de chefs» (Nietzsche 1993i, *Par-delà le bien et le mal*, p. 647) pour renverser les valeurs et se défaire de l'erreur démocratique. Dans le jargon sartrien, les chefs sont les «salauds», c'est-à-dire ceux qui refusent l'évidence de

---

12 Sur la vision du monde où le Beau se confonde avec le nécessaire pour former un univers d'idéalités cf. le bonheur de Nietzsche (Nietzsche 1993 h, *Le Gai Savoir*, p. 80) et le dégoût de Roquentin (Sartre 2009c : N, p. 147) devant la mer.

13 Ce texte est précédé par un *Chant de la contingence*, dont le titre évoque « Le chant de la mélancolie » d'*Ainsi parlait Zarathoustra*. Juliette Simont rend compte de façon remarquable de la référence de Sartre à l'*Empédocle* de Nietzsche (cf. Simont 2016b, p. 79–100).

leur propre contingence : «Tout membre de la classe dominante est homme de droit divin» (Sartre 2003 : MatR, p. 137). Dans «L'Enfance d'un chef» (1939), «celui qui sait commander, celui dont la nature a fait un ‹ seigneur ›» (Nietzsche 1993f, *La Généalogie de la morale*, p. 827) vanté par Nietzsche, est méprisé par Sartre, car le *salaud* s'arroge le droit d'humilier, d'exploiter, de tuer. Si bien que, dans «Érostrate» (1939), Paul Hilbert est le personnage caricatural de l'homme seul, qui regarde les hommes d'un balcon du sixième étage, car il se croît «de l'espèce des revolvers, des pétards et des bombes» (Sartre 2009a : Ero, p. 269). L'aphorisme d'*Ecce homo* est connu, «je ne suis pas un homme, je suis de la dynamite» (Nietzsche 1993c, *Ecce homo*, p. 1191), mais cet antihéros sartrien est un pseudo-surhomme narcissique et impuissant qui n'accomplit pas son crime anti-humaniste et se fait passer à tabac.

L'expérience de la mobilisation, de l'emprisonnement et de la vie collective au Stalag XII D introduisent une profonde crise existentielle et éthique dans la vie de Sartre. Sa seule certitude est qu'il ne mourra pas avant d'avoir fait son œuvre, à l'instar de Nietzsche : «Ma passion et ma compassion – qu'importe d'elles ? Est-ce que je recherche *le bonheur* ? Je recherche mon *œuvre* !» (Nietzsche 1993a, *Ainsi parlait Zarathoustra*, p. 545 ; cf. Sartre 2010a : Carn, p. 361). Va-t-il dès lors faire sienne la devise chérie par Nietzsche, «deviens ce que tu es» ? Non, dans les *Carnets de la drôle de guerre* (1939–1940), il écrit qu'il s'était placé en philosophie «sous le signe de Descartes parce que Descartes est un penseur à explosions. [...] On ne peut devenir que ce qu'on n'est pas encore. Donc on ne peut pas devenir ce qu'on est» (Sartre 2010a : Carn, p. 368). C'est ici que Sartre, lecteur de Nietzsche, force la pensée de Nietzsche, voire la fracture. À la volonté de puissance affirmative nietzschéenne se substitue une faiblesse ontologique de l'homme, un «*manque* par nature» (Sartre 2010a : Carn, p. 516) qui pourtant est totalement responsable de sa vie. C'est bien en tant qu'il est manque que le pour-soi fait paraître l'en-soi, donc la corrélation est irréductible. Il est ainsi particulièrement éclairant de rapprocher les déclarations de Nietzsche sur le mensonge – où il affirme que nous avons parfois besoin d'un «masque» pour vivre (Nietzsche 1993i, *Par-delà le bien et le mal*, p. 594) – et l'élaboration de Sartre d'une morale valorisant l'authentique assomption de la situation. Accepter de n'être pas afin d'exister authentiquement, vouloir la défaite ontologique de la contingence permet de gagner la liberté, tel est le ressort du «qui perd gagne ». En ce sens, il faut vouloir cette tension vers l'être même avec l'échec qu'elle comporte. Ce n'est pas en acceptant ce qui lui arrive, mais en assumant la tragique ambiguïté de sa condition que l'homme manifeste l'existence comme un bonheur et le monde comme source de joie. Il en découle une dialectique (pas encore méthodique) entre l'aspect objectif et l'aspect subjectif. Puisqu'exister c'est *se faire* manque d'être et se jeter dans le monde pour le revêtir de signification humaine, «on peut considérer comme des sous-hommes ceux qui s'emploient à retenir ce mouvement originel» (Beauvoir 2013 : PMA, p. 56), ce perpétuel mouvement de néantisation (de soi et du monde) qui fonde la liberté de la conscience. On comprend que Nietzsche, pour Sartre et Beauvoir, est un *sous-homme*, un *salaud* qui suscite le mépris, car il élude la passion de la liberté et de ses exigences, mais il n'en demeure pas moins responsable. Sa ruse c'est la doctrine du Retour Éternel, une tentative d'imprimer au devenir le caractère de l'être, alors que la liberté fait passer l'être à l'existence dans un mouvement sans cesse dépassé[14]. Comme le souligne Sartre, la liberté est condamnée à se choisir perpétuellement. Quant à son autodiagnostic, «la guerre a vraiment divisé ma vie en deux» (Sartre 1976 : ASD, p. 180), peut-on penser qu'il y ait un Sartre d'avant la guerre, dandy nietzschéen, penseur anarchiste de la liberté individuelle, et un Sartre d'après la guerre, chantre de l'engagement politique ? Oui, mais Sartre continue à s'intéresser à la pensée de Nietzsche.

Libre et responsable devant les hommes : *Ecce homo*

C'est dans le sens d'un enracinement résolu dans l'ici-bas terrestre que Sartre parle d'un «humanisme nietzschéen» : «La saveur de la pensée nietzschéenne vient de ce qu'elle est profondément et uniquement terrestre. Nietzsche est un athée qui tire durement et logiquement toutes les conséquences de son athéisme» (Sartre 1947 : NouvM, p. 178). À la fois fidèle et déloyal envers Nietzsche, Sartre encadre *L'Être et le néant* (1943) avec deux références au philosophe au marteau, l'une explicite au début de l'ouvrage – «Nous nous sommes une fois dépris de ce que Nietzsche appelait ‹ l'illusion des arrière-mondes ›» (Sartre 2012a : EN, p. 12) –, l'autre implicite à la fin précisant que l'ontologie a révélé «l'origine et la nature de la *valeur*» (Sartre 2012a : EN, p. 674), soit le désir constitutif de l'existence humaine comme désir d'être Dieu[15]. Dès lors, Sartre tire les conséquences de l'idée fondamentale de contingence sans pour autant prédire aucun surhomme, car son cri, «Dieu, s'il existe, est contingent» (Sartre

---

14 La volonté de puissance et la pensée du retour éternel s'impliquent l'une l'autre, cf. Nietzsche 1993a, *Ainsi parlait Zarathoustra*, p. 454–455 et Nietzsche 1993 h, *Le Gai Savoir*, p. 202. Sur la critique du libre-arbitre cf. Nietzsche 1993 g, *Le Crépuscule des idoles*, p. 979–980.

15 Cf. Nietzsche 1993a, *Ainsi parlait Zarathoustra*, p. 305–306 et Nietzsche 1993 f, *La Généalogie de la morale*, p. 800–801.

2012a : EN, p. 117), annonce l'homme[16]. Voilà ce qui reste une fois rangé «l'impossible Salut au magasin des accessoires» (Sartre 2010b : Mots, p. 139). En d'autres termes, Sartre enracine les valeurs dans la liberté du pour-soi et son manque d'être. Ainsi, faut-il reconnaître le caractère subjectif des valeurs. Aussi, l'identification de la conscience à la liberté suppose une théorie de l'imaginaire et une philosophie de la temporalité. Par le fait qu'il est libre, l'homme fait exister le monde comme horizon de ses possibles, car il se modifie en modifiant le monde. En effet, l'hypothèse de l'irréalité de la valeur permet de dépasser l'aporie causée par la dichotomie entre faits et valeurs. Dans ce cadre, la notion de «situation» montre l'enchevêtrement entre l'irréel et le donné, les valeurs et les actions, ainsi que la résistance des choses et d'autrui. Notons que l'objection selon laquelle la liberté sartrienne serait dépourvue de toute limite oublie que transcendance et facticité font couple : la liberté s'exerce dans un monde donné sans qu'elle l'ait choisi. Comme le révèle l'angoisse éthique, les valeurs trouvent leur fondement ontologique dans la réalité-humaine en tant que libre transcendance qui dépasse ce qui est vers ce qui n'est pas et auquel la liberté confère du même coup une valeur. Si la liberté est l'unique fondement des valeurs, alors rien ne justifie l'adoption de telle ou telle échelle de valeurs et l'homme, en tant qu'être par qui les valeurs existent, est injustifiable.

Tout en soutenant qu'aucune des dimensions temporelles n'a de priorité ontologique sur les autres, Sartre met l'accent sur le présent, qui est le «creux de non-être indispensable à la forme synthétique totale de la temporalité» (Sartre 2012a : EN, p. 177), le moment de la liberté, du choix et de l'action. On songe à la leçon de Zarathoustra – «Je t'ai appris à dire ‹ Aujourd'hui ›» (Nietzsche 1993a, *Ainsi parlait Zarathoustra*, p. 459) –, mais il faut préciser qu'« [aujourd'hui] n'existe que comme engagement et parti pris» (Beauvoir 2013 : PMA, p. 98). La nature du «pour-soi» est «de devoir être ‹ un creux toujours futur ›» (Sartre 2012a : EN, p. 163), donc on ne peut éviter de vivre le présent à travers un projet et aucun projet n'est purement contemplatif. Le circuit de l'ipséité, comme possibilisation transcendant le présent existant, est la racine même du temps : le choix d'une quelconque conduite, parce qu'elle vise ce qui n'est pas encore, décide du futur qui constitue le présent comme tel, et confère au donné la signification du passé. Impossible de se placer sur le terrain d'une objectivité inhumaine et par conséquent fausse ; aucun surplomb, aucun survol n'est possible, on ne sort pas de l'humain comme on ne s'évade pas de ce monde. Nous voyons donc quel est le vice fondamental de Nietzsche selon Sartre. Beauvoir appelle «esthétique» cette attitude d'oisif : «celui qui l'adopte prétend

16 Cf. l'annonce du surhomme, Nietzsche 1993a, *Ainsi parlait Zarathoustra*, p. 293–294.

n'avoir avec le monde d'autre rapport que celui d'une contemplation détachée ; hors du temps, loin des hommes, il se pose en face de l'histoire, à laquelle il ne croit pas appartenir, comme un pur regard» (Beauvoir 2013 : PMA, p. 94). On remarque aussitôt qu'une telle attitude égalise toutes les situations : en fait elle reste une position de repli, une manière poétique de fuir le présent.

Quoi qu'il en soit, la pièce *Les Mouches* (1943) est truffée de références nietzschéennes, comme l'a montré de façon décisive Jean-François Louette, mais Sartre défend des thèses inverses. Oreste aux pieds légers[17] est désireux de se lester d'un lourd forfait[18] ; il se décide pour le crime, car il connaît le secret des Dieux et des rois depuis que la liberté a fondu sur lui comme la foudre[19] ; il ressent moins de la vengeance que de la joie[20], parce qu'il sait que son acte est bon et accepte d'inventer son chemin[21]. «La vie humaine commence de l'autre côté du désespoir» (Sartre 2005c : Mou, p. 66)[22]. Pour Sartre, comme pour Nietzsche, la politique suppose l'oubli de Dieu : «J'ai perdu quelqu'un qui m'était cher» (Sartre 2005a : DBD, p. 498 ; cf. Nietzsche 1993 h, *Le Gai Savoir*, p. 131–132), dira Gœtz avant de se rallier à la révolte populaire, mais Oreste demeure trop léger car son geste n'entraîne qu'un salut personnel. *A posteriori*, Sartre jugea sévèrement son héroïsme aristocratique, qui montre l'aporie éthique et politique à laquelle conduit le nietzschéisme en refusant de proposer une table de valeurs nouvelles bien définies[23]. En somme, Oreste est un exemple d'engagement négatif[24], soit il faut inverser sa conduite pour exercer la liberté. Cette stratégie est employée aussi dans *Huis clos* (1944), où Sartre, suivant la leçon de Zarathoustra, s'attache à liquider le mythe chrétien de l'enfer par la dérision et la perversion[25]. La célèbre formule, «L'enfer, c'est les Autres », montre que les autres sont l'enfer que si on se met dans leur dépendance. On aurait donc tort,

17 Cf. Sartre 2005c : Mou, p. 12 et Nietzsche 1993a, *Ainsi parlait Zarathoustra*, p. 314.

18 Cf. Sartre 2005c : Mou, p. 40 et Nietzsche 1993c, *Ecce homo*, p. 1123.

19 Cf. Sartre 2005c : Mou, p. 53 et Nietzsche 1993a, *Ainsi parlait Zarathoustra*, p. 512.

20 Cf. Sartre 2005c : Mou, p. 51 et Nietzsche 1993a, *Ainsi parlait Zarathoustra*, p. 359–361.

21 Cf. Sartre 2005c : Mou, p. 53, 65 et Nietzsche 1993a, *Ainsi parlait Zarathoustra*, p. 436–437. Cf. aussi Louette 1996, p. 13–70, qui développe amplement ces thèmes.

22 Cf. Nietzsche 1993i, *Par-delà le bien et le mal*, p. 578–579, où l'auteur reprend l'image de l'océan inexploré de la connaissance, de l'angoisse et de la hardiesse des aventuriers qui s'y risquent.

23 Cf. Gerassi 2011, p. 209. En 1940, Sartre avait déjà dénoncé la conduite d'Oreste, cf. Sartre 2010a : Carn, p. 623. Cf. Nietzsche 1993a, *Ainsi parlait Zarathoustra*, p. 322–324.

24 L'expression est de Jean-François Louette.

25 Cf. Louette 1996, p. 90 sq., où l'auteur montre qu'en mettant en scène la luxure Sartre torture le public chrétien dont Nietzsche a montré le blâme contre le sexe (cf. Nietzsche 1993d, *Humain, trop humain*, p. 520–523).

dit Beauvoir, à « assimiler l'existentialisme à un solipsisme qui exalterait, comme Nietzsche, la seule volonté de puissance ». Cela reviendrait à concevoir l'individu comme un tyran qui chercherait à imposer à autrui les valeurs dont il est créateur. Une telle conception débouche sur « un conflit des volontés adverses, enfermées dans leur solitude », mais ne concerne pas la morale existentialiste qui, loin d'être une doctrine formelle, indique un contenu à cette liberté qu'elle veut engagée : « se vouloir libre, c'est aussi vouloir les autres libres » (Beauvoir 2013 : PMA, p. 91–92). C'est cet aspect positif de la moralité que souligne le *faire* de la liberté, autrement dit avoir à être son être pour le pour-soi signifie *vouloir* la liberté, pour soi-même et pour les autres hommes, et tenter de la conquérir.

Pour un élitisme des égaux

Reconquérir l'Histoire, le social, la politique, ce sont les chemins de la liberté. Mathieu, le protagoniste de la trilogie romanesque, ne se laisse pas tenter par des morales mensongères dont l'erreur est de placer les valeurs hors de l'existence humaine alors que c'est la pluralité des hommes qui les fait surgir. C'est ce qu'énonce l'humanisme décrit dans la fameuse conférence de 1945 qui engendre une philosophie de l'action, de la lutte et de la solidarité. Délaissé sur terre, l'homme se définit « par rapport à un engagement » (Sartre 2012b : EH, p. 67). De par ce souci de l'immanence Sartre prend ses distances à l'égard des rêveries de lévitations et de lointains nietzschéens (cf. Nietzsche 1993c, *Ecce homo*, p. 1148–1149). Dans ce cadre, le désespoir auquel l'homme est condamné constitue sa chance pour se faire humain, car l'individu se fait en choisissant sa morale et le bon choix est celui qui conserve la liberté comme condition de tous les choix. Ceci est le socle de l'éthique sartrienne et définit les luttes avec tous ceux qui ne sont pas justifiés et veulent créer leur monde (les ouvriers, les colonisés, etc.). Sans doute a-t-il fallu l'expérience de la guerre, pour que Roquentin apparaisse à Sartre comme privé de volonté, car « la volonté a besoin du monde et de la résistance des choses » (Sartre 2010a : Carn, p. 316) ; il faut « du plomb dans la tête » pour que Mathieu se reproche son mépris en observant ses camarades « inhumains, trop humains, au-delà et en deçà du désespoir : des hommes » (Sartre 2009b : MAme, p. 1305, 1205). D'où l'exigence de se faire historique, ou l'appartenance objective à son époque : « Nous ne voulons rien manquer de notre temps : peut-être en est-il de plus beaux, mais c'est le nôtre ; nous n'avons que *cette* vie à vivre, au milieu de *cette* guerre, de *cette* révolution peut-être » (Sartre 1948a : PréTM, p. 13). À l'inverse, Nietzsche écrit : « Quelques-uns naissent d'une façon posthume. [...] Je serais en contradiction absolue avec moi-même, si je m'attendais aujourd'hui déjà à trouver des oreilles, *à trouver des mains* pour *mes* vérités» (Nietzsche 1993c, *Ecce homo*, p. 1145). Faut-il réduire Sartre aux prises de position sociales et politiques de l'intellectuel total qui se veut le porte-voix du peuple à venir ? La question posée, on le voit, contient déjà sa réponse : de fait, c'est en assumant le conflit qu'on annonce l'humain, c'est-à-dire la liberté comme principe d'une auto-définition infinie de l'homme comme singularité irréductible. «Il faut donc vouloir l'homme. Non pas le découvrir mais l'inventer» (Sartre 1983 : CM, p. 23).

On sait que Nietzsche s'oppose aux thèses marxistes[26] et que l'enquête généalogique se veut une histoire naturelle des origines de la morale[27]. Or, tout au long des *Cahiers pour une morale* (1947–1948), Sartre trace l'histoire de la morale en termes existentialo-marxistes afin de cerner le moment de la morale, «*le moment de l'homme*» (Sartre 1983 : CM, p. 430). À l'*ens causa sui* se substitue une logique de l'action montrant que l'existence d'autrui comme liberté définit ma situation et est en même temps la condition de ma propre liberté. Dans *Vérité et existence* (1948), Sartre montre que la vérité se révèle à l'action et au projet qui l'anime, elle «disparaît avec l'homme» (Sartre 2008b : VérE, p. 19). Cet homme n'est pas un individu isolé, car la vérité comme réalité concrète est la manifestation des dévoilements du monde de tous les hommes. On comprend du coup le sens de la formule de Ponge adoptée par Sartre, «l'homme est l'avenir de l'homme» : puisque chacun dépend des autres, ce sont les autres hommes qui m'ouvrent l'avenir. Aussi est-il clair que le concept de solidarité rend compte de la formule «rien dans les mains, rien dans les poches» : la liberté étant en sursis, il faut renoncer non seulement aux arrière-mondes nietzschéens mais aussi à ceux de l'avenir historique. Cela ne signifie pas seulement que la liberté de chacun est porteuse du sens du monde, mais qu'on fait *don* du monde à la liberté des autres sans l'imposer.

Dans *Saint Genet* (1952), Sartre s'attache à montrer que la séparation abstraite du Bien et du Mal exprime l'aliénation de l'homme. «Ou la morale est une faribole ou c'est une totalité concrète qui réalise la synthèse du Bien et du Mal. [...] On comprendra, j'espère, qu'il ne s'agit nullement d'un ‹ au-delà › nietzschéen du Bien et du Mal mais plutôt d'une *Aufhebung* hégélienne» (Sartre 2011 : SG, p. 211–212 note). De fait, «en cas d'*impossibilité*, le choix du Bien conduit à renforcer l'impossible, il faut choisir le Mal pour trouver le Bien ». Parier pour le Mal quand le Bien est aliéné, c'est «l'élément *tragique* de l'Histoire» (Sartre 1983 : CM, p. 420–421). Mais la Beauté comme signature du Mal sombre dans

26 Cf. Nietzsche 1993i, *Par-delà le bien et le mal*, p. 708–709 et Nietzsche 1993 g, *Le Crépuscule des idoles*, p. 1015.

27 Cf. Nietzsche 1993i, *Par-delà le bien et le mal*, p. 631–632 et Nietzsche 1993f, *La Généalogie de la morale*, p. 773–774.

une théâtralité de gestes inefficaces. «La vérité de Nietzsche, c'est le professeur de philosophie dans un État bourgeois et militaire» (Sartre 1983 : CM, p. 108). Plus encore, Sartre épingle l'antinomie du Retour Éternel par lequel Nietzsche s'est voué à devenir ce qu'il est déjà et à détruire ce qu'il ne peut empêcher d'être. Si l'on peut vouloir ce qui n'est pas – agencer des moyens et transformer le présent en vue d'une fin future –, vouloir devenir ce que l'on est déjà en niant le déroulement temporel de l'acte c'est faire tomber la volonté dans l'imaginaire et réduire la *praxis* à l'impuissance, «car l'acte transforme le possible en réel, mais Nietzsche [...] dissout la totalité de l'être en possibilité» (Sartre 2011 : SG, p. 386–387). La question est ici politique avant d'être morale : il faut aboutir à ce que toute oppression soit abolie. Selon Sartre, tout système de valeurs est complice d'un système d'exploitation et d'oppression. À l'instar de Marx, il pense qu'il faut dépasser les inégalités produites par l'économie capitaliste. Le modèle d'une révolte éthique, où l'homme transforme non seulement lui-même mais aussi le monde, est donné par Gœtz, d'abord «bouffon» (Sartre 2005a : DBD, p. 423) du Mal puis faussaire du Bien, qui se libère de la supercherie du Bien et du Mal en intégrant la violence dans la lutte pour l'émancipation des opprimés.

On retrouve dans la *Critique de la raison dialectique* (1960) cette dénonciation des morales des valeurs et du devoir. Plus encore, selon Sartre, Nietzsche est pire que Kant, parce que «la Valeur n'est pas l'aliénation de la fin ou de l'objectivité réalisée, c'est celle de la *praxis* elle-même» (Sartre 1985 : CRD, p. 356–357, note). On l'a vu, à l'opposé de l'angoisse éthique, l'esprit de sérieux consiste à dissimuler sa responsabilité éthique, c'est-à-dire à tenir les valeurs comme des données transcendantes. Tout se passe comme s'il s'agissait de refonder la morale sur la subjectivité humaine afin de montrer que le respect des normes se base sur la liberté de faire ce qu'il y a à faire comme exigence de l'humain. Dans les années 1964–1965, Sartre s'adonne à sa morale commençant par «éliminer les *morales impératives* (Kant, Nietzsche, etc.)» qui tendent «à remanier les ‹ tables des valeurs › ou les impératifs, en objectivant sous une forme morale (et de ce fait universelle) des impulsions subjectives et originales» (Sartre 2013 : DetL, p. 735). Bien que le manuscrit sur la morale ait été interrompu pour l'étude sur Flaubert, dans *L'Idiot de la famille* (1972) Sartre ne se contente pas de critiquer l'idée de surhomme, «haine de l'homme inversée », mais dénonce les liens entre la pensée nietzschéenne, l'idéologie du Progrès et la violence bourgeoise. «On montre plus clairement que l'homme est présentement un ratage en indiquant ce qu'il *aurait pu être* ou ce qu'il sera plus tard [...] : l'homme est un moyen de produire le surhomme ; son rôle est de se supprimer pour que cette nouvelle espèce surgisse» (Sartre 1988 : IF, p. 264). Comme le montre la psychanalyse existentielle, on ne distingue pas entre faire et être pour la réalité-humaine, car l'homme se réalise en s'engageant dans le monde et le projet s'incarne dans des conduites définies. C'est pourquoi il faut l'aimer pour ce qu'il est, avec ses vices et défauts. Pour le dire avec les mots des *Mains sales* (1948), ceux qui s'intéressent à ce que les hommes pourront devenir ne sont que des destructeurs : ils ne veulent pas changer le monde, ils veulent le faire sauter (cf. Sartre 2005b : MS, p. 333). Il apparaît clairement ici le sens de la chute des *Mots*, à l'opposé de l'arrogance nietzschéenne. Puisque le problème de l'homme, comme *personne*, est celui de tous, c'est-à-dire de l'*humanité*, «la violence se fera contre les élites (ainsi pissait Zarathoustra) au nom de l'égalité et la supériorité absolue [...] : un = tous» (Sartre 2010c : CLut, p. 919–920). Contre les élites, c'est-à-dire aussi contre soi, enfant d'élite, jusqu'à «briser des os dans sa tête» (Beauvoir 2008 : EBS, p. 13). Avec cet ouvrage, la boucle est bouclée.

À défaut d'avoir épuisé le rapport de Sartre à Nietzsche, contentons-nous ici de relever que, bien qu'il soit resté aussi fondamental en 1972 qu'il l'avait été en 1920, il est de part en part critique. On pourrait conclure avec l'aphorisme d'*Aurore* : «Nous autres chercheurs, comme tous les conquérants, tous les explorateurs, tous les navigateurs, tous les aventuriers, nous sommes d'une moralité audacieuse et il nous faut trouver bon de passer, somme toute, pour méchants» (Nietzsche 1993b, Aurore, p. 1160). En définitive, Sartre semble-t-il le méchant de tous les philosophes. Tout repose sur une stratégie, avouée par la prosopopée qu'il prête au Tintoret : «Les anciens, les nouveaux, je les prends tous et je les bats sur leur propre terrain» (Sartre 1980b : SéqV, p. 327)[28].

Bibliographie

Beauvoir, Simone de (2008) : *La Cérémonie des adieux suivi de Entretiens avec Jean-Paul Sartre (août-septembre 1974)*. Paris.

Beauvoir, Simone de (2013) : *Pour une morale de l'ambiguïté*. Paris.

Beauvoir, Simone de (2018a) : «Mémoires d'une jeune fille rangée». Dans : *Mémoires*. Jean-Louis Jeannelle/Éliane Lecarme-Tabone (éd.), t. 1. Paris, p. 1–338.

Beauvoir, Simone de (2018b) : «La Force de l'âge». Dans : *Mémoires*. Jean-Louis Jeannelle/Éliane Lecarme-Tabone (éd.), t. 1. Paris, p. 353–922.

Dassonneville, Gautier (2018) : «Liste des emprunts de Jean-Paul Sartre à la Bibliothèque des Lettres de l'École Normale Supérieure (1924–1928)». Dans : *Études sartriennes*, No. 22, p. 247–299.

Gerassi, John (2011) : *Entretiens avec Sartre*. Paris.

Louette, Jean-François (1996) : *Sartre* contra *Nietzsche*. Grenoble.

28 Je remercie Hadi Rizk pour la lecture critique de ce texte.

Nietzsche, Friedrich (1993a) : «Ainsi parlait Zarathoustra ». Dans : Friedrich Nietzsche : *Œuvres*, tome 2. Paris, p. 267–545.
Nietzsche, Friedrich (1993b) : «Aurore ». Dans : Friedrich Nietzsche : *Œuvres*, tome 1. Paris, p. 957–1211.
Nietzsche, Friedrich (1993c) : «Ecce homo ». Dans : Friedrich Nietzsche : *Œuvres*, tome 2. Paris, p. 1105–1198.
Nietzsche, Friedrich (1993d) : «Humain, trop humain ». Dans : Friedrich Nietzsche : *Œuvres*, tome 1. Paris, p. 417–956.
Nietzsche, Friedrich (1993e) : «L'Antéchrist ». Dans : Friedrich Nietzsche : *Œuvres*, tome 2. Paris, p. 1031–1103.
Nietzsche, Friedrich (1993f) : «La Généalogie de la morale ». Dans : Friedrich Nietzsche : *Œuvres*, tome 2. Paris, p. 739–889.
Nietzsche, Friedrich (1993 g) : «Le Crépuscule des idoles ». Dans : Friedrich Nietzsche : *Œuvres*, tome 2. Paris, p. 931–1029.
Nietzsche, Friedrich (1993 h) : Le Gai Savoir «. Dans : Friedrich Nietzsche : *Œuvres*, tome 2. Paris, p. 1–265.
Nietzsche, Friedrich (1993i) : «Par-delà le bien et le mal ». Dans : Friedrich Nietzsche : *Œuvres*, tome 2. Paris. p. 547–737.
Sartre, Jean-Paul (1947) : «Un nouveau mystique». Dans : Jean-Paul Sartre : *Situations, I.* Paris, p. 143–188.
Sartre, Jean-Paul (1948a) : «Présentation des *Temps modernes*». Dans : Jean-Paul Sartre : *Situations, II.* Paris, p. 7–30.
Sartre, Jean-Paul (1948b) : «Qu'est-ce que la littérature ?». Dans : Jean-Paul Sartre : *Situations, II.* Paris, p. 55–330.
Sartre, Jean-Paul (1976) : «Autoportrait à soixante-dix ans». Dans : Jean-Paul Sartre : *Situations, X.* Paris, p. 133–226.
Sartre, Jean-Paul (1980a) : «Paul Nizan». Dans : Jean-Paul Sartre : *Situations, IV.* Paris, p. 130–188.
Sartre, Jean-Paul (1980b) : «Le séquestré de Venise». Dans : Jean-Paul Sartre : *Situations, IV.* Paris, p. 291–346.
Sartre, Jean-Paul (1983) : *Cahiers pour une morale.* Paris.
Sartre, Jean-Paul (1985) : *Critique de la raison dialectique*, t. I. Paris.
Sartre, Jean-Paul (1988) : *L'Idiot de la famille. Gustave Flaubert de 1821 à 1857*, t. 3. Paris.
Sartre, Jean-Paul (1990a) : «Carnet Midy». Dans : *Écrits de jeunesse.* Michel Contat/Michel Rybalka (éd.). Paris, p. 443–497.
Sartre, Jean-Paul (1990b) : «Er l'Arménien». Dans : *Écrits de jeunesse.* Michel Contat/Michel Rybalka (éd.). Paris, p. 293–334.
Sartre, Jean-Paul (1990c) : «La Semence et le Scaphandre». Dans : *Écrits de jeunesse.* Michel Contat/Michel Rybalka (éd.). Paris, p. 140–187.
Sartre, Jean-Paul (1990d) : «Une défaite». Dans : *Écrits de jeunesse.* Michel Contat/Michel Rybalka (éd.). Paris, p. 204–286.
Sartre, Jean-Paul (1990e) : «Une lettre sur les étudiants d'aujourd'hui». Dans : *Écrits de jeunesse.* Michel Contat/Michel Rybalka (éd.). Paris, p. 433–434.
Sartre, Jean-Paul (2003) : «Matérialisme et révolution». In : *Situations, III.* Paris, p. 103–166.
Sartre, Jean-Paul (2005a) : *Le Diable et le Bon Dieu.* Dans : *Théâtre complet.* Michel Contat (éd.). Paris, p 375–501.

Sartre, Jean-Paul (2005b) : *Les Mains sales.* Dans : *Théâtre complet.* Michel Contat (éd.). Paris, p. 245–354.
Sartre, Jean-Paul (2005c) : *Les Mouches.* Dans : *Théâtre complet.* Michel Contat (éd.). Paris, p. 1–70.
Sartre, Jean-Paul (2008a) : *Lettres au Castor et à quelques autres*, t. 2. Simone de Beauvoir (éd.) Paris.
Sartre, Jean-Paul (2008b) : *Vérité et existence.* Paris.
Sartre, Jean-Paul (2009a) : «Érostrate». Dans : *Œuvres romanesques.* Michel Contat/Michel Rybalka (éd.). Paris, p. 262–278.
Sartre, Jean-Paul (2009b) : *La Mort dans l'âme.* Dans : *Œuvres romanesques.* Michel Contat/Michel Rybalka (éd.). Paris, p. 1135–1457.
Sartre, Jean-Paul (2009c) : *La Nausée.* Dans : *Œuvres romanesques.* Michel Contat/Michel Rybalka (éd.). Paris, p. 1–210.
Sartre, Jean-Paul (2010a) : *Carnets de la drôle de guerre.* Dans : *Les Mots et autres écrits autobiographiques.* Jean-François Louette (éd.). Paris, p. 143–651.
Sartre, Jean-Paul (2010b) : *Les Mots.* Dans : *Les Mots et autres écrits autobiographiques.* Jean-François Louette (éd.). Paris, p. 1–139.
Sartre, Jean-Paul (2010c) : »*Cahier Lutèce*» Dans : *Les Mots et autres écrits autobiographiques.* Jean-François Louette (éd.). Paris, p. 907–935.
Sartre, Jean-Paul (2011) : *Saint Genet, comédien et martyr.* Paris.
Sartre, Jean-Paul (2012a) : *L'Être et le néant.* Paris.
Sartre, Jean-Paul (2012b) : *L'Existentialisme est un humanisme.* Paris.
Sartre, Jean-Paul (2013) : «Détermination et liberté». Dans : *Écrits de Sartre.* Michel Contat/Michel Rybalka (éd.). Paris, p. 735–745.
Sartre, Jean-Paul (2014) : *Lettres au Castor et à quelques autres*, t. 1. Simone de Beauvoir (éd.) Paris.
Sartre, Jean-Paul (2016) : «Empédocle». Dans : *Études sartriennes*, No. 20, p. 27–50.
Simont, Juliette (2016) : «Empédocle chez Nietzsche et Sartre. Impuissance de la volonté et volonté de puissance». Dans : *Études sartriennes*, No. 20, p. 79–100.

Camus und Nietzsche im Überblick

Erminio Maglione

Albert Camus et la philosophie de Nietzsche

Abstract: Albert Camus and Nietzsche's Philosophy. The aim of this article is to trace, in general, the main stages of the reception of Nietzsche in Albert Camus's philosophy. The analysis intends to highlight the originality of Camus's existentialism and how this theory was influenced by the German thinker. However, it also contains several criticisms of Nietzsche. The French philosopher tries, in fact, to detach himself from the Nietzschean philosophy through the ideas of «pensée de Midi» and «affirmative negation» linked to his theorizing on «good nihilism». The latter draws its inspiration from the philosophy of the late Nietzsche (1886–1889) and, in particular, from reflections on *active nihilism*. This is where we find the most original Camus, the one who finally takes leave of the Nietzsche of *The Birth of Tragedy* to develop his own conception of «pensée de Midi». The «good nihilism» then assumes the form of a nihilism where the negation is not an *exclusive negation*, which removes the otherness of the opposite thesis, but an *inclusive negation* that contemplates the exercise of the negative in its tension with the opposite polarity. The *revolt* represents the apogee of this thought of contradiction. The «affirmative negation» describes a limit that cannot be crossed: the value of human life. The *revolt* is therefore the place where the distancing from Nietzsche and his conception of absolute nihilism takes place.

> Si j'avais à écrire ici un livre de morale, il aurait cent pages et 99 seraient blanches. Sur la dernière, j'écrirais : «Je ne connais qu'un seul devoir et c'est celui d'aimer». Et, pour le reste, je dis non. Je dis non de toutes mes forces.
> (Camus, *Carnets*, Septembre 1937)

1 Nietzsche vivant

Camus fut philosophe car il fut artiste et il fut artiste car il fut philosophe[1]. C'est peut-être précisément là que réside sa plus grande dette envers Nietzsche. Il

1 Pour un cadre historique de la figure d'Albert Camus (1913–1960) sont importantes les biographies suivantes : Lottman 1980 ; Todd 1996 ; Tanase 2010. Pour le Camus philosophe cf. :

https://doi.org/10.1515/9783110760163-005

nous apprend que la vision théorique est toujours une vision esthétique et que, pour spéculer sur les formes, il faut avant tout les avoir vues, imaginées, et mis en ordre mentalement dans un dessin spécifique. Certes, Camus fut un des plus sublimes « simplificateurs du monde», pour utiliser les mots de *Richard Wagner à Bayreuth* (1876), puisqu'il réussit à restituer, malgré sa *Weltanschauung* puissamment tragique, la beauté palpitante du monde en devenir (cf. Nietzsche 1967, WB 4, p. 20)[2]. Il fut pour cela, l'un des plus grands héritiers de Nietzsche au vingtième siècle : comme ce dernier, il fut un *tragique*, pensa comme un *Grec* et élabora une philosophie qui fit de la sagesse de la mesure son point de référence[3]. Camus comme Nietzsche pensa comme un *Grec* car il *vit* et *ressentit* comme un Grec. Il entra en contact avec l'auteur qui sera central pour le reste de sa vie dans les années trente et sa lecture fut filtrée par les publications du Mercure de France menées par Henri Albert[4]. Il entra donc immédiatement en contact avec le Nietzsche subversif et agitateur, oscillant entre anarchisme et tentations réactionnaires. Pour l'écriture du juvénile *Essai sur la Musique* (1932) il choisit en effet de s'appuyer, par exemple, sur *La Philosophie de Nietzsche* (1898) par Henri Lichtenberger. Il ne fut pas non plus étranger à l'interprétation du philosophe allemand comme penseur dionysiaque et vitaliste, comme peut clairement en témoigner la production de la deuxième moitié des années trente. En somme il respira l'air du Nietzsche subversif et révolutionnaire typique de la première phase de la réception française (1877–1914)[5].

L'idée de proposer un Nietzsche *accessible*, qui puisse aider concrètement l'homme dans sa recherche sur les questions fondamentales de l'existence, découla de cette influence. Roger Grenier parla d'un réel fil rouge nietzschéen au sein de la production de Camus, d'une voie maîtresse authentique qui pourrait être parcourue pour reconstruire l'ensemble de son parcours intellectuel (cf. Grenier 1992).

Jean Grenier, professeur de Camus à l'époque de l'hypokhâgne en Algérie, l'initia aux auteurs qui laisseront en lui une trace indélébile et qui le nourriront

---

Bachman 1998 ; Barilier 1977 ; Bartfeld 1988 ; Del Vecchio 1979 ; Grenier 1987 ; Hengelbrock 1982 ; Montano 2003 ; Quilliot 1980.

2 La définition de «simplificateur du monde [*Vereinfacher der Welt*]» est ici renvoyée par Nietzsche au compositeur Wagner. Les abréviations des œuvres sont celles qui sont utilisées dans la langue allemande.

3 Sur l'influence de la philosophie de Nietzsche sur Camus cf. : Rosenthal 1977 ; Weyembergh 1998 (en particulier cf. les p. 41–61 et 75–84) ; Merlio 2009, p. 284–312 ; Novello 2000, p. 128–141.

4 En général, pour la réception de Nietzsche en France cf. : Le Rider 1999 ; Boudot 1970 ; Pinto 1995.

5 Sur la première période de la réception de Nietzsche en France cf. : Gonzi 2012.

dans sa vocation existentialiste : évidemment au côté d'Augustin, Pascal, Schopenhauer et la mystique, il y avait aussi Nietzsche. Dorénavant son attention se focalisera toujours plus sur ce dernier, comme on peut le déduire des *Carnets*, remplis de notes de lectures nietzschéennes. Le témoin de ce grand intérêt est résumé aussi dans la documentation solide que Camus, au cours des années, recueillit. De l'analyse des archives de Francine Faure, deuxième femme de Camus, il ressort en outre la volonté d'être mis au courant des tendances les plus récentes sur les sujets nietzschéens. À partir de l'analyse de la bibliothèque l'on peut voir que le philosophe privilégia les éditions du Mercure de France, desquelles il possédait au moins onze œuvres de Nietzsche. Parmi ces titres, certains ont été très annotés : *La Volonté de puissance*, *L'Origine de la tragédie*, *Le Crépuscule des idoles*, *Le Cas Wagner*, *Aurore* et *Ecce homo*[6]. Camus s'arrêta de façon majeure donc sur la philosophie du Nietzsche tardif, celui des années 1886–1889, laissant de côté à la fois le Nietzsche juvénile (1872–1876) ainsi que celui de la période intermédiaire, la soi-disant phase «des lumières» (1878–1882).

Ayant choisi de s'attarder sur cette dernière étape, Camus faisait un choix décisif : il se rapprochait *philosophiquement* de Nietzsche insistant sur la période qui, selon Colli et Montinari, est la plus dense théoriquement (cf. Colli, Montinari 1964, p. 423). Ce ne fut donc pas le Nietzsche *Kulturkritiker* qui intéressa le plus Camus, ni le moraliste ou le supposé «politique», mais le philosophe, le Nietzsche théoricien. A partir de cette réflexion, il conçut son Nietzsche *effectif*, qui devait évaluer sa teneur non seulement sur le plan logico-spéculatif mais sur celui pratico-existentiel ; *libertaire* puisqu'il eut le courage de penser l'homme *en dehors de* la dialectique hégélienne, sans le brider dans un processus progressif se concluant en une synthèse. Le penseur allemand, en définitive, accompagna chaque principale phase de la production camusienne : de la réflexion des années trente autour de la question du tragique et du dionysiaque, passant par la spéculation sur l'absurde des années quarante, jusqu'à la dernière phase où à partir des *Lettres à un ami allemand* (1945), la vision deviendra plus critique et détachée. Ce dernier élément est fondamental car, en plus de mettre en évidence l'originalité de la proposition camusienne, c'est ce qui le rend un *nietzschéen de gauche*, un nietzschéen en mesure de penser l'échec, capable de critiquer et déconstruire son grand maître sans y adhérer de manière soumise.

---

6 Sur cet aspect cf. l'important article, philologiquement très approfondi, par Arnold 1979, p. 95–99.

2 La sagesse du tragique

Dans l'analyse du rapport entre Camus et Nietzsche nous avons choisi de nous concentrer sur les œuvres de la période allant des années 1936 aux années 1951 (année de la publication de *L'Homme révolté)*, avec une attention particulière sur la production essayiste. Selon nous, en effet, à cette période la réflexion de Camus peut être saisie dans toute la clarté de sa connotation *philosophique*. Analysant ce *cas* spécifique du nietzschéisme français, nous aurons la possibilité d'accéder à deux questions fondamentales : celle de la *pensée négative* comme *philosophie tragique et nihiliste* – développée surtout dans *Le mythe de Sisyphe* (1942) – et celle qui concerne le *diagnostic sur le nihilisme contemporain*, porté à maturation dans l'essai de 1951.

Les deux problèmes sont évidemment étroitement liés mais, le deuxième aspect, est probablement le plus intéressant car, connecté à cette définition du phénomène nihiliste, nous trouvons aussi la tentative originale de résolution et de dépassement de ce problème fondamental[7]. La pensée négative de Camus puise ses racines dans sa conception du dionysiaque. Il fut entraîné vers ces thématiques à la suite de deux rencontres : celle avec Jean Grenier, dont il lut *Les Îles* (1933) et avec l'écrivain André de Richaud, dont il aima le roman *La Douleur* (1931) (cf. Grenier 1933 et Richaud 1931). Avec le premier il découvrit la passion pour le dionysiaque et la fascination vers la beauté éphémère, avec le second le sens du tragique.

Le roman autobiographique de Richaud est la transposition de ce que vivait l'auteur en 1915 : Thérèse Delombre, veuve d'un officier français tombé au front, tombe follement amoureuse d'un prisonnier de guerre allemand et est submergée par une passion incontrôlable qui lui fait oublier toute convention et toute pudeur[8]. Comme on peut le voir, il y a de nombreux éléments en filigrane qui seront typiques du tragique camusien : la scission, la tension incontrôlable, la corporéité qui ternit et prévaut sur le raisonnement et les sentiments, ce dernier véritable protagoniste d'œuvres comme *L'Étranger* (1942) et *Caligula* (première version en 1938). En ce qui concerne *Les Îles*, dans la préface écrite pour l'ancien

7 Camus, à travers sa réflexion sur Nietzsche, représente l'un des principaux analystes du thème du nihilisme dans le contexte de la philosophie contemporaine. Pour une analyse historico-philosophique du phénomène nihiliste cf. Riedel 1978, p. 371–411 ; Volpi 2004.

8 Rappelant l'effet que la lecture du roman de Richaud, recommandé par Grenier en 1932, avait sur lui, Albert Camus écrivit en 1951 : « Je n'ai jamais oublié son beau livre, qui fut le premier à me parler de ce que je connaissais : [...] tout cela pouvait donc se dire ! » (Camus 2008d : Gide, p. 881–882).

maître en 1959 à l'occasion de la réédition du texte, Camus comparera sa propre expérience de lecture avec le *choc* produit par *Les Nourritures terrestres* de Gide sur la génération précédente d'intellectuels. Dans l'œuvre de Jean Grenier, la «heureuse barbarie» gidienne s'est retransformée à travers l'immanence solaire des terres algériennes, acquérant une fraîcheur sans précédent et un nouvel effet revigorant :

> la vérité du monde était dans sa seule beauté, et dans les joies qu'elle dispensait. Nous vivions ainsi dans la sensation, à la surface du monde, parmi les couleurs, les vagues, la bonne odeur des terres. (Camus 2008f : Îles, p. 621)

Si donc le point de départ de Gide et Grenier est le même – l'hédonisme de l'immanence pratiqué à travers l'immédiateté de la sensation, la joie sensuelle du corps qui doit se perdre dans l'Autre pour se retrouver, la jouissance de la surface comme lieu où la beauté se révèle – les conclusions auxquelles les deux arrivent sont décidément divergentes. En fait, contrairement à Gide, cette *jouissance sensuelle* n'est pas inattaquable, ce n'est pas une possession éternelle, comme une sorte d'ataraxie hédoniste, mais elle apparaît toujours menacée par la possibilité de la fin, par la mort, par la fugacité et la finitude de l'homme et des choses[9].

Ces thématiques convergeront dans *l'Envers et l'endroit* (1937), œuvre centrée sur la *bifacéite* de la vie, partagée entre *l'endroit* – la joie de vivre, la passion du corps, l'éruption de la beauté naturelle – et *l'envers* représenté par l'absurdité de l'existence, la mort, la souffrance, la maladie[10]. Apparaît ici, à travers la tension insoluble entre la «mort pour tous» et toute la «lumière du monde» (Camus 2006b : EnvEnd, p. 46), l'image des hommes clairvoyants inspirée de la *Naissance de la tragédie* (cf. Nietzsche 1972a, GT 17, p. 111–112), ceux qui sont en mesure de regarder le monde pour ce qu'il est, sans espoir ou désespoir. Cette réflexion domine la prose de *Noces* (1939), œuvre dans laquelle le dionysisme de Camus est thématisé de façon plus ponctuelle. L'accès à la dimension initiatique du regard sans illusion s'acquiert, selon Camus, en se fondant dans la Nature. Cette communion avec la terre est instructive car elle aide à se perdre, à abandonner l'emprise du *logos* sur l'existant. Le dionysisme camusien est alors en

9 Cet enseignement de Grenier aura des influences incalculables sur l'œuvre camusienne : «Il nous fallait des maîtres plus subtils et qu'un homme, par exemple né sur d'autres rivages, amoureux lui aussi de la lumière et de la splendeur des corps, vînt nous dire, dans un langage inimitable, que ces apparences étaient belles, mais qu'elles devaient périr et qu'il fallait alors les aimer désespérément. [...] Les Îles venaient, en somme, de nous initier au désenchantement ; nous avions découvert la culture» (Camus 2008e : DerInt, p. 621–622).

10 Camus attrapa la tuberculose pendant l'été 1930.

première place dans ce processus de nullification, d'abandon provisoire du contrôle de soi, qui permet d'aller à l'origine passant par le *pré-logique*.

Il y a ici le Nietzsche privilégié par Camus lors de sa jeunesse, celui wagnérien et schopenhauerien, qui réfléchit au dépassement du *principium individuationis* dans *La Naissance de la tragédie* (cf. Nietzsche 1972a, GT 8 et 17, p. 61 et p. 111–112). La première acceptation du négatif chez Camus est donc cette déclinaison panico-dionysiaque vue comme *anéantissement*, nullification de l'esprit humain, *dénuement*. Nous aimerions définir cette position théorique avec le nom de *primitivisme*, c'est à dire la nécessité de réduire au stade de la simplicité biologique, de l'élémentaire, pour apprendre à ne pas *subir* la chair mais à la *penser.* C'est un retour à la simplicité de l'*immanence*, à la pureté de l'absence de médiation, à l'immédiate innocence du simple subsister et jouir du corps. Dans cette fusion avec la nature, la pensée est finalement reconduite au corps. Le corps pour Camus et Nietzsche est le *simple*, la première substance, précisément car, sans elle, nous ne pourrions être, nous ne pourrions ressentir, *donc* penser. « Nous sommes parce que nous ressentons » est l'axiome du primitivisme camusien : *sum, ergo cogito*, comme dirait le Nietzsche lecteur hérétique de Descartes dans *Le Gai Savoir* (Nietzsche 1965b, FW 276, p. 159)[11].

La seule vérité indiscutable, claire, est la vérité du corps et c'est d'elle qu'il faut partir. L'importance de la corporéité est dans le fait d'être un lieu de frontière, dans le fait d'enseigner à la pensée sa limite, ce qui permet de ne pas tomber dans l'*hybris* du *logos*. Ce *dénuement* est alors le début de la pensée négative, le point de départ de chaque recherche sur l'homme.

À partir de ces hypothèses, Camus réfléchit probablement à la théorie du retour éternel comme une stratégie pour remettre au centre de la réflexion de la philosophie occidentale la corporéité. L'hypothèse du retour éternel, de ce point de vue, sert à conférer à la vie le poids qui lui est propre. C'est là le sens de l'être comme retour du même chez Nietzsche : la réalité, se répétant, ne fait que souligner son propre poids affirmant sa propre existence. L'éternel retour des choses est donc le paroxysme de l'immanence, c'est la corporéité qui triomphe dans sa constante réaffirmation. L'éternel retour restitue alors à l'instant son absence de répétitions, sa densité, dans son éternel retour sur soi : c'est cela *le plus grand poids*, le plus important paradoxe créateur de la métaphysique du Nietzsche mature[12].

---

11 Sur les implications théoriques de la notion de corps chez Nietzsche cf. Deleuze 1962.

12 Sur le puissant impact existentiel du «plus grand poids [*das grösste Schwergewicht*]», écrit Nietzsche: «Wenn jener Gedanke über dich Gewalt bekäme, er würde dich, wie du bist, verwandeln und vielleicht zermalmen; die Frage bei Allem und Jedem ‹ willst du diess noch einmal und noch unzählige Male? › würde als das grösste Schwergewicht auf deinem Handeln liegen!

Dans la vision tragique de l'homme, l'aporie donc représente le trait marquant de l'existence. Cependant elle est, en conformité avec Nietzsche, une philosophie absolument optimiste car l'individu lucide, qui a saisi l'absurde comme sens de l'existence, accepte le monde même dans son intime contradiction, gardant les yeux rivés sur l'être dans sa nudité, libéré des camouflages de la métaphysique. Pour Camus la beauté réside dans le fait de se consacrer au monde dans sa contradiction et son étrangeté. La beauté, le dépouillement de tout voile, c'est l'exposition de l'existence dans son absurdité. Se présenter comme une contradiction est le caractère du tragique selon Camus et, cette contradiction, coïncide avec la conscience indiscutable qu'il n'y aura aucune conciliation entre notre exigence de clarté et l'impénétrabilité du monde, hostile et muet[13]. Le sens de la *Weltanschauung* tragique se traduit donc dans le refus nihiliste d'un système transcendant qui justifie la condition humaine. Le tragique est contradictoire car il se fonde sur l'absurde, lui-même contradictoire : c'est pour cette raison que Camus fait clairement appel, dans cette phase, au Nietzsche de *La Naissance de la tragédie* (cf. Nietzsche 1972a, GT 7, pp. 55–56). Le philosophe cependant ne subit pas le contrecoup de l'absurde de manière passive mais, au contraire, il l'affronte par le biais de la *négation*, de la révolte à l'égard de ce qu'il n'accepte pas.

L'expression du vitalisme sauvage de l'univers camusien est en effet, en plus de Sisyphe, Caligula. C'est une autre figure de la lutte infatigable face à l'absurde : il choisit de se rebeller à l'égard d'un ordre qui ne le satisfait pas en décidant de l'épuiser de l'intérieur à travers la folie. Il combat l'absurde avec les armes de l'absurde, en se transformant en *catalyseur* de ce dernier. En fait, l'empereur romain connaît la liberté illimitée qui le mènera à la folie *après* la mort de sa bien-aimée sœur Drusilla, *après* la percée de la forme la plus violente et la plus absolue de l'absurde, la mort – «Ah ! mes enfants. Je viens de comprendre la vertu du pouvoir. Il va de pair avec la liberté d'esprit. Aujourd'hui et pour tout le temps qui va venir, ma liberté n'a pas de limites» (Camus 2006a : Cal, p. 397).

Si le tragique camusien se fonde alors sur la lucidité consciente de la mort, il n'est pas séparé néanmoins du bonheur dérivant du renoncement à toute foi

---

Oder wie müsstest du dir selber und dem Leben gut werden, um nach Nichts m e h r z u v e r l a n g e n, als nach dieser letzten ewigen Bestätigung und Besiegelung?» (Nietzsche 1965b, FW 341, p. 202).

13 Ce clivage entre l'homme et le monde, entre le désir de compréhension absolue et l'absence de réponses, est fortement souligné par Camus : «L'absurde est essentiellement un divorce. Il n'est ni dans l'un ni dans l'autre des éléments comparés. Il naît de leur confrontation» (Camus 2006c : MS, p. 239).

dans le transcendant. Cet homme privilégierait toujours de revivre depuis le début ce qu'il a déjà vécu. L'homme de Camus, comme celui de Nietzsche, aime follement l'instant car il sait qu'après ce dernier, il n'y a rien d'autre qui vaille la peine de vivre[14]. C'est probablement une des composantes de la philosophie de Camus la plus connotée sur le plan de l'existentialisme : il est tendu, du début à la fin, vers une transcendance qu'il admet inatteignable et, pour cela, fait exister cette transcendance dans l'horizon du sensible. Il reste l'aspiration à l'unité mais, en même temps, on reconnaît l'impossibilité d'un même désir et on substitue l'espoir dans la transcendance avec une projectualité immanente. Il reste le pur projet humain, une action qui reconnaît l'impossibilité de son propre accomplissement.

Camus, en tant qu'existentialiste, imagine la transcendance comme une figure de l'immanence, comme un trait de cette dernière. Camus s'estimait en accord avec les prémisses des philosophes qu'il jugeait comme des précurseurs de l'existentialisme – Pascal, Nietzsche, Kierkegaard et Chestov[15] – ; lui-même, selon nous, peut se compter parmi les existentialistes pour une forte raison philosophique : Camus choisit comme objectif centrale de sa propre réflexion l'existant concret saisi dans ses caractéristiques ontologiques (la temporalité, le rapport avec la transcendance, l'angoisse, le désespoir, l'absurde, la révolte, exc.) : ceci représente le noyau fondamental de toutes visions existentialistes[16].

Malgré cela, pourtant, Camus chercha toujours à maintenir un détachement critique à l'égard de l'existentialisme : il critiqua Sartre et sa doctrine de l'angoisse, selon lui dangereusement convergente avec les positions d'auteurs comme Kierkegaard, Chestov, Jaspers et Heidegger. La polémique majeure est tournée vers l'horizon de désespoir infranchissable présent dans ce type de réflexion. À juste titre il estime que son vitalisme est inconciliable avec cet horizon théorique. Les critiques soulevées par le *Mythe de Sisyphe* sont connues au sujet de l'existentialisme professé par des figures comme Léon Chestov et Benjamin Fondane. C'est une conception existentielle théologisant et irrationnelle qui aboutissait à une révolte contre la raison à travers une expérience de l'absurde débouchant sur la foi. Camus retient que c'est un réel «suicide phi-

---

14 Sur le lien fondamental entre nécessité et *amor fati* cf. NF-1881–1882, 15[20].

15 «Si les prémisses de l'existentialisme se trouvent, comme je le crois, chez Pascal, Nietzsche, Kierkegaard et Chestov, alors je suis d'accord avec elles» (Camus 2008e: DerInt, p. 663).

16 Un autre élément qui témoigne du point de conjonction entre la réflexion de Camus et celle de la philosophie de l'existence opérée par des auteurs comme Nietzsche et Kierkegaard est, par exemple, son intérêt pour un écrivain-philosophe comme R. W. Emerson. Pour ce thème, nous aimerons renvoyer à Maglione 2020, p. 143–162.

losophique», une faille injustifiable à l'espoir[17]. L'homme tragique, le sage, est pour l'Algérien l'homme sans espoir, le pèlerin de l'immanence, qui assume la mort comme inévitable et, malgré cela, il aime la vie de toutes ses forces et combat pour elle. Sisyphe est heureux car il a regardé la mort en face et, malgré tout, il a choisi la vie, a compris l'absurde et l'a accepté. C'est pour cela que Sisyphe est l'image du philosophe tragique et dionysiaque qui, ayant accepté la logique de l'*amour fati* comme destin, mène sa propre existence sans se soucier de l'absurdité qui lui pèse. L'absurde est le moteur de l'existence et Sisyphe décida, d'une manière lucide, de supporter cette énigme :

> Sisyphe enseigne la fidélité supérieure qui nie les dieux et soulève les rochers. Lui aussi juge que tout est bien. Cet univers désormais sans maître ne lui paraît ni stérile ni futile. Chacun des grains de cette pierre, chaque éclat minéral de cette montagne pleine de nuit, à lui seul, forme un monde. La lutte elle-même vers le sommet suffit à remplir un cœur d'homme. (Camus 2006c : MS, p. 304)

3 Le « bon nihilisme »

Si la découverte de l'absurde est le point de départ de la pensée négative de Camus, c'est important de souligner qu'elle ne représente certainement pas une *impasse* mais une ouverture à la possibilité de la projectualité humaine. La raison de l'homme *clairvoyant* ne peut conclure en effet à une absence d'un sens propre du monde, à la négation absolue de toute valeur. Ce serait un geste irrationnel : voir le nihilisme comme un *nihilisme absolu* signifie trahir la logique interne du nihilisme même.

Il est vrai que la négation absurde représente la *pars destruens* d'un mouvement de renversement dont l'objectif essentiel est d'éliminer tout mensonge métaphysique mais, tout cela, s'avère être orienté par un *telos* positif : la création d'une morale de l'immanence, entièrement basée sur l'homme et sur le relatif. Ce parcours tendant à une positivité, n'est pas une synthèse dialectique, mais un mouvement dans la tension du relatif qui est souligné à travers la double évocation à Nietzsche et Descartes : Camus pratique le *doute méthodique* à travers la

17 Cf. le paragraphe du *Mythe* intitulé précisément *Le suicide philosophique* dans Camus 2006c : MS, p. 238–253 : «Je prends la liberté d'appeler ici suicide philosophique l'attitude existentielle. Mais ceci n'implique pas un jugement. C'est une façon commode de désigner le mouvement par quoi une pensée se nie elle-même et tend à se surpasser dans ce qui fait sa négation. Pour les existentiels, la négation c'est leur Dieu. Exactement, ce dieu ne se soutient que par la négation de la raison humaine» (Camus 2006c : MS, p. 247).

négation des idoles métaphysiques comme fit Nietzsche écrivant, ainsi dans *L'Homme révolté*, «le *Discours de la méthode* de son temps» (Camus 2008b : HR, p. 117). Nier signifie douter, annihiler méthodiquement tout ce qui est étranger à l'immanence et qui n'est pas immédiatement clair et évident. Camus est un nihiliste cartésien qui a corrigé Descartes avec Nietzsche : il croit fermement en la possibilité de la raison comme instrument pour bien vivre parmi les hommes, comme seule voie pour conquérir le bonheur. Le *logos* n'est pas omniprésent, il n'est pas infaillible, mais il peut faire beaucoup : il peut critiquer et nier, c'est peut-être l'un des principaux enseignements que Camus hérite du penseur allemand, avec la grande confiance dans les possibilités de la raison typique de Descartes aussi, fait évidemment les distinctions nécessaires[18].

La théorisation camusienne sur la catégorie du «bon nihilisme» se développe dans ce contexte et puise son inspiration dans la philosophie du Nietzsche tardif (1886–1889) et, en particulier, dans les réflexions sur le nihilisme *actif*[19]. C'est là que l'on trouve, selon nous, le Camus le plus original, celui qui prend finalement congé du Nietzsche de *La Naissance de la tragédie* pour développer, de façon originale, sa propre conception de la *pensée de Midi*. Le «bon nihilisme» assume alors la forme d'un nihilisme où la négation n'est pas une *négation exclusive*, qui retire l'altérité de la thèse opposée, mais une *négation inclusive* qui contemple l'exercice du négatif dans la tension de la polarité opposée[20]. La

---

18 Par exemple, sur l'importance de la critique dans le domaine philosophique, en référence à l'expérience de son maître, cf. ce que Nietzsche a déclaré dans *Schopenhauer als Erzieher* (1874): «Die einzige Kritik einer Philosophie, die möglich ist und die auch etwas beweist, nämlich zu versuchen, ob man nach ihr leben könne, ist nie auf Universitäten gelehrt worden: sondern immer die Kritik der Worte über Worte» (Nietzsche 1972b, SE 8, p. 447). Un lecteur important qui s'attarda davantage sur les points de contact que sur ceux de divergence entre Nietzsche et Descartes fut, en Italie, Giuliano Campioni. Est important à cet égard son livre : Campioni 2001. Pour ce thème, nous aimerons renvoyer à Maglione 2018, p. 377–389.

19 Le terme «bon nihilisme» est utilisé dans le premier projet de quatrième de couverture du *Mythe* envoyé à l'éditeur Gallimard en septembre 1942. Sur cet aspect cf. Todd 1996, p. 303. Pour la spéculation nietzschéenne sur le nihilisme *passif* et *actif* dans le cadre théorique du *nihilisme complet* cf. NF-1887–1888, 9[35]. L'ouverture du fragment est célèbre: «*Nihilism:* es fehlt das Ziel; es fehlt die Antwort auf das ‹ Warum? › was bedeutet Nihilism? – *daß die obersten Werthe sich entwerthen*» (Nietzsche 1964, NL 1887, 9[35]; Nietzsche 1971, p. 12).

20 Sur le nihilisme comme point de départ et comme doute méthodique dans le contexte de la recherche philosophique, cf. l'entretien de Camus avec Rauhut datant de septembre 1951 : Rauhut 1963 – À cet égard, on peut dire que le point de départ théorique de Nietzsche et de Camus est le même. Pour Nietzsche comme nihiliste parfait, comme *nihiliste classique*, cf. le § 3 du NF-1887–1888, 11[411] dans lequel le philosophe allemand se définit «als der erste vollkommene Nihilist Europas, der aber den Nihilismus selbst schon in sich zu Ende gelebt hat, – der ihn hinter sich, unter sich, außer sich hat...» (Nietzsche 1971, NL 1887, 11[411], p. 393).

révolte représente l'apogée de cette pensée de la contradiction. C'est une «négation affirmative», selon Camus, une tension entre la fidélité à la terre, l'acceptation de l'existence, et le refus de la souffrance et de la mort (cf. Camus 2008b : HR, p. 71 et Camus 2008c : RemRév, p. 325). Elle affirme qu'il y a une limite qui ne peut être franchie : la valeur de la vie humaine[21].

La *révolte* est donc le lieu où s'effectue la prise de distance face à Nietzsche et la période où cette réflexion s'opère de manière structurée est comprise entre 1944 et 1950. Au cours de ces années-là, Camus se remet à étudier intensément Nietzsche aussi en raison de l'instrumentalisation dont il a été l'objet par les nazis. Il faut être, selon *L'Homme révolté*, «les avocats de Nietzsche» par le biais d'une relecture attentive de ces écrits qui met en lumière, aux côtés des possibles dangers, la leçon positive de sagesse (Camus 2008b : HR, p. 125). Camus fait remarquer une profonde incohérence dans l'interprétation nietzschéenne : son point de départ se basait sur la révolte, sur la pensée qui à travers la négation nihiliste se libère des mensonges métaphysiques, mais ses conclusions, liées à l'acceptation intégrale de la logique du *nihilisme positif*, se transforment en une trahison de la révolte elle-même, atteignant un *nihilisme absolu*. Cette pensée s'élabore dans les *Lettres à un ami allemand* (1945) pour atteindre son apogée dans *l'Homme révolté*. Le nihilisme camusien est un *nihilisme relatif*, un «bon nihilisme», car il s'oppose au nihilisme absolu nietzschéen comme négation indiscriminée de la vie humaine.

Le *nihilisme absolu* nietzschéen est en effet un acte de démesure et, pour cela, un acte d'infidélité à l'égard de la finitude du savoir de l'homme qui a choisi d'être fidèle à la terre. Ce passage de la pensée du philosophe allemand se prête à l'époque à l'instrumentalisation et à la méprise dans le sens de l'homicide indiscriminé systématiquement perpétré par les idéologies totalitaires. La transvaluation des valeurs (*Umwertung aller Werte*) représente le point culminant du mouvement nihiliste, son accomplissement parfait selon Camus. Après avoir niée les valeurs antiques, elle culmine dans l'*amor fati* qui est l'adhésion de l'*Übermensch* à la spirale du devenir à travers l'acceptation inconditionnée de tout ce qui arrivera. Cette adhésion *totale* à l'existence se transforme cependant en un consensus démesuré au monde et donc, aussi, au *mal* et à la *mort*. Camus essaie alors d'adapter l'*amor fati* en une pratique de la mesure et du soin de l'homme. Accepter le monde dans sa *bifacité* ne signifie pas en effet y *consentir* totalement ; il faut alors remplacer cette affirmation sans freins par des forces équilibrées de la *négation affirmative*. Cette dernière, effectivement, ne dit pas

21 Sur le caractère central de la négation affirmative dans la philosophie de Camus cf. Gay-Crosier 2009, p. 29–44.

seulement *oui* mais *oui-et-non*, affirmant l'existence d'une limite – la vie humaine précisément – dans l'équilibre des opposés[22].

L'expression de cette composition toujours provisoire, constamment remise en question, est la *pensée de Midi* proposée dans *l'Homme révolté*. Ici Camus, évoquant la beauté grecque comme composition dans la mesure, fonde sur l'esthétique la *philosophie de la limite*, qui se reconnectant au savoir grec antique, souligne de nouveau, malgré les profondes critiques, son rapport *essentiel* avec Nietzsche. Camus écrit à ce sujet, dans son recueil d'essais intitulé *L'Été* (1954), une réflexion importante qui mérite d'être rapportée dans son intégralité et qui représente, en général, un élément fondamental de clarification des questions rapidement abordées dans ce travail :

> Nous avons exilé la beauté, les Grecs ont pris les armes pour elle. Première différence, mais qui vient de loin. La pensée grecque s'est toujours retranchée sur l'idée de limite. Elle n'a rien poussé à bout, ni le sacré, ni la raison, parce qu'elle n'a rien nié, ni le sacré, ni la raison. Elle a fait la part de tout, équilibrant l'ombre par la lumière. Notre Europe, au contraire, lancée à la conquête de la totalité, est fille de la démesure. [...] Ulysse peut choisir chez Calypso entre l'immortalité et la terre de la patrie. Il choisit la terre, et la mort avec elle. Une si simple grandeur nous est aujourd'hui étrangère. (Camus 2008a: Eté, p. 597 et p. 600)[23]

Bibliographie

Arnold, James A. (1979) : «Camus lecteur de Nietzsche». Dans : *Revue des Lettres modernes*, IX, pp. 95–99.

Bachman, John V. (1998) : *Camus' Rebellious Thought*. New York.

Barilier, Étienne (1977) : *Albert Camus. Philosophie et littérature*. Lausanne.

Bartfeld, Fernande (1988) : *L'Effet tragique. Essai sur le tragique dans l'œuvre de Camus*. Paris, Genève.

Boudot, Pierre (1970) : *Nietzsche et les écrivains français de 1930 à 1960*. Paris.

Campioni, Giuliano (2001) : *Les lectures françaises de Nietzsche*. Paris.

Camus, Albert (2006a) : «Caligula (version de 1941)». Dans : Albert Camus : *Œuvres complètes*, tome I (1931–1944). Jacqueline Lévi-Valensi (ed.). Paris, pp. 389–442.

---

22 Avec les paroles de Camus : «Si la révolte pouvait fonder une philosophie [...] ce serait une philosophie des limites, de l'ignorance calculée et du risque. Celui qui ne peut tout savoir ne peut tout tuer. [...] La révolte [...] ne vise qu'au relatif et ne peut promettre qu'une dignité certaine assortie d'une justice relative. Elle prend le parti d'une limite où s'établit la communauté des hommes. Son univers est celui du relatif» (Camus 2008b : HR, p. 309–310). Sur cette question cf. Novello 2000, p. 137–139.

23 Intéressant pour approfondir la relation entre Camus et la Grèce classique, avec une attention particulière à l'esthétique de cette dernière, cf. Rey 2000 ; Modler 2000.

Camus, Albert (2006b) : «L'envers et l'endroit». Dans : Albert Camus : *Œuvres complètes*, tome I (1931–1944). Jacqueline Lévi-Valensi (ed.). Paris, pp. 29–72.
Camus, Albert (2006c) : «Le Mythe de Sisyphe». Dans : Albert Camus : *Œuvres complètes*, tome I (1931–1944). Jacqueline Lévi-Valensi (ed.). Paris, pp. 217–315.
Camus, Albert (2008a) : «L'été». Dans : Albert Camus : *Œuvres complètes*, tome III (1949–1956). Raymond Gay-Crosier (ed.). Paris, pp. 565–623.
Camus, Albert (2008b) : «L'Homme révolté». Dans : Albert Camus : *Œuvres complètes*, tome III (1949–1956). Raymond Gay-Crosier (ed.). Paris, pp. 61–324.
Camus, Albert (2008c) : «Remarque sur la révolte (L'Existence, 1945)». Dans : Albert Camus : *Œuvres complètes*, tome III (1949–1956). Raymond Gay-Crosier (ed.). Paris, pp. 325–337.
Camus, Albert (2008d) : «Rencontres avec André Gide». Dans : Albert Camus : *Œuvres complètes*, tome III (1949–1956). Raymond Gay-Crosier (ed.). Paris, pp. 881–885.
Camus, Albert (2008e) : «Dernière interview (Décembre 1959)». Dans : Albert Camus : *Œuvres complètes*, tome IV (1957–1959). Raymond Gay-Crosier (ed.). Paris, pp. 661–664.
Camus, Albert (2008f) : «Préface aux ‹ Îles ›». Dans : Albert Camus : *Œuvres complètes*, tome IV (1957–1959). Raymond Gay-Crosier (ed.). Paris, pp. 621–624.
Colli, Giorgio, Montinari, Mazzino (1964) : «Notizie e note ai Frammenti postumi 1887–1888». Dans : *Opere di Friedrich Nietzsche*, VIII/2. Giorgio Colli/Mazzino Montinari (ed.). Milano.
Del Vecchio, Marcello (1979) : *La Fenomenologia dell'assurdo: Albert Camus.* Firenze.
Deleuze, Gilles (1962) : *Nietzsche et la philosophie.* Paris.
Gay-Crosier, Raymond (2009) : «De l'*homo faber* à l'*homo ludens* : défense et illustration de la pensée de midi». Dans : Dolorès Lyotard (ed.) : *Albert Camus contemporain.* Villeneuve d'Ascq, pp. 29–44.
Gonzi, Alice (2012) : *Zarathustra a Parigi. La ricezione di Nietzsche nella cultura francese del primo Novecento.* Roma.
Grenier, Jean (1933) : *Les Îles.* Paris.
Grenier, Roger (1987) : *Albert Camus, soleil et ombre.* Paris
Grenier, Roger (1992) : «Camus : Sisyphe et les autres». Dans : *Le Magazine littéraire*, n. 298, pp. 75–76.
Hengelbrock, Jürgen (1982) : *Albert Camus. Ursprünglichkeit der Empfindung und Krisis des Denkens.* Freiburg, München.
Le Rider, Jacques (1999) : *Nietzsche en France, de la fin du XIXe au temps présent.* Paris.
Lottman, Herbert R. (1980) : *Albert Camus.* Paris.
Maglione, Erminio (2018) : «Nietzsche interprete di Descartes nella lettura di Giuliano Campioni». Dans : *Giornale Critico di Storia delle Idee. Rivista internazionale di filosofia*, n. 1, pp. 377–389.
Maglione, Erminio (2020) : «Albert Camus e il Trascendentalismo americano». Dans : *Studi Sartriani*, XIV, pp. 143–162.
Merlio, Gilbert (2009) : «Sisyphos und der Übermensch. Auf den Spuren Nietzsches bei Camus». Dans : Thorsten Valk (ed.) : *Friedrich Nietzsche und die Literatur der klassischen Moderne.* Berlin, pp. 284–312.
Modler, Karl (2000) : *Soleil et mesure dans l'œuvre de Camus*, Paris.
Montano, Aniello (2003) : *Camus : un mistico senza Dio*, Padova.

Nietzsche, Friedrich (1965a) : «Frammenti postumi (1881–1882)» [NF-1881–1882]. Dans : Friedrich Nietzsche : *Opere di Friedrich Nietzsche*, volume V, tome 2. Giorgio Colli/Mazzino Montinari (ed.). Milan, pp. 277–511.
Nietzsche, Friedrich (1965b) : « La gaia scienza». Dans : Friedrich Nietzsche : *Opere di Friedrich Nietzsche*, volume V, tome 2. Giorgio Colli/Mazzino Montinari (ed.). Milan, pp. 11–276.
Nietzsche, Friedrich (1967) : «Richard Wagner a Bayreuth». Dans : Friedrich Nietzsche : *Opere di Friedrich Nietzsche*, volume IV, tome 1. Giorgio Colli/Mazzino Montinari (ed.). Milan, pp. 1–80.
Nietzsche, Friedrich (1971) : «Frammenti postumi (1887–1888)» [NF-1887–1888]. Dans : Friedrich Nietzsche : *Opere di Friedrich Nietzsche*, volume VIII, tome 2. Giorgio Colli/Mazzino Montinari (ed.). Milan, pp. 1–418.
Nietzsche, Friedrich (1972a): «La nascita della tragedia». Dans : Friedrich Nietzsche : *Opere di Friedrich Nietzsche*, volume III, tome 1. Giorgio Colli/Mazzino Montinari (ed.). Milan, pp. 1–163.
Nietzsche, Friedrich (1972b) : «Schopenhauer come educatore». Dans : Friedrich Nietzsche : *Opere di Friedrich Nietzsche*, volume III, tome 1. Giorgio Colli/Mazzino Montinari (ed.). Milan, pp. 357–457.
Novello, Samantha (2000) : « Nihilismo e pensiero tragico. La ricezione di Nietzsche nei saggi di Albert Camus». Dans : *Il Ponte*, LVI, pp. 128–141.
Pinto, Louis (1995) : *Les Neveux de Zarathustra. La réception de Nietzsche en France.* Paris.
Quilliot, Roger (1980) : *La Mer et les Prisons. Essai sur Albert Camus.* Paris.
Rauhut, Franz (1963) : « Du nihilisme à la ‹ mesure › et à l'amour des hommes». Dans : *La Revue des Lettres Modernes*, n. 90–93, IV, p. 17–40.
Rey, Pierre-Louis (2000) : *Camus. Une morale de la beauté.* Paris.
Richaud, André de (1931) : *La Douleur.* Paris.
Riedel, Manfred (1978) : « Nihilismus». Dans : Otto Brunner, Werner Conze, Reinhart Kosellek (ed.) : *Geschichtliche Grundbegriffe. Historisches Lexikon zur politisch-sozialen Sprache in Deutschland*, 4. Stuttgart, pp. 371–411.
Rosenthal, Bianca (1977) : *Die Idee des Absurden bei Friedrich Nietzsche und Albert Camus.* Bonn.
Tanase, Virgil (2010) : *Camus.* Paris.
Todd, Olivier (1996) : *Albert Camus, une vie.* Paris.
Volpi, Franco (2004) : *Il nichilismo.* Roma-Bari.
Weyembergh, Maurice (1998) : *Albert Camus ou la mémoire des origines.* Bruxelles.

Der Nihilismus und seine Überwindung

Alfred Betschart

Nietzsche, Sartre und der Vorwurf des Nihilismus

Abstract: Nietzsche, Sartre, and the Allegation of Nihilism. In the 19th and 20th centuries, the allegation of nihilism was a weapon frequently used in the fight against those denouncing the reigning morality. The ideas of Nietzsche and Sartre, too, fell prey to this performative use of the term „nihilism". Were Nietzsche and Sartre indeed nihilists? A comparison of their moral theories based on Kierkegaard's model of the three stages shows that Nietzsche's and Sartre's moral theories share important characteristics that qualify to call Nietzsche and Sartre nihilists, in particular their criticism of the reigning morality and their moral relativism.

Der Nihilismusvorwurf in Ostberlin und Zürich

Im Dezember 1965 findet das berüchtigte 11. Plenum des ZKs der SED statt, bekannt unter der Bezeichnung „Kahlschlag-Plenum", weil es beschloss, die in der ersten Hälfte der 1960er Jahre erfolgten zaghaften Lockerungen im Kulturbereich schlagartig rückgängig zu machen. An die Kulturschaffenden und Jugendlichen gewandt hält Erich Honecker, nach Walter Ulbricht die Nummer zwei der SED, in seinem Bericht ans ZK fest:

> Wollen wir die Arbeitsproduktivität und damit den Lebensstandard weiter erhöhen, woran doch alle Bürger der DDR interessiert sind, dann kann man nicht nihilistische, ausweglose und moralzersetzende Philosophien in Literatur, Film, Theater, Fernsehen und Zeitschriften verbreiten. („Bericht des Politbüros" 1965, S. 123)

Mit nihilistisch und moralzersetzend meint er „die Einflüsse der kapitalistischen Unkultur und Unmoral, wie sie in der amerikanischen Sex-Propaganda und der Verherrlichung des Banditentums zum Ausdruck kommt" („Bericht des Politbüros" 1965, S. 118). Die DDR sei schließlich ein sauberer Staat. Zusammen mit Begriffen wie Dekadenz, Subjektivismus, Pornographie und Skeptizismus bildet der Ausdruck „Nihilismus" ein von der SED häufig benutztes Begriffsfeld im Kampf gegen unliebsame kulturelle Phänomene (siehe Weißgerber 2010).

Ziemlich genau zwölf Monate später kommt dasselbe Arsenal ideologischer Angriffswaffen im sog. Zürcher Literaturstreit auf der anderen Seite des Eisernen Vorhangs zum Einsatz. Professor Emil Staiger von der Universität Zürich, einer der

https://doi.org/10.1515/9783110760163-006

bekanntesten Neugermanisten der damaligen Zeit, holt zum Rundschlag gegen die modernen Schriftsteller aus, die

> behaupten, die Kloake sei ein Bild der wahren Welt, Zuhälter, Dirnen und Säufer Repräsentanten der wahren, ungeschminkten Menschheit, so frage ich: In welchen Kreisen verkehren sie? [...] Beachten wir lieber, wann und wo eine Trümmerliteratur gedeiht, in welchen Lagen sich nihilistische Stimmung auszubreiten pflegt. [...] Der Nihilismus ist, in erstaunlich vielen Fällen, ein Luxusartikel. (Staiger 1967, S. 91)

Die Kritik ist groß. Max Frisch antwortet Staiger in einem Beitrag mit dem Titel *Endlich darf man es wieder sagen:* „Ich nenne es ein Standgericht [...] So, erinnern wir uns, verfuhr man schon immer, wenn die Rede war von entarteter Kunst. [...] [D]eine grundlegende Rede, gehalten in Moskau, würde offiziös nicht minder begrüsst als in Zürich" (Frisch 1966, S. 105, 107). Der Angriff auf den Nihilismus, die Dekadenz und die Entartung der Schriftsteller und Kunstschaffenden ist etwas, dies erkennt Frisch sehr treffend, was die Konservativen aller Strömungen eint, ob nun linkskonservative Kommunisten, Rechtsbürgerliche oder Nationalsozialisten.

Der Begriff des Nihilismus als Keule im Kulturkampf hat seit den Zeiten von Honecker und Staiger fast ausgedient. Benutzt wird er nur noch von religiösen Fundamentalisten wie Abu Bakr Rieger, einem deutschen islamistischen Intellektuellen (Rieger 2020), oder von dessen katholischem Bruder im Geiste, Joseph Ratzinger, dem späteren Papst Benedikt XVI., einem der Spiritus rectores der Enzyklika *Fides et ratio* (Johannes Paul II. 1998), in der der Vatikan mit dem Nihilismus und Relativismus der westlichen Philosophie abrechnet. In den Jahren zwischen 1945 und 1965, in denen der Begriff des Nihilismus ein sehr beliebtes Argument in den Auseinandersetzungen um die ‚richtige' Kultur darstellt, werden mit diesem Vorwurf fortschrittliche Musiker, bildende Künstler und vor allem Schriftsteller bekämpft, von Jean Genet bis Peter Weiss, die den Menschen als verzweifelndes, aber auch sexuelles und mordendes Wesen darstellen.

Der Begriff des Nihilismus und sein performativer Gebrauch

Die Nihilismuskeule wird auch gegen Nietzsche und Sartre geschwungen. György Lukács wirft 1954 in *Die Zerstörung der Vernunft*, jenem Werk, in dem Lukács nach Theodor W. Adornos Worten seine eigene Vernunft zerstörte (Adorno 1961, S. 153), Nietzsche Relativismus, Pessimismus und Nihilismus vor (Lukács 1954, S. 13). François de Menthon, der französische Chefankläger am Nürnberger Prozess und spätere katholische Politiker, benutzt zwar nicht den Begriff „Nihilismus", wirft aber in diesem Sinne Nietzsche während des Nürnberger Prozesses vor, mit seiner Kritik an den traditionellen Werten ein Ahnherr des Nationalsozialismus gewesen zu sein („Nuremberg Trial Proceedings Vol. 5" 1946, S. 376–377). Auch Hermann Rauschning bringt in seinen zwischen 1938 und 1954 publizierten Büchern[1] immer wieder Nihilismus, Nationalsozialismus und Nietzsche miteinander in Verbindung.

In besonderem Ausmaß trifft die Nihilismuskeule in den Nachkriegsjahren Sartre, was aufgrund des Inhalts seiner literarischen Werke, in denen die Menschen nicht nur scheitern, sondern auch alle möglichen Arten von sexuellen Beziehungen pflegen, wenig erstaunt. Unter Sartres ideologischen Kritikern finden sich Kommunisten ebenso sehr wie Bürgerliche, der Kommunist Lukács gleichermaßen wie der oldenburgische Geistesaristokrat Karl Jaspers. Schon 1944 sieht sich Sartre mit dem Vorwurf konfrontiert, nihilistische Lehren zu vertreten (Sartre 2000: PropE, S. 113). Besonders bemerkenswert ist der Beitrag *Was ist Existentialismus?* von 1951 von Karl Jaspers, der seinen mehrjährigen Kampf gegen den Nihilismus schon in der Nr. 1 seiner Zeitschrift *Die Wandlung* vom November 1945 eingeläutet hatte. In seinem Essay *Was ist Existentialismus?* macht er nicht nur deutlich, was ihn am Existentialismus besonders stört, er stellt auch Sartre und Nietzsche in eine Linie:

> Wenn ein Jüngling in Paris auffällig gekleidet ist, sich erotische Zügellosigkeiten erlaubt, im Café lebt und wunderliche Redensarten macht, sagt man, er sei ein Existentialist; geschah dasselbe vor einem halben Jahrhundert in Berlin, so sagte man, er sei ein Nietzscheaner. (Jaspers 1968, S. 497)

Der Begriff des Nihilismus wird zwar gelegentlich auch in einem konstativen Sinn gebraucht, so bei Martin Heidegger, für den das Wesen des Nihilismus in der Seinsvergessenheit besteht. In der neueren Literatur haben vor allem Elisabeth Kuhn (1992) und Eicke Brock (2015) die konstative Bedeutung von ‚Nihilismus' in Nietzsches Werk aufgearbeitet. Diesem konstativen Gebrauch kommt in der Geschichte des Wortes ‚Nihilismus' jedoch eine eher marginale Stellung zu. Seit den Zeiten der Französischen Revolution (Goudsblom 1980, S. 3–10; Slocombe 2006, S. 10) bis in die 1960er Jahre bei Honecker und Staiger ist dessen Historie wesentlich eine der performativen Nutzung. Dieter Arendt bezeichnet ‚Nihilismus' zurecht als Scheltwort des Idealismus (Arendt 1970, S. 9). Steht in Friedrich Heinrich Jacobis Brief an Johann Gottlieb Fichte 1799 die Verurteilung von dessen

---

1 *Die Revolution des Nihilismus. Kulisse und Wirklichkeit im Dritten Reich* (1938), *Masken und Metamorphosen des Nihilismus – Der Nihilismus des XX. Jahrhunderts* (1954).

angeblichem Atheismus im Vordergrund („das Nichts oder einen Gott", Jacobi 1799, S. 48), so in Ivan Turgenevs *Väter und Söhne* (1862) die Ablehnung der traditionellen Moral („Ein Nihilist ist ein Mensch, der sich vor keiner Autorität beugt", Turgenjew 1949, S. 36). Fëdor Dostoevskijs Feststellung „wenn es keinen Gott gibt, ist alles erlaubt"[2] verbindet wiederum die beiden Auffassungen von Nihilismus als Atheismus und Ablehnung der herrschenden Moral im Sinne Jacobis resp. Turgenevs.

Während Sartre am Begriffsfeld des Nihilismus vollkommen desinteressiert ist (siehe Daigle 2002, S. 7, 118), kommt diesem Terminus in Nietzsches Denken eine zentrale Stellung zu, vor allem in der Zeit von Herbst 1885 bis Herbst 1888 (Tongeren 2018, S. 26). In seinem ausgeprägt performativen Gebrauch steht Nietzsche im Einklang mit den Russen Turgenev und Dostoevskij. Dieser Bezug zur russischen Kultur und Geschichte kommt auch darin zum Ausdruck, dass Nietzsche erstmals ein Wort aus dem Begriffsfeld ‚Nihilismus' in seinen Notizen vom Sommer 1880 benutzt, nur wenige Monate nach dem fehlgeschlagenen Attentat von ‚Nihilisten' auf Mikhail Loris-Melikov, einen der damals bedeutendsten russischen Politiker. Bemerkenswert ist bei Nietzsches performativem Gebrauch von ‚Nihilismus', dass er zusammen mit ‚Décadence' und ‚Skepsis' ein Begriffsfeld benutzt, das in den 1960er Jahren weitgehend identisch zur Anwendung kommt. Sein eigenes ambivalentes Verhältnis zum Nihilismus bringt Nietzsche in einem Fragment von 1887 in folgender Selbstbeschreibung zum Ausdruck: „als der erste vollkommene Nihilist Europas, der aber den Nihilismus selbst schon in sich zu Ende gelebt hat, – der ihn hinter sich, unter sich, außer sich hat..." (Nietzsche, eKGWB, NF-1887, 11[411]).

Stellen wir uns die Frage nach den Ursachen dieses Unterschieds im Gebrauch des Worts ‚Nihilismus' zwischen Nietzsche und Sartre, so ist die Antwort wohl am ehesten in deren jeweiligem persönlichen Hintergrund zu finden. Sartre wächst in einer linksliberalen Familie auf, in der die Männer nicht viel von Religion halten und die Frauen primär wegen der schönen Musik in die Kirche gehen. Ihn, den Nominalisten, der sich ohnehin nicht um Bezeichnungen kümmert, stört der Vorwurf des Nihilismus nicht. Ganz anders verläuft die Konstitution von Nietzsches Persönlichkeit. Nietzsche wächst in einem frommen evangelischen

2 Dieser oft zitierte Ausdruck findet sich allerdings nicht *tel quel* in Dostoevskijs *Die Brüder Karamasow.* Im fünften Buch der Erzählung *Der Großinquisitor* steht zwar mehrfach der Ausdruck „Alles ist erlaubt", jedoch ohne die erste Hälfte des Ausdrucks. Ivan stimmt jedoch Mitjas Aussage zu, die im russischen Originaltext lautet: „Без бога-то и без будущей жизни? Ведь это, стало быть, теперь всё позволено, всё можно делать?" (Die Brüder Karamasow, Teil 4, Buch 11, Kapitel IV; Dostoevskij 1880): „Ohne Gott und ohne zukünftiges Leben? Ja heißt dies dann, jetzt ist alles erlaubt, alles kann man tun?" (Übersetzung A.B.).

(Frauen-)Haushalt mit einer langen Tradition an Pfarrern auf. Die Konfrontation mit der ‚anderen' Welt, in der die Religionskritik von Ludwig Feuerbach, David Friedrich Strauß und Bruno Bauer omnipräsent ist, muss für ihn prägend gewesen sein. Einerseits sind die alten Gewissheiten verschwunden – in den Wissenschaften und der Moral, wie die Adoption des Assassinen-Motivs „‚Nichts ist wahr, alles ist erlaubt'" (Nietzsche, eKGWB, Za-IV-Schatten; GM-III-24) zeigt. Andererseits gibt er die Suche nach neuen festen Werten – die letztlich in der Herrenmoral endet – nie auf. Entsprechend ambivalent ist Nietzsches Haltung gegenüber dem Nihilismus (s.o.: Nietzsche, eKGWB, NF-1887, 11[411]).

Grundsätzlich finden sich bei Nietzsche und Sartre beide Bestandteile des Vorwurfs des Nihilismus, sowohl jener des Atheismus wie jener der Ablehnung der traditionellen Moral. Beide Philosophen sind bekennende Atheisten. Seinen Angaben gemäß spricht Nietzsche nie das Tischgebot an der Pforta (Nietzsche, eKGWB, NF-1879, 42[68]) und in *Die fröhliche Wissenschaft* lässt er den tollen Menschen ausrufen: „Wohin ist Gott? rief er, ich will es euch sagen! Wir haben ihn getödtet" (Nietzsche, eKGWB, FW-125). Ähnlich stellt auch Zarathustra fest: „Dieser alte *Gott* nämlich lebt nicht mehr: der ist gründlich *todt*." (Nietzsche, eKGWB, Za-IV-Dienst; siehe auch eKGWB, Za-IV-Menschen-2). Sartre wiederum sagt über sich, dass er schon mit zwölf Jahren den Glauben an Gott verlor (Sartre 1996: Carn, S. 260; Sartre/Beauvoir 1986: EBS, S. 487, 553–554), verlorengegangen auf dem Schulweg (Sartre 1972: Mots, S. 142). Seine Philosophie stuft er in die Kategorie des atheistischen Existentialismus ein (Sartre 2000a: EH, S. 148).

Da sich im Laufe der historischen Entwicklung des performativen Gebrauchs des Terminus „Nihilismus" das Gewicht immer mehr vom Atheismus auf die Kritik an den herrschenden Moralvorstellungen verschiebt, will ich mich im Folgenden auf den moralischen Aspekt des Nihilismusbegriffs konzentrieren[3]. Trotzdem gestaltet sich dieses Vorhaben nicht einfach, denn Nietzsche und Sartre gehen in ihren philosophischen Tätigkeiten von unterschiedlichen Selbstverständnissen aus. Nietzsche sah sich nicht als klassischen Philosophen, sondern eher als „Arzt der Cultur" (siehe u. a. Nietzsche, Bf. an Erwin Rohde, ca. 22. März 1877, BVN-1873,300), der mit einer Anamnese beginnt, hernach eine Diagnose erstellt und eine Therapie empfiehlt (Kuhn 1992, S. 198–236). Diesem Ansatz steht Sartre grundsätzlich kritisch gegenüber. In den 1950er und 1960er Jahren wirft er in analogen Fällen den Marxisten und Michel Foucault – in *Questions de méthode* (Sartre 1964: QM, S. 28) resp. *Sartre répond* (Sartre 1966: Arc66, S. 95) – vor, dass

3 Siehe auch Sommer (2006, S. 250, 256), wo Andreas Urs Sommer den Nihilismus bei Nietzsche zwar im Sinne des Assassinen-Motivs als Negation jeglicher moralischen und metaphysischen Wahrheit definiert, aber dann doch festhält, dass die Frage der Moral die primäre des Nihilismus ist.

sie übersähen, dass es der Mensch ist, der die Geschichte macht. Noch gewichtiger ist Sartres Selbstverständnis als eines sehr klassischen, weil systemischen Philosophen. Philosophie ist für Sartre nicht nur systematisches, sondern wesentlich auch systemisches Denken. Ähnlich wie in Kants *Logik* (Kant 1800, S. 25) steht bei Sartre die Anthropologie – mit der Psychologie von *L'être et le néant* und der Sozialontologie der *Critique de la raison dialectique* – im Zentrum seines philosophischen Denkens. Diese Anthropologie beruht einerseits auf Sartres Ontologie und Epistemologie und bildet andererseits die Basis der Ethik und Metaethik und darauf aufbauend von Sartres Ästhetik und dessen politischer Philosophie der 1970er Jahre. Nietzsches auf Aphorismen beruhende Kulturphilosophie bildet geradezu das Gegenstück zu einer Philosophie in Sartres Verständnis. Wenig erstaunlich bezeichnet Sartre schon im Alter von 19 Jahren in seinem *Carnet Midy* Nietzsche als einen Dichter, der das Pech hat, für einen Philosophen gehalten zu werden (Sartre 1990: CMid, S. 471)[4].

Moral aus ästhetischer Sicht – die Neigungen und das subjektiv-individuelle Gute

Der Vergleich von Nietzsches und Sartres Moraltheorien erfolgt im Folgenden in lockerer Anlehnung an Kierkegaards Drei-Stadien-Modell mit dem ästhetischen, dem ethischen und dem religiösen Stadium. Jedem dieser drei Stadien lässt sich eine typische Sichtweise auf die Moral zuordnen:

- in der ästhetischen Sicht auf Moral stehen die Neigungen als Ausdruck des subjektiv-individuellen Guten im Zentrum,
- in der ethischen Sicht auf Moral liegt der Fokus auf dem Entscheidungskriterium im Fall von Konflikten zwischen den Neigungen und
- in der religiösen Sicht auf Moral geht es um die Frage nach der Möglichkeit und Bedeutung einer objektiven Moral.

4 In *Aller et retour* hält Sartre 1944 nochmals fest: „Bekanntlich war Nietzsche kein Philosoph." (Sartre 2000c: AllR, S. 102) Die Beurteilung Nietzsches als Dichter darf aber, zumindest für den Fall von 1924, nicht nur negativ verstanden werden, denn Sartre selbst will zu jener Zeit nicht Philosoph, sondern Schriftsteller werden. Nahm er sich Nietzsche sogar als Vorbild? In *Une défaite*, einem 1927 geschriebenen Roman über Nietzsche, Wagner und Cosima, erscheint Sartre in der Figur des Frédéric (Friedrich Nietzsche), weshalb er von seinen Freunden den Spitznamen *pauvre Frédéric* erhält (Beauvoir 2000: FA, S. 35).

Heute von eher untergeordneter Bedeutung ist die ästhetische Sichtweise, in der Geschichte der Moraltheorien lange eine sehr gewichtige, wenn nicht sogar die entscheidende. Ευδαιμονία ist der zentrale Begriff der antiken Ethik – wie immer diese Glückseligkeit definiert wird, ob in einem epikureischen, stoischen, kynischen oder kyrenäischen Sinn. Die Gegenposition hierzu wird durch Kant vertreten, der die Neigungen im Rahmen seiner deontologischen Ethik für ethisch irrelevant hält. Mit dieser Haltung bringt Kant geradezu paradigmatisch jene lustfeindliche Haltung des deutschen Bürgertums zum Ausdruck, die letztlich im Gehorsam gegenüber Führer und Drittem Reich kulminiert[5]. Mit Ausnahme von Foucault, der die antike Ethik in den 1970er Jahren zu rehabilitieren versucht, dominiert in Politik und Kultur seit zwei Jahrhunderten eine deontologische Moral, insbesondere im deutschsprachigen Raum.

Was Nietzsche und Sartre in Bezug auf den ästhetischen Blickwinkel eint, ist die große Bedeutung, die beide ihm zuerkennen. Beide gehören mit den Utilitaristen, aber auch Freud, zu den relativ wenigen modernen Vertretern, für die der ästhetische Blickwinkel zentral ist. Unabhängig davon, welcher Begriff hierbei im Zentrum steht, ob Trieb, Begierde, Bedürfnis oder Nutzen, was sie alle eint, ist der damit verbundene Relativismus im Sinne von *de gustibus et coloribus non est disputandum*. Um zum Ausdruck zu bringen, dass es nicht nur eine, sondern viele Moralen gibt, benützt Nietzsche das Wort Moral immer wieder im Plural. Ähnlich versteht auch Sartre die Moralen im Plural als Ausdruck unterschiedlicher sozialer Systeme, Klassen, Milieus oder Gesellschaften (Sartre 2005a: MorH, S. 307; Sartre 2015: RacE, S. 16, 36). Schon in *L'être et le néant* bringt Sartre seinen Relativismus deutlich zum Ausdruck, wenn er festhält, dass alle menschlichen Aktivitäten äquivalent sind, ob man sich nun einsam betrinkt oder Völker regiert (Sartre 1995: EN, S. 1071). Kants Reich der Zwecke wird via Marxens Reiche der Freiheit und der Notwendigkeit in den 1970er Jahren bei Sartre zur Grundlage einer anarchistischen politischen Philosophie, in der Menschen mit ähnlichen Neigungen in separaten Gemeinschaften zusammenleben. Bei Nietzsche lässt sich ein ähnlicher Zug erkennen, denn für ihn wie alle anderen Relativisten gilt, dass die subjektive ευδαιμονία ihren Ursprung nur in etwas Individuellem haben kann. Zusammen mit den Utilitaristen stehen Nietzsche und Sartre in der Tradition des Sophisten Protagoras und der Δισσοί λόγοι. Im Gegensatz zu den Vertretern der ionisch-attischen Philosophie von Thales bis Platon ist der thrakischen Philosophie – Leukippos, Demokritos und vor allem Protagoras – die Vorstellung

5 Siehe Adolf Eichmanns Aussage während des Prozesses in Jerusalem, dass er nur seine Pflicht erfüllt habe und den Gesetzen gehorchte. Er sei sein Leben lang den Moralvorschriften Kants gefolgt (Arendt, 1996, S. 231–235).

eines *unum, verum, bonum, pulchrum* fremd, das die Grundlage für jene Vorschriften bildet, was andere als wahr, gut und schön zu halten haben.

Doch so nah sich Nietzsche und Sartre in der grundsätzlichen Haltung in Bezug auf die ästhetische Sichtweise sind, es gibt auch unübersehbare Differenzen. Sie zeigen sich schon in der Begrifflichkeit. Einer der für Nietzsche in diesem Zusammenhang zentralen Begriffe ist jener des ‚Willens zur Macht', verstanden hier weder in einem politischen Sinne wie zuzeiten des Dritten Reiches noch in einem metaphysischen wie bei Heidegger, sondern als das Streben hin zum Leben in all seinen Ausprägungen. In *Jenseits von Gut und Böse* schreibt Nietzsche: „Vor Allem will etwas Lebendiges seine Kraft auslassen – Leben selbst ist Wille zur Macht." (Nietzsche, eKGWB, JGB-13) Ähnlich heißt es in der *Götzen-Dämmerung:* „jede gesunde Moral ist von einem Instinkte des Lebens beherrscht" (Nietzsche, eKGWB, GD-Moral-4). Wenn Nietzsche anders als Schopenhauer nicht vom Willen zum Leben, sondern vom Willen zur Macht spricht, dann hat dies seine Gründe, wie er im *Zarathustra* schreibt: „Vieles ist dem Lebenden höher geschätzt, als Leben selber" (Nietzsche, eKGWB, Za-II-Ueberwindung). In einem Fragment von 1881 hält er sogar fest: „Es giebt keinen Selbsterhaltungstrieb!" (Nietzsche, eKGWB, NF-1881, 11[108]). Leben ist mehr als nur Selbsterhaltung und Erhaltung der Gattung: „Das Leben ist bloß ein E i n z e l f a l l des Willens zur Macht" (Nietzsche, eKGWB, NF-1888, 14[121]). So fragt Nietzsche in *Jenseits von Gut und Böse:* „Ist Leben nicht Abschätzen, Vorziehn, Ungerechtsein, Begrenzt-sein, Different-sein-wollen?" (Nietzsche, eKGWB, JGB-9) und in *Zur Genealogie der Moral* lobt er die vornehmen Rassen für „ihre Gleichgültigkeit und Verachtung gegen Sicherheit, Leib, Leben, Behagen, ihre entsetzliche Heiterkeit und Tiefe der Lust in allem Zerstören, in allen Wollüsten des Siegs und der Grausamkeit" (Nietzsche, eKGWB, GM-I-11). Im *Zarathustra* preist er sogar das Böse: „Denn das Böse ist des Menschen beste Kraft. ‚Der Mensch muss besser und böser werden' – so lehre ich. Das Böseste ist nöthig zu des Übermenschen Bestem." (Nietzsche, eKGWB, Za-IV-Menschen-5). Damit spurt er Freud vor, der in seinem Werk *Jenseits des Lustprinzips* von 1920 dem Lebenstrieb Eros den Todestrieb Thanatos entgegensetzt.

Nietzsches Relativierung des Selbsterhaltungstriebs und sein Lob des Bösen finden ihre Parallelen bei Sartre und Beauvoir. Richtet sich Nietzsches Kritik vor allem gegen den mit Charles Darwin und Herbert Spencer aufkommenden Biologismus, so jene von Sartre und Beauvoir gegen die Reduktion des Menschen auf den Sexualtrieb im Zusammenhang mit der Freudschen Psychoanalyse. Nach der kurzen Periode von 1944 bis 1949, in der Sartre und Beauvoir eine Verbindung ihrer relativistischen existentialistischen Moral mit dem objektiv Guten anstrebten, kehren sie mit *Faut-îl brûler Sade?* und *Saint Genet* wieder zu ihrer alten Auffassung zurück, wonach der Mensch in seinem Leben nicht nur das Gute, sondern auch das Böse wählen kann – das Böse hier verstanden als Kategorien einer sozialen Moral und nicht eines individuellen moralischen Urteils. Der lebende Mensch bedarf sogar des Bösen, denn wie Sartre, der Bewunderer von Baudelaires *Les fleurs du mal*, es in *Saint Genet* ausdrückt: „das Gute ohne das Böse ist das Parmenideische Sein, das heißt der Tod" (Sartre 1986: SG, S. 293). Entsprechend kann Beauvoir behaupten, dass es unter den humanistischen Philosophien allein der Existentialismus ist, der ähnlich den Religionen das Böse wirklich in Betracht zieht (Beauvoir 1997: PMA, S. 98–99).

Auf eine wesentliche Differenz zwischen Nietzsche und ihm in Bezug auf die ästhetische Sichtweise verweist Sartre in seinen Tagebüchern von 1939/40:

> Die Absurdität des ‚Willens zur Macht' bei Schopenhauer oder Nietzsche besteht darin, daß man, wenn man ihn als eine Kraft auffaßt, niemals verstehen kann, daß er sich durch *Begierden* oder *Willensäußerungen* ausdrückt. (Sartre 1996: Carn, S. 450)

Im Zentrum steht hier die Frage nach dem Verhältnis zwischen der Freiheit des Subjekts und dessen Neigungen. Ob einer Sklaven- oder Herrenmensch ist, ist für Nietzsche eine Frage des Typus oder Charakters[6] – eines Wortpaars, dessen sich Nietzsche seit seinen frühen Werken um 1870 bedient. Nicht unähnlich Freuds Modell scheint der Mensch bei Nietzsche von Trieben regiert zu sein. Selbst im Intellekt sieht Nietzsche nur das blinde Werkzeug eines Triebs (Nietzsche, eKGWB, M-109; JGB-158). Seit den Anfängen zeigt er sich gegenüber der Freiheit des Willens (Nietzsche, eKGWB, Bf. an Wilhelm Pinder, Ende März – Anfang April1859, BVN-1859,62) reserviert, die er im *Antichrist* sogar als imaginäre Ursache bezeichnet (Nietzsche, eKGWB, AC-15). Entsprechend heißt es – in essentialistischer Manier – in *Die fröhliche Wissenschaft:* „‚Du sollst der werden, der du bist.'" (Nietzsche, eKGWB, FW-270).

Sartres existentialistische Version lässt sich stattdessen eher zusammenfassen als ‚Du bist der, den du sein willst' (siehe Sartre 2000a: EH, S. 149–150). Für Sartre ist die Willensfreiheit ein Eckstein seiner Philosophie. Anders als Nietzsche und Freud und in Linie mit der Psychologie von Alfred Adler und Ludwig Binswanger, die später auch von Viktor Frankl und der amerikanischen humanistischen Psychologie übernommen wird, erklärt sich bei Sartre die Wahl der Neigungen nicht durch Typus oder Charakter eines Menschen oder durch dessen Triebe, sondern durch dessen Entwurf oder Lebensplan, eine präreflexive Urwahl. Diese ist das Ergebnis eines auf Willensfreiheit beruhenden, willkürlichen Aktes

6 Andreas Urs Sommer spricht in seinem Kommentar zu *Jenseits von Gut und Böse* in diesem Zusammenhang von psychischen Dispositionen, zuerst von Individuen, dann von ganzen Gesellschaften, die (anders als bei Sartre, A.B.) nicht wie Weltanschauungen beliebig gewechselt werden können (Sommer 2016, S. 124).

(*acte gratuit*). Der Mensch findet seine Werte und seinen Lebensplan nicht, sondern erfindet sie (Sartre 2000a: EH, S. 158). „Wenn Gott nicht existiert, haben wir allein über den Sinn des SEINS zu entscheiden" (Sartre 2005b: CM, S. 847). Sartre hält zeitlebens an dieser absoluten ontologischen Freiheit fest. Das Skandalon, dass auch der Sklave diese absolute Freiheit besitzt, findet sich nicht nur in *L'être et le néant*, sondern auch in der *Critique* (Sartre 1995: EN, S. 944; 1967: CRD, S. 612).

Sartre will damit nicht die Bedeutung von Trieben, Genen oder der Erziehung leugnen und noch weniger, dass die Möglichkeiten, aus denen ein Mensch wählen kann, beschränkt sind. Schon 1946 hält Sartre fest: „Frei sein heißt nicht, tun können, was man will, sondern wollen, was man kann" (Sartre 2000b: LibCar, S. 128). Zehn Jahre später führt Sartre seine theoretischen Vorstellungen zur Konstituierung und der darauffolgenden Personalisierung des Menschen weiter aus: der Mensch ist nicht das, wozu er gemacht wurde; der Mensch ist das, was er daraus macht, wozu er gemacht wurde (Sartre 1964: QM, S. 28; 1966: Arc66, S. 95). Für Sartre stehen entsprechend nie die Triebe im Vordergrund, sondern, in zeitlicher Reihenfolge, zuerst die Begierden und später die Bedürfnisse (Sartre 1995: EN, S. 185; 2015: RacE, S. 72), in denen die Triebe zwar ihren Niederschlag finden, aber immer vermittelt durch den Entwurf des Individuums.

Moral aus ethischer Sicht – das Entscheidungskriterium im Fall von ethischen Konflikten

So nahe sich Nietzsche und Sartre in der Beurteilung der Moral aus dem ästhetischen Blickwinkel stehen, so groß ist ihr Gegensatz in Bezug auf die ethische Sichtweise. In dieser geht es um die für die moderne Ethik zentrale Frage, wie Konflikte, intra- und intersubjektive, zwischen verschiedenen Neigungen zu lösen sind. Die antiken Ethiken versuchten dies mittels der Tugendlehre. Bei Kant, für den dieser Blickwinkel, der in der Ethik entscheidende ist, werden die Tugenden durch ein vernunftbasiertes Kriterium abgelöst, den kategorischen Imperativ. Eine strukturell verwandte Antwort findet sich später in John Rawls' *A Theory of Justice*, in der die Frage nach der Gerechtigkeit über jener nach dem Guten steht. Für die Utilitaristen hingegen ist die Entscheidung eher eine mathematische Frage nach der intersubjektiven Maximierung des totalen Nutzens. Einen ganz anderen Weg verfolgen die Marxisten, die nicht wie die vorhergehenden Ethiken auf ein formales, inhaltlich offenes Kriterium, sondern auf ein materielles setzen: gut ist, was der proletarischen Revolution und der Schaffung einer kommunistischen Gesellschaft dient.

Vergleichen wir Nietzsches Auffassung mit diesen Ethiken, so steht sie strukturell wohl der marxistischen am nächsten. Wie diese setzt auch Nietzsche auf ein materielles Kriterium: gut ist, was der Herrenmoral entspricht, dessen Vertreter der Übermensch ist (Nietzsche, eKGWB, Za-IV-Menschen-2; EH-Bücher-1). Dieses materielle Kriterium steht allerdings in Widerspruch zum Relativismus der Moral in der ästhetischen Dimension. Auch wenn das materielle Kriterium durch einen Rekurs auf das Sein – sei es auf ein entsprechendes Geschichtsmodell wie im Marxismus oder bei Nietzsche auf eine Genealogie der Moral – nicht einem *deus ex machina* gleich eingeführt wird, dem Problem des Sprungs vom Sein ins Sollen können weder Nietzsche noch die Marxisten entgehen. Dies gilt auch für Nietzsches Analyse der herrschenden Sklavenmoral als einer Perversion, die den ursprünglichen Gegensatz von Gut und Schlecht aufgrund von Ressentiments in einen Gegensatz von Gut und Böse verwandelt hat (Nietzsche, eKGWB, JGB-260; GM-I-11). Eine Aussage über historisches Sein kann aus der Perspektive von Sartres Philosophie grundsätzlich nicht eine Aussage bezüglich ethischen Sollens rechtfertigen – was auch für Nietzsches Forderung nach einer Umwertung der Werte im Sinne der Rückkehr zu einer Herrenmoral gelten muss.

Dass der Gegensatz zwischen Sartre und Nietzsche kein totaler ist, zeigt sich in Sartres und Beauvoirs Versuch in den Jahren 1944 bis 1949, mit dem Respekt vor der Freiheit des Andern ein materielles Kriterium in ihre Ethik einzuführen. „Ich kann meine Freiheit nur zum Ziel machen, indem ich auch die der anderen zum Ziel mache“, heißt es in Sartres *L'existentialisme est un humanisme* (Sartre 2000a: EH, S. 172). Zur Zeit der Abfassung der *Cahiers pour une morale* spielt Sartre sogar mit dem Gedanken der Notwendigkeit einer allgemeinen Konversion, um umfassende praktische Freiheit in einer Welt ohne gegenseitige Unterdrückung zu realisieren. Später wird Sartre diesen Versuch einer Ethik mit deutlich kantianischer Schlagseite als idealistisch, „völlig mystifiziert“ verurteilen (Sartre/Astruc/Contat 1978: SFilm, S. 66). Das zunehmende politische Engagement, bei dem sich Sartre mit den Widrigkeiten der konkreten Welt auseinandersetzen muss, und insbesondere seine und Beauvoirs Beschäftigungen mit Genet und Sade führen 1948 – 51 zur Aufgabe der Beschäftigung mit einer normativen Ethik, wie sie sein Denken von *L'existentialisme est un humanisme* bis zu den *Cahiers pour une morale* kennzeichnet.

Sartre kehrt wieder zu seiner ursprünglichen existentialistischen Ethik zurück – nur noch die Einheit von Zwecken und Mitteln lässt er aus seiner Zeit der normativen Ethik gelten[7]. Mit *Les mouches* wird die existentialistische Ethik

---

7 Mit ihrer Version der Verantwortungsethik (Zwecke und Mittel bilden eine Einheit, die das Subjekt vor dem konkreten Andern verantworten muss) lehnen Sartre und Beauvoir sowohl die

überwiegend als eine Ethik der Freiheit verstanden, was sie in ästhetischer Sicht auch durchaus ist. Sprechen wir jedoch über Moral aus einer ethischen Sicht, so steht ihr Charakter als Diskursethik im Vordergrund. Der Mensch muss sich vor seinen Mitmenschen – und zwar den konkreten, nicht den abstrakten – für seine Taten ver-antworten, sie rechtfertigen. „Der Mensch ist dazu verurteilt, frei zu sein. Verurteilt, weil er sich nicht selbst erschaffen hat, und dennoch frei, weil er, einmal in die Welt geworfen, für all das verantwortlich ist, was er tut" (Sartre 2000a: EH, S. 155). Diese Verantwortlichkeit gegenüber dem Andern – und nicht, dass die andern die Hölle sind – ist die Kernaussage seines Dramas *Huis clos* (*Geschlossene Gesellschaft*). Diesen Aspekt der Diskursethik betont Sartre nochmals in seiner Metaethik der 1960er Jahre. In seinem Vortrag am Gramsci-Institut in Rom 1964 geht Sartre sogar so weit, den Begriff der Ethik für solche diskursethischen Situationen zu reservieren, während der Begriff der Moral für die entfremdeten, durch Religion und Tradition überlieferten Moralsysteme steht (Sartre 2015: RacE, S. 36–49).

Ethisches Handeln ist für Sartre immer ein Handeln im Angesicht des Andern. Auch wenn Sartre Dostoevskijs Satz „Wenn Gott nicht existiert, ist alles erlaubt" liebt, so hält er doch schon in seinen Tagebüchern 1939/40 fest, dass die Existenz Gottes für die Ethik irrelevant ist, denn Moral ist eine Angelegenheit unter Menschen (Sartre 1996: Carn, S. 306). In *L'existentialisme est un humanisme* geht er noch einen Schritt weiter: „[S]elbst wenn Gott existierte, würde das nichts ändern" (Sartre 2000a: EH, S. 176) – der Mensch müsse sich immer noch vor dem andern ver-antworten. Hiermit wird einmal mehr deutlich, wie sehr Sartre nicht dem Protestantismus, sondern der katholischen Lehre nahesteht, wo die Sünden dem Priester und nicht Gott gegenüber bekennt werden.

Die für Nietzsche so bedeutende Unterscheidung zwischen Herren und Sklaven ist für Sartre nicht akzeptabel. Der letzte Satz in seinem autobiographischen Roman *Les mots* lautet: „Ein ganzer Mensch, gemacht aus dem Zeug aller Menschen, und der soviel wert ist wie sie alle und soviel wert wie jedermann" (Sartre 1972: Mots, S. 145; siehe auch Sartre/Beauvoir 1986: EBS, S. 315–323). Während die ersten drei Viertel dieser Aussage eher an Nietzsches Elitismus und

---

kommunistische Zweckethik (der Zweck rechtfertigt die Mittel) wie eine Gesinnungsethik in Camus' Art ab (keine Mittel, die den Zwecken widersprechen). Der Streit zwischen Sartre und Camus von 1952 beruht letztlich auf diesem Konflikt zwischen Verantwortungs- und Gesinnungsethik und lässt sich in deren Werken bis 1944 zurückverfolgen: Camus: *Lettres à un ami allemand* 1943–1945, *Ni victimes, ni bourreaux* 1946, *Les justes* 1949, *L'homme révolté* 1951; Beauvoir: *Idéalisme moral et réalisme politique* und *Les bouches inutiles* 1945; Sartre: *La responsabilité de l'écrivain* und *Qu'est-ce que la littérature?* 1946/47, *Cahiers pour une morale* 1947/48, *L'engrenage* 1946/48, *Les mains sales* 1948, *Le diable et le bon dieu* 1951.

den Übermenschen gemahnen (Louette 1996, S. 9), endet der Satz darin, dass auch dieser Übermensch nicht mehr wert ist als jedermann. Sartre ist ein Urdemokrat, für den alle Menschen gleichberechtigt sind und nicht diskriminiert werden dürfen. Mit seinem Engagement für Juden, Schwarze, Frauen, Homosexuelle und viele andere Minderheiten legt er hierfür zeitlebens konkretes Zeugnis ab.

Während in theoretischer Hinsicht Nietzsche und Sartre in der ethischen Dimension weitauseinander liegende Positionen beziehen, weist Sartres Position unter einem praktischen Blickwinkel durchaus gewisse elitäre Züge auf, die ihn wieder in die Nähe von Nietzsches Übermenschen bringen. Walter Kaufmann, der so viel für die Einführung von Nietzsche wie Sartre in den USA tut, hat nicht unrecht, dass die Ethik von *Les mouches*, personifiziert durch Orest, den Übermenschen, im Gegensatz zur Ethik von *Huis clos* nietzscheanisch ist (Kaufmann 1964, S. 70). In seiner Zeit am Lycée schließt Sartre mit Nizan einen Pakt der Übermenschen, bei dem sie sich die Namen der bretonischen Götter Bor'hou und R'hâ geben. An der ENS predigt Sartre eine nietzscheanische Moral der Freude, wie er in seinem Kriegstagebuch festhält (Sartre 1996: Carn, S. 279). Auch die Figuren von Roquentin in *La nausée* und Hilbert in *Érostrate* atmen den Geist des Übermenschen aus, einen Geist, den Sartre noch während der *drôle de guerre* zeigt – er nennt seine Kameraden seine Akoluthen – und von dem ihn wohl erst das Gefangenenlager befreit.

Seit den dreißiger Jahren vertritt Sartre eine Authentizitätsmoral mit Authentizität als höchstem Ideal, das nur wenige erreichen können. Er lässt zwar gewisse Ausnahmen zu – bspw. für Juden und Homosexuelle –, doch in *L'existentialisme est un humanisme* kämpft er vehement gegen die elitären *salauds*, die Schweinehunde, die in inauthentischer Haltung ihre Existenz für notwendig halten, und gegen die *lâches*, die Feiglinge, die Massen, die die totale Freiheit nicht wahrhaben wollen (Sartre 2000a: EH, S. 172). Die authentischen Menschen sind Sartres Übermenschen. Wenn er in seinem übrigen Werk über Verantwortung spricht, entsteht oft der Eindruck, dass seinen Forderungen nur der kritische Intellektuelle entsprechen kann, aber nicht der Proletarier am Förderband. Auch wenn Sartre den Begriff des Übermenschen kaum benutzt, ist er in seinem Werk indirekt sehr präsent. Um ihn nicht zu verwenden, nimmt Sartre eine geschickte Transposition vor: der Übermensch wird in seinen Texten zum Menschen, der (gewöhnliche) Mensch zum ‚Untermenschen'. Es ist ein Terminus, dessen sich Sartre von den *Cahiers pour une morale* bis zu *L'espoir maintenant* immer wieder bedient.

Sartres Verständnis von Übermenschentum steht jedoch nicht in der Tradition des Nazis mit ihrem Vernichtungswahn, sondern in jener progressiver deutscher und vor allem französischer Sozialisten. Zu diesen zählen Kurt Eisner, der Anführer der Novemberrevolution in München 1918, Lucien Herr, der wichtigste sozialistische Intellektuelle Frankreichs zu Beginn des 20. Jh., den Sartre noch selbst in seinem ersten Jahr an der ENS als Bibliothekar erlebt, und Charles Andler, der Nietzsche-Biograph und gute Bekannte von Sartres Großvater. In diesem Zusammenhang ist auch Jean Jaurès zu erwähnen, der bedeutendste französische Sozialist zu Beginn des 20. Jh., der in drei Vorträgen über *Nietzsche et le socialisme* im Februar 1902 in Genf die Ansicht vertritt, wonach die ganze Menschheit zu Übermenschen in Nietzsches Sinn werden müsse (zit. nach Rebérioux 1961, S. 14). In seinen Gesprächen mit Beauvoir hält Sartre entsprechend fest, dass die Existenz von Übermenschen und Genies nicht im Gegensatz zur Idee der Gleichheit steht, denn sie seien auch nur Menschen (Sartre/Beauvoir 1986: EBS, S. 317). Es ist ein Verständnis, das – wie es in der russischen Philosophie heißt (Stellino 2015, S. 19) – den christlichen богочеловек (bogočelovek, бог=-Gott; человек=Mensch) durch den человекобог (čelovekobog) ersetzt, den christlichen Gottmenschen durch den Feuerbachschen Menschengott. Jeder Mensch soll ein Übermensch werden.

Moral aus religiöser Sicht – Möglichkeit und Bedeutung von objektiver Moral

So bedeutend die Unterschiede zwischen Sartre und Nietzsche in Bezug auf die Moral unter einem ethischen Blickwinkel sind, so nahe stehen sie sich wiederum in einer religiösen Sichtweise. Hier steht nicht mehr die Frage nach den durch die Vernunft zu vertretenden ethischen Regeln im Fokus, sondern das Verhältnis der Moral zu den durch Religion, aber auch Tradition überkommenen Regeln. Was Religion und Tradition vereint, ist ihre Ansicht, dass es eine „objektive" Moral mit klaren, festen Regeln gibt. Kierkegaard stellt sich dieser Frage in *Furcht und Zittern* in seinen Äußerungen zum Problem, ob Abraham richtig handelt, als er sich anschickt, Isaak zu töten. Während sich Kierkegaard auf die Seite des religiösen Gebots stellt, vertritt Kant in *Die Religion innerhalb der Grenzen der bloßen Vernunft* die gegenteilige These. Nicht nur kann nach Kant ein Befehl, der der Moralität widerspricht, nicht von Gott stammen, vielmehr muss der Mensch in jedem Fall der Stimme der Vernunft folgen, denn bei der von ihm gehörten Gottesstimme kann er nie sicher sein, ob sie wirklich von Gott kommt. Für Kant ist klar, dass die religiöse Sichtweise hinter die ethische zurückzutreten hat (Schröder 2002, S. 65–80). Auch in der Diskussion zwischen den liberalen Vertretern einer Theorie der Gerechtigkeit in der Nachfolge von Rawls und den Kommunitaristen und in den Gesprächen zwischen Jürgen Habermas und Ratzinger geht es zentral um die

Frage nach der eigenständigen Berechtigung der (traditionell-)religiösen Sicht in der Diskussion um Moral. Von den Konservativen aller Provenienzen wird die Bedeutung einer ‚objektiven' Moral hervorgehoben, ob nun deren Advokaten zu den politisch konservativen Traditionalisten oder den Vertretern von theistischen oder politischen Religionen (wie jenen des Kommunismus) gehören. Da sich der Vorwurf des Nihilismus zentral um die Gültigkeit von ‚objektiven' Moralvorstellungen dreht, kommt der (traditionell-)religiösen Sicht die entscheidende Rolle hinsichtlich der Frage zu, ob Nietzsche und Sartre selbst Nihilisten im performativen Sinn des Wortes sind.

Nietzsche selbst bezeichnet sich immer wieder, von *Menschliches, Allzumenschliches* über *Jenseits von Gut und Böse* bis zu *Götzen-Dämmerung* und *Ecce Homo*, als (ersten) Immoralisten (Nietzsche, eKGWB, MA-I-Vorrede2–1; JGB-226; GD-Moral-3; EH-UB-2). In seinen für seine Vorstellungen über Moral zentralen Werken *Jenseits von Gut und Böse* und *Zur Genealogie der Moral* entwickelt er als Kulturphilosoph seine zentrale These einer historisch langfristigen Entwicklung von einer Herrenmoral zu einer auf dem Ressentiment und dem asketischen Ideal beruhenden Sklavenmoral. Gestützt auf die Herrschaft der Priester – zuerst jene der Juden, dann jene der Christen, abgelöst wiederum durch jene der Philosophen – wird der Gegensatz von Gut und Schlecht der Herrenmoral durch jenen von Gut und Böse der Sklavenmoral ersetzt. Dadurch übernehmen die Sklaven die Macht von den Herren. Das Christentum und auch große Teile der Philosophie sind nihilistisch, weil sie den Gegensatzbegriff des Lebens als höchstes Gut verehren (Nietzsche, eKGWB, NF-1888, 14[225]; NF-1888, 14[135]). Für Nietzsche ist die bisherige Moral lebensfeindlich (Nietzsche, eKGWB, NF-1886, 5[97]), was auch für den Sozialismus mit seinem *„Heerden-Blödsinn* des ‚Guten Wahren Schönen'" (Nietzsche, eKGWB, NF-1885, 34[177]; Kuhn 1992, S. 197–198) gilt, in dem sich in versteckter Form die christlichen Ideale wiederfinden (Nietzsche, eKGWB, NF-1888, 12[1]). In historischer Perspektive stellt Nietzsche seit der vorsokratischen Zeit eine Entwicklung Richtung Dekadenz und Entwertung aller Werte fest, die nur durch gelegentliche positive Momente und Perioden unterbrochen wurde. Zu diesen zählt er die Renaissance und den französischen Adel der Barockzeit. Nach Nietzsche sind wir in einer Zeit des Nihilismus angekommen, in der sich die obersten Werte entwerten (Nietzsche, eKGWB, NF-1887, 9[35]) und als „Arzt der Cultur" fordert er – in seiner Therapie ähnlich radikal wie in seiner Anamnese und Diagnose – nichts weniger als die Umwertung der Werte.

Schon aus theoretischen Gründen kann sich Sartre in gewissen Punkten Nietzsche nicht anschließen. Sartre hat gegenüber allen historisierenden Spekulationen große Vorbehalte. Da es für ihn letztlich die Subjekte sind, die in ihrer Freiheit Geschichte machen (Sartre 1964: QM, S. 28), verpönt er den historischen Determinismus, sowohl in seinen Vergangenheitsanalysen wie in seinen Prognosen für die Zukunft. Dies gilt für den Marxismus und Foucaults Analysen ebenso sehr wie für Nietzsches Auffassungen bezüglich Moral. Auch die Begriffe der Entwertung wie der Umwertung können aus prinzipiellen Gründen keinen Platz in Sartres Vokabular finden, da jeder Mensch immer wertbasiert handelt und alle Weltanschauungen mit den dazugehörenden Werten gleich wertvoll sind. Weder kann es bei Sartre eine Entwertung von Werten geben noch ist eine Forderung nach einer Umwertung begründbar. Sartres Position in Bezug auf diesen dritten religiösen Blickwinkel basiert, anders als bei Nietzsche, nicht auf einer kulturphilosophischen Analyse, sondern auf seiner Ontologie. Das An-sich enthält keine Werte; diese treten erst durch das Für-sich, das Bewusstsein, in die Welt ein. Und da es unzählige Individuen mit ihrem jeweiligen Für-sich gibt, gibt es viele unterschiedliche und auch gleichwertige Standpunkte. Oder, wie Sartre es unter Bezugnahme auf Dostoevskijs Satz „Wenn Gott nicht existiert, ist alles erlaubt" in *L'existentialisme est un humanisme* ausdrückt: es gibt kein *a priori* Gutes (Sartre 2000a: EH, S. 154), denn Existenz geht Essenz voraus – ginge die Essenz der Existenz voraus, gäbe es keine Moral (Sartre 2005b: CM, S. 73).

Nietzsches und Sartres Nihilismus und seine Folgen

Wo sich Nietzsche und Sartre treffen, ist in der Kritik der herrschenden Moral. Nietzsches Verurteilung des Ressentiments findet seine Parallele bei Sartre in der Missbilligung der Reue resp. der Gewissensbisse, die in Sartres Orestie-Drama durch die Fliegen personifiziert werden. Die Sklavenmoral, die Moral, die Argos in *Les mouches* beherrscht, ist eine fast die ganze Gesellschaft beherrschende Moral der *mauvaise foi*, der Unwahrhaftigkeit, der Bösgläubigkeit. Das heute vorherrschende Verständnis von Nietzsche als wegweisendem Kritiker der Moral in westlicher Tradition, wie es Peter Berkowitz konstatiert (Berkowitz 1995, S. ix), wird seinen Werken sicher gerechter als die frühere Auffassung von Nietzsche als einem konservativen Kulturkritiker. Während sich die Kritik der herrschenden Moral bei Nietzsche im Wesentlichen in der historischen Analyse zeigt, nimmt sie bei Sartre vor allem die Form eines literarischen und politischen Kampfes gegen die Vertreter der herrschenden Moral an. Für Sartre bilden Diskussionen um Moral ein entscheidendes Mittel im Kampf gegen Herrschaft. Als er sich in der ersten Hälfte der 1970er Jahre auf die Seite der Pseudo-Maoisten der Gauche Prolétarienne schlägt, begründet er dies primär mit den ausgeprägt moralischen Aspekten ihres politischen Kampfes. Über seinen Einsatz gegen Militarismus, (Neo-)Kolonialismus und Rassismus hinaus kämpft er zeitlebens gegen die herrschenden

Vorstellungen betreffend Autorität und Ehre, Geld und insbesondere Familie und Sexualität. Er setzt sich für Juden und Schwarze ein und lehnt alle Ehrungen und Formen der Respektbezeugungen bis hin zum Nobelpreis ab. In seinen literarischen Werken beschreibt er alle möglichen Varianten von Verhalten, die gegen die traditionelle Moral verstoßen, von unverheirateten Paaren, außerehelichen Verhältnissen und *ménages à trois* über Homosexualität und Voyeurismus bis zu Inzest und Abtreibung – wobei er, seiner Philosophie getreu, diese nie in einem befürwortenden Sinne beschreibt, sondern immer nur wertfrei als von anderen Menschen getroffene Wahl.

Wird Nihilismus in der in seiner Geschichte als Vorwurf erlangten Bedeutung der Ablehnung der herrschenden religiösen und traditionellen Werte verstanden, dann trifft dieser Vorwurf offensichtlich sowohl auf Nietzsche wie auf Sartre zu. Beider Kritik an der herrschenden Moral hat ihre Wurzeln in einer entsprechenden Ähnlichkeit in der ästhetischen Sichtweise. Auch wenn Nietzsches und Sartres Auffassungen von Moral vom ethischen Standpunkt aus deutliche Differenzen erkennen lassen, so überwiegen doch die Gemeinsamkeiten. Beide zählen zu den bedeutendsten Kritikern der herrschenden Moral des 19. und des 20. Jahrhunderts. In seinem oben zitierten Beitrag *Was ist Existentialismus?* erkennt Jaspers dies ganz richtig. Sowohl Nietzsches wie Sartres Kritik ist fundamental und trifft die respektablen Vertreter der Elite, auch der intellektuellen Elite, schwer. Nietzsche wie Sartre müssen dafür einen hohen Preis bezahlen. Ihr Denken ist trotz überwältigenden Erfolgs aus den akademischen Curricula weitgehend ausgeschlossen. Nietzsches aphoristischer Stil hat die Integration seines Denkens in die Akademia ebenso behindert wie Sartres politische Haltung. Die Rezeptionsgeschichte zeigt jedoch, dass dies nicht die wesentlichen Gründe gewesen sein können. Heideggers Nazi-Vergangenheit verhindert nicht dessen Aufnahme in den erlauchten Kreis der unvergänglichen Philosophen ab den 1960er Jahren. Blaise Pascal und Camus, beide eher Moralisten als Philosophen, stoßen trotz ihres Stils auf keine großen Vorbehalte.

Jaspers' Einwände geben schon eher einen Hinweis. Wie im Rahmen einer Konferenz vor zwei Jahren an der Universität Wien festgestellt werden konnte, deren Vorträge unter dem Titel *Sartre and the International Impact of Existentialism* (Betschart/Werner 2020) erschienen sind, schließen sich selbst dem Existentialismus gegenüber positiv eingestellte Philosophen zumindest teilweise dem Vorwurf des Nihilismus an. Der deutsch-britische Philosoph Fritz Heinemann, der viel für die Verbreitung des Existentialismus in Großbritannien tut, schreibt von *scènes amoureuses* in *L'être et le néant*, die ein Skandal in der Geschichte der Metaphysik seien (Heinemann 1971, S. 189). Und seine amerikanischen Kollegen, Kaufmann und Alfred Stern, der erste deutscher, letzterer österreichischer Herkunft, sprechen davon, dass *L'être et le néant* D.H. Lawrences *Lady Chatterley's*

Lover in den Schatten stelle resp. dass *L'être et le néant* teilweise pornographisch sei (Kaufmann 1975, S. 45; Stern 1967, S. 250). Die Vermutung liegt nahe, dass es bei Sartre wie bei Nietzsche weder die politische Haltung, mit der sie identifiziert werden, noch der Stil sind, die den Grund für den weitgehenden Ausschluss aus den höheren akademischen Kreisen bilden. Der Grund scheint eher in der von beiden geübten Kritik an den herrschenden moralischen Vorstellungen zu liegen. Viele Philosophen verstehen sich offensichtlich – genauso wie Politiker und viele anderen Mitglieder der Elite – als Wächter einer moralischen Anständigkeit im Sinne der traditionellen und z.T. auch religiösen Moral. Nichts belegt diese Feststellung so sehr wie die Ähnlichkeit in der Argumentation von Honecker und Staiger trotz des tiefen Grabens zwischen ihren politischen Ansichten.

Bibliographie

Adorno, Theodor W. (1961): *Noten zur Literatur II.* Frankfurt a. M.

Arendt, Dieter (1970): „Einleitung". In: Dieter Arendt (Hrsg.): *Nihilismus. Die Anfänge von Jacobi bis Nietzsche.* Köln, S. 9–106.

Arendt, Hannah (1996): *Eichmann in Jerusalem. Ein Bericht über die Banalität des Bösen.* München.

Beauvoir, Simone de (1997): „Für eine Moral der Doppelsinnigkeit". In: Simone de Beauvoir: *Soll man de Sade verbrennen?* Reinbek b. Hamburg, S. 77–192.

Beauvoir, Simone de (2000): *In den besten Jahren.* Reinbek b. Hamburg.

„Bericht des Politbüros an die 11. Tagung des Zentralkomitees der Sozialistischen Einheitspartei Deutschlands" (1965). http://www.argus.bstu.bundesarchiv.de/dy30bho/mets/dy30bho_2067/index.htm?target=midosaFraContent&backlink=/dy30bho/index.htm-kid-d5ceedf0-6b4b-482f-83ac-4192751495e7&sign=2067, besucht am 11. 4. 2020.

Berkowitz, Peter (1995): *Nietzsche. The Ethics of an Immoralist.* Cambridge (Mass.).

Betschart, Alfred/Werner, Juliane (2020): *Sartre and the International Impact of Existentialism.* Cham.

Brock, Eike (2015): *Nietzsche und der Nihilismus.* Berlin.

Daigle, Christine (2005): *Le nihilisme est-il un humanisme? Étude sur Nietzsche et Sartre.* Saint-Nicolas.

Dostoevskij, Fëdor (1880): *Братья Карамазовы* (*Die Brüder Karamasow*). https://klassika.ru/read.html?proza/dostoevskij/karams.txt&page=142, besucht am 18. 4. 2021.

Frisch, Max (1966): „Endlich darf man es wieder sagen". In: Walter Höllerer (Hrsg.): *Der Zürcher Literaturstreit.* Sprache im Technischen Zeitalter, Nr. 22. Stuttgart April/Juni 1967, S. 104–109 (24. Dez. 1966).

Goudsblom, Johan (190): *Nihilism and Culture.* Oxford.

Heinemann, Heinz (1971): *Existenzphilosophie lebendig oder tot?* Stuttgart.

Jacobi, Friedrich Heinrich (1799): *Jacobi an Fichte.* Hamburg.

Jaspers, Karl (1968): „Was ist Existentialismus?". In: Hans Saner (Hrsg.): *Aneignung und Polemik.* München, S. 497–501 (1951).

Johannes Paul II. (1998): „Enzyklika Fides et Ratio". http://www.vatican.va/content/john-paul-ii/de/encyclicals/documents/hf_jp-ii_enc_14091998_fides-et-ratio.html, besucht am 17.12.2020.

Kant, Immanuel (1800): *Logik*. Akademieausgabe. http://www.korpora.org/Kant/aa09/, Bd. 9, S. 1–150, besucht am 3.6.2020.

Kaufmann, Walter (1964): „Nietzsche between Homer and Sartre: Five Treatments of the Orestes Story". In: *Revue Internationale de Philosophie*, Bd. 18, Nr. 67 (1), S. 50–73.

Kaufmann, Walter (1975): *Existentialism from Dostoevsky to Sartre*. Rev. and exp. ed. New York.

Kuhn, Elisabeth (1992): *Friedrich Nietzsches Philosophie des europäischen Nihilismus*. Berlin.

Louette, Jean-François (1996): *Sartre contra Nietzsche*. Grenoble.

Lukács, Georg (1954): *Die Zerstörung der Vernunft. Irrationalismus und Imperialismus*. Bd. 2. Darmstadt 1974.

„Nuremberg Trial Proceedings Vol. 5" (1946). https://avalon.law.yale.edu/imt/01-17-46.asp, besucht am 17.12.2020.

Rebérioux, Madeleine (1961): „La philosophie de Nietzsche et le socialisme". In: *Bulletin de la Société d'études jauresiennes*. Nr. 3, Okt. 1961, S. 6–14.

Rieger, Abu Bakr (2020): „Islam und Nihilismus". http://www.abubakrrieger.de/islam-und-nihilismus/, besucht am 11.4.2020.

Sartre, Jean-Paul (1964): *Marxismus und Existentialismus*. Reinbek b. Hamburg.

Sartre, Jean-Paul (1966): „Jean-Paul Sartre répond". In: *L'Arc*, Nr. 30, S. 87–96.

Sartre, Jean-Paul (1967): *Kritik der dialektischen Vernunft*, Bd. 1. Reinbek b. Hamburg.

Sartre, Jean-Paul (1972): *Die Wörter*. Reinbek b. Hamburg.

Sartre, Jean-Paul (1986): *Saint Genet*. Reinbek b. Hamburg.

Sartre, Jean-Paul (1990): „Carnet Midy". In: Michel Contat/Michel Rybalka (Hrsg.): *Écrits de Jeunesse*. Paris, S. 443–497.

Sartre, Jean-Paul (1995): *Das Sein und das Nichts: Versuch einer phänomenologischen Ohtologie*. Reinbek b. Hamburg.

Sartre, Jean-Paul (1996): *Tagebücher. Les carnets de la drôle de guerre September 1939 – März 1940*. Reinbek b. Hamburg.

Sartre, Jean-Paul (2000a): „Der Existentialismus ist ein Humanismus". In: Jean-Paul Sartre: *Der Existentialismus ist ein Humanismus*. Reinbek b. Hamburg, S. 145–192.

Sartre, Jean-Paul (2000b): „Die cartesianische Freiheit". In: Jean-Paul Sartre: *Der Existentialismus ist ein Humanismus*. Reinbek b. Hamburg, S. 122–144.

Sartre, Jean-Paul (2000c): „Hin und zurück". In: Jean-Paul Sartre: *Der Existentialismus ist ein Humanismus*. Reinbek b. Hamburg, S. 55–112.

Sartre, Jean-Paul (2000d): „Zum Existentialismus. Eine Klarstellung". In: Jean-Paul Sartre: *Der Existentialismus ist ein Humanismus*. Reinbek b. Hamburg, S. 113–121.

Sartre, Jean-Paul (2005a): „Morale et histoire". In: *Les Temps Modernes*, Juillet-Octobre, Nr. 632–633–634, S. 268–414.

Sartre, Jean-Paul (2005b): *Entwürfe für eine Moralphilosophie*. Reinbek b. Hamburg.

Sartre, Jean-Paul (2015): „Les racines de l'éthique. Conférence à l'Institut Gramsci, mai 1964". In: *Études sartriennes*, 19. Bruxelles, S. 11–118.

Sartre, Jean-Paul/Astruc, Alexandre/Contat, Michel (1978): *Sartre: Ein Film von Alexandre Astruc und Michel Contat*. Reinbek b. Hamburg.

Sartre, Jean-Paul/Beauvoir, Simone de (1986): „Gespräche mit Jean-Paul Sartre. August-September 1974". In: Simone de Beauvoir: *Die Zeremonie des Abschieds.* Reinbek b. Hamburg, S. 167–568.
Schröder, Winfried (2002): *Moralischer Nihilismus. Typen radikaler Moralkritik von den Sophisten bis Nietzsche.* Stuttgart-Bad Cannstatt.
Slocombe, Will (2006): *The (Hi)Story of a Difficult Relationship from Romanticism to Postmodernism.* New York.
Sommer, Andreas Urs (2006): „Nihilism and Skepticism in Nietzsche". In: Keith Ansell Pearson (Hrsg.): *A Companion to Nietzsche.* Malden (MA).
Sommer, Andreas Urs (2016): *Kommentar zu Nietzsches* Jenseits von Gut und Böse. *Historischer und kritischer Kommentar zu Friedrich Nietzsches Werken.* Hrsg. v. d. Heidelberger Akademie der Wissenschaften. Bd. 5/1 (NK 5/1). Berlin, Boston.
Staiger, Emil (1967): „Literatur und Öffentlichkeit". In: Walter Höllerer (Hrsg.): *Der Zürcher Literaturstreit.* Sprache im Technischen Zeitalter, Nr. 22. Stuttgart, S. 90–97 (17. Dez. 1966).
Stellino, Paolo (2015): *Nietzsche and Dostoyevsky. On the Verge of Nihilism.* Bern.
Stern, Alfred (1967): *Sartre. His Philosophy and Existential Psychoanalysis.* New York.
Tongeren, Paul van (2018): *Friedrich Nietzsche and European Nihilism.* Newcastle.
Turgenjew, Iwan (1949): *Väter und Söhne.* Zürich.
Weißgerber, Ulrich (2010): *Giftige Worte der SED-Diktatur. Sprache als Instrument von Machtausübung und Ausgrenzung in der SBZ und der DDR.* Münster.

Eike Brock

Camus' Nietzsche. *Amor fati* und Revolte

Abstract: Camus's Nietzsche. *Amor fati* and Revolt. Camus truly is a Nietzschean philosopher; but being a Nietzschean philosopher rules out discipleship. Therefore, Camus develops an appreciative, but also at the same time critical assessment of Nietzsche's philosophy. This paper is about Camus's engagement with Nietzsche's philosophy in his two major essays *The Myth of Sisyphus* and *The Rebel*. In the process, I will pay close attention to Camus's changing estimation of Nietzsche's philosophical concept of *Amor fati*. Finally, I will ask the (open) question whether we can find a truly Camusian version of *Amor fati* in *The Plague*.

> Ich hatte immer das Gefühl, auf hohem Meer zu leben, bedroht,
> im Herzen eines königlichen Glückes. (Camus 2003, S. 144)

1 Einleitung

Der französische Philosoph Michel Onfray beklagt in seinem Buch über das Leben und Werk Albert Camus', *Im Namen der Freiheit*, dass es in der Rezeptionsgeschichte Camus' ein wichtiges Kapitel gebe, das bis dato viel zu kurz geraten sei: „*Camus als nietzscheanischer Philosoph* – das ist eine allzu selten erzählte Geschichte." (Onfray 2015, S. 75) Wirklich gibt es da eine Menge zu erzählen, zumal – und das stellt der Untertitel von Onfrays Buch im französischen Original[1] klar heraus: *La vie philosophique d'Albert Camus* – dieses Kapitel von zwei Denkern handelt, bei denen Philosophie nicht etwas ist, was man eben tut, so wie andere eben anderes tun (etwa Schwimmen, Reiten, Lesen), sondern eine Lebensform. Philosophie ist für Nietzsche und Camus ein Unternehmen von so gravierender existenzieller Bedeutung, dass ihr Leben und ihre Philosophie ein unzertrennliches Amalgam bilden: Ihr Leben durchpulst ihre Philosophie und ihre Philosophie durchwirkt ihr Leben so, dass ihr Leben philosophisch und ihre Philosophie lebendig ist.

Nun ergeht sich Onfray nicht einfach im Klagen, sondern er gibt sein Bestes, die Lage zu verbessern, indem er seine Version der Geschichte vom nietzschea-

1 Die deutsche Übersetzung des Untertitels „*Leben und Philosophie des* Albert Camus" suggeriert hingegen eine Trennung von Leben und Philosophie, was schon grundsätzlich problematisch ist (Nietzsche etwa schreibt in JGB 6, KSA 5, S. 19, ihm habe sich „jede grosse Philosophie" zuletzt als „das Selbstbekenntnis ihres Urhebers" offenbart), in Camus' Fall aber ganz bestimmt fehlgeht.

https://doi.org/10.1515/9783110760163-007

nischen Philosophen Camus erzählt. Außerdem beschränkt er sich nicht auf das Erzählen, sondern er tut zudem, was Philosophen tun, d. h. er macht sich daran, den Gegenstand seiner Betrachtung so genau wie möglich zu bestimmen. Onfray arbeitet also mit begrifflichen Mitteln heraus, was es bedeutet, ein nietzscheanischer Philosoph zu sein. Sein erster Annährungsversuch besteht dabei in einer Negation, einer Abgrenzung um genau zu sein: Er bestimmt, was ein nietzscheanischer Philosoph gerade *nicht* ist, nämlich ein Epigone Nietzsches oder ein antichristlicher Nachbeter von Nietzsches säkularem Evangelium, dessen kategorischer Imperativ in den Worten Zarathustras lautet: „[B]leibt der Erde treu und glaubt Denen nicht, welche euch von überirdischen Hoffnungen reden!" (Nietzsche, Za I Vorrede 3, KSA 4, S. 15) Das ist eine Lektion, die Leserinnen und Leser Nietzsches spätestens seit *Also sprach Zarathustra* begriffen haben sollten, denn Zarathustra erklärt wiederholt, weder Jünger und Gläubige um sich scharen zu wollen, noch (solche) Schüler, die den Worten ihres Lehrers sklavisch ergeben sind: „Man vergilt einem Lehrer schlecht, wenn man immer nur Schüler bleibt" (Nietzsche, Za I Tugend 1, KSA 4, S. 101) ist genauso ein pädagogischer Hauptsatz Zarathustras wie folgende, im wahrsten Sinne des Wortes wegweisende Ermahnung: „Und ungern nur fragte ich stets nach Wegen – das ging immer gegen den Geschmack. Lieber fragte und versuchte ich die Wege selber. [...] ‚Das nun ist mein Weg – wo ist der eure?', so antwortete ich denen, welche mich nach dem Wege fragten. Den Weg nämlich, den gibt es nicht." (Nietzsche, Za III Schwere 2, KSA 4, S. 245) Ein nietzscheanischer Philosoph – das lässt sich nun als positive Bestimmung des gesuchten Gegenstandes verbuchen – nimmt sich Zarathustras Worte zu Herzen und macht sich daran, seinen eigenen Weg zu suchen, wobei ihm Nietzsche stets ein Wegbegleiter bleibt. Ein nietzscheanischer Philosoph weiß mithin genau, wessen Schüler er ist, nämlich der Schüler oder die Schülerin eines Lehrers, „der lehrt, dass man sich von ihm lösen muss, er [der Lehrer, EB] fertigt eine Karte der Wirklichkeit an, aber er zeichnet nicht den Weg vor. Er überlässt seinem Schüler die Navigation innerhalb eines Universums, das er gemeinsam mit diesem kartiert hat." (Onfray 2015, S. 76)

Mit Onfrays Version der Geschichte des nietzscheanischen Philosophen Camus, die kurz gesagt darauf hinausläuft, dass Nietzsche und Camus ein ideales Schüler-Lehrer-Paar im skizzierten Sinne abgeben, ist das Nietzsche-Kapitel der Camus-Rezeption allerdings nicht geschlossen. Die Geschichte ist einfach zu reichhaltig, als dass man schnell damit fertig werden könnte. Wollte man sie wirklich zu Ende erzählen, würde das Camus-Nietzsche-Kapitel unter der Hand zu einem umfangreichen Buch anwachsen – das dann vermutlich selbst wiederum kommentiert, kritisiert und ergänzt würde. Vor diesem Hintergrund sollte klar sein, dass ich mit diesem Text nicht den Anspruch erhebe, die „allzu selten erzählte Geschichte" (s. o., Onfray 2015, S. 75) zu Ende zu erzählen. Immerhin möchte ich aber dafür werben, sie weiter und neu zu erzählen, denn sie ist erzählenswert. Den folgenden Beitrag verstehe ich also als eine Art Basisarbeit, bei der ich mich zunächst darauf konzentrieren möchte, Camus' Nietzsche-Rezeption in seinen zentralen philosophischen Essays *Der Mythos des Sisyphos* und *Der Mensch in der Revolte* nachzuzeichnen. Im Zuge dessen wird sich zeigen, dass Nietzsches Versuch, das im Grunde doch absurde Dasein in der allumfassenden philosophischen Umarmung des *Amor fati* zu rechtfertigen, in beiden Essays den Kern von Camus' Auseinandersetzung mit Nietzsche bildet. Im *Sisyphos* ist der *Amor fati* sogar der Mittelpunkt des ganzen Essays. Denn hier ist es Sisyphos selbst, der Ja und Amen zu seinem maximal absurden Leben sagt. In *Der Mensch in der Revolte* revoltiert Camus dann jedoch (auch) gegen seine frühere Sisyphos-Philosophie, indem er Nietzsches *Amor fati* kritisiert. Am Ende des Beitrags komme ich in beinahe sträflicher Kürze auf Camus' Roman *Die Pest* zu sprechen. Dabei interessiert mich abermals der Gedanke des *Amor fati*. Insbesondere dieser letzte Teil soll eine Einladung sein, die Geschichte von Camus als nietzscheanischem Philosophen weiterzuerzählen, indem ich als offene Frage eine Szene aus *Die Pest* ins Spiel bringe, die ich für eine literarische Darstellung des *Amor fati* halte, in welcher der Schüler Camus dem Lehrer Nietzsche die Ehre erweist, mit dem Lehrer über den Lehrer hinauszugehen. Meine Frage lautet: Kann man diese Szene als genuin camussche Version des *Amor fati* lesen, in der sowohl die Philosophie des Sisyphos als auch die der Revolte zu ihrem Recht kommen.

2 Nietzsches Präsenz in *Der Mythos des Sisyphos*

Was bedeutet es, ein Mensch in der Welt zu sein? Und wie findet der Mensch einen bzw. seinen (je eigenen) Platz in der Welt? Das Mensch-Welt-Verhältnis ist womöglich *das* zentrale Thema der Existenzphilosophie und des Existenzialismus. Klar ist, dass es für den Menschen „keinerlei Möglichkeit [gibt], dieser Welt zu entfliehen" (Beauvoir 2007, S. 125). Es gibt für ein Wesen, das sich seiner selbst und der Welt bewusst ist, aber auch keine Möglichkeit, gleichsam glatt in der Welt aufzugehen. Der Mensch ist in der Welt, strebt intentional ständig auf sie zu (auf Dinge in der Welt) und ist dabei, wie Sartre nicht müde wird aufzuzeigen, doch immer von ihr getrennt. Der Mensch ist das differenzierende und negierende Wesen. So kommt durch ihn ein „Riß in das Sein" (Sartre 2006: EN, S. 170), der auch durch den Menschen selbst geht (vgl. Sartre 2006: EN, S. 42). Wie also, fragt Sartre, „müssen der Mensch und die Welt sein, damit das Verhältnis zwischen ihnen möglich ist"? (Sartre 2006: EN, S. 50) In *Der Mythos des Sisyphos* stellt

Camus die gleiche Frage. Freilich geht er bei der Beantwortung dieser Frage weniger systematisch vor als Sartre. Das Thema brennt ihm existenziell unter den Nägeln, darum will er möglichst rasch zu einer Antwort vorstoßen. Man versteht diese Dringlichkeit am besten vor dem Hintergrund des Ergebnisses seiner Reflexion über das Verhältnis des Menschen zur Welt: Camus stellt fest, dass das menschliche Leben absurd ist, weil es sich durch eine tiefe Sehnsucht nach Sinn und Verständnis seiner selbst auszeichnet, welcher die Welt nicht entspricht. Auch Camus konstatiert einen Riss im Sein, der unüberbrückbar zwischen dem menschlichen Sinn- und Klarheitsbegehren und der Welt klafft. Bleibt dann tatsächlich, wie Camus zu Beginn des Essays in den Raum stellt (vgl. Camus 2002: MS, S. 11), nur der Selbstmord? Mitnichten. Camus will, wie vor ihm Nietzsche, ein Fürsprecher des Lebens sein. Als ein solcher zieht er folgenden Schluss: Wenn das Leben nicht ohne das Absurde zu haben ist, muss man eben absurd leben. Es kommt dann darauf an, „Lebensregeln für diesen Zustand" (Camus 2002: MS, S. 56), d. h. für ein Leben im Zeichen des Absurden, zu ermitteln. Als oberste Regel gilt, das Absurde nicht zu verleugnen: „Leben heißt das Absurde leben lassen. Es leben lassen heißt vor allem ihm ins Auge sehen." (Camus 2002: MS, S. 72) Absurd zu leben bedeutet, mit offenem Visier einem unbesiegbaren Feind zu begegnen, ohne sich vor dessen Übermacht in den Staub zu werfen. Mit viel Pathos vertritt Camus eine Philosophie der Auflehnung.[2] Indem wir uns gegen das Unabänderliche stemmen, erfüllen wir unser Dasein mit einer vom Trotz angefeuerten, ungeheuren Intensität und Leidenschaft. Camus glaubt sogar an die wertbildende Kraft der Auflehnung: „Diese Auflehnung gibt dem Leben seinen Wert." (Camus 2002: MS, S. 73)

Um seiner Philosophie der Auflehnung schärfere Kontur zu verleihen, greift Camus auf den antiken Mythos zurück und präsentiert Sisyphos sowohl als Arche- und Prototyp wie auch als Idealgestalt des absurden Helden. Was Sisyphos für dieses Amt qualifiziert, ist nicht allein die legendäre Strafe bzw. die spezifische ‚Qual', die er erleiden muss: die ewige Wiederholung einer zutiefst absurden Tätigkeit. Sondern auch sein leidenschaftliches, von Götterverachtung geprägtes Vorleben disponiert ihn, die Rolle des absurden Helden zu besetzen (vgl. Camus 2002: MS, S. 156). Für seine Taten, die Camus in Form von drei Anklagepunkten zusammenfasst: „[s]eine Verachtung der Götter, sein Hass auf den Tod und sein

---

2 „Sie ist eine ständige Konfrontation des Menschen mit seiner eigenen Dunkelheit. Sie ist der Anspruch auf eine unmögliche Transparenz. Sie stellt die Welt in jeder Sekunde in Frage. [...] Sie ist die ständige Anwesenheit des Menschen bei sich selbst. Sie ist kein Sehnen, sie ist ohne Hoffnung. Diese Auflehnung ist nichts als die Gewissheit eines erdrückenden Schicksals, weniger die Resignation, die es begleiten sollte" (Camus 2002: MS, S. 72–73).

leidenschaftlicher Lebenswille" (Camus 2002: MS, S. 156)[3], haben die Götter Sisyphos hart bestraft. Er ist dazu verurteilt, einen Felsbrocken immer wieder aufs Neue auf den Gipfel eines Hügels hinauf zu wälzen, nur um dann erleben zu müssen, wie dieser gewaltige Stein von dort aus wieder ins Tal hinunterrollt. Dieses genauso mühsame wie sinnlose[4] Geschäft, bei dem „sein ganzes Sein sich abmüht, ohne etwas zu vollenden", ist ihm ewig aufgegeben. „Das ist der Preis für die Leidenschaften dieser Welt." (Camus 2002: MS, S. 156) Trotzdem behauptet Camus: „Wir müssen uns Sisyphos als einen glücklichen Menschen vorstellen." (Camus 2002: MS, S. 160) Dieses merkwürdige Glück begründet Camus mit der Klarsicht des Verurteilten. Während Sisyphos den Hang dem Felsen nach hinuntersteigt – „Auf diesem Rückweg, während dieser Pause, interessiert mich Sisyphos" (Camus 2002: MS, S. 157) –, bleibt ihm Zeit, seine Situation zu bedenken. Er erkennt ihre Ausweglosigkeit. Man sollte meinen, gerade diese hoffnungsraubende Einsicht würde ihn seine Strafe als Strafe erfassen lassen und so sein Leiden auf die Spitze treiben. Doch Sisyphos bringt die Kraft auf, sein grausames Schicksal zu verachten: „Die Klarsichtigkeit, die Ursache seiner Qual sein sollte, vollendet zugleich seinen Sieg. Es gibt kein Schicksal, das durch Verachtung nicht überwunden werden kann." (Camus 2002: MS, S. 158) Vom Gipfel des Hügels aus betrachtet Sisyphos sein Schicksal, den am Fuß des Hügels auf ihn wartenden Felsblock, und erhebt sich darüber, indem er es verachtet. Aber noch ist sein Glück nicht vollkommen. Erst wenn er in einer fulminanten Wendung das Verachtete zu lieben beginnt, dadurch dass er dem Schicksal das Moment der Fremdbestimmtheit nimmt, indem er es sich zu eigen macht, ist sein Triumph perfekt. Wie absurd es auch scheinen mag – Sisyphos bejaht sein Schicksal. Er macht es zu seinem Schicksal. Er entreißt es den Händen der Götter, indem er es an sich reißt. Dieser Aneignungsakt des Sisyphos hat auch eine prometheische Dimension. Die Auflehnung von Camus' Sisyphos ist nicht nur ein Akt der Selbstermächtigung eines bestimmten Individuums, sondern eine Tat, die auf die gesamte menschliche Sphäre ausstrahlt. Wenn Sisyphos glücklich sein kann, dann kann es auch der Rest der Menschheit; so zum Beispiel Ödipus:

> „Ich finde, daß alles gut ist", sagt Ödipus, und dieses Wort ist heilig. [...] Es vertreibt aus dieser Welt einen Gott, der in sie eingedrungen war mit der Unzufriedenheit und mit dem

---

3 Zu den Taten selbst vgl. Camus 2002: MS, S. 155.

4 Anzumerken ist allerdings, dass es sich bei dieser sinnlosen Tätigkeit um eine Strafe für ein bestimmtes Vergehen handelt. Mag also die Tätigkeit als solche auch sinnlos oder absurd sein, so verdankt sie sich doch nicht reiner Willkür, sondern hat ihren Platz innerhalb eines Systems (des Mythos), in dem sie Sinn ergibt. Im Rahmen des Mythos gilt: Wer sich den Göttern widersetzt, wird bestraft.

> Gefallen an sinnlosen Schmerzen. Es macht aus dem Schicksal eine menschliche Angelegenheit, die unter Menschen geregelt werden muss. Darin besteht die verborgene Freude des Sisyphos. Sein Schicksal gehört ihm. Sein Fels ist seine Sache. (Camus 2002: MS, S. 159)

Dass Camus sich mit diesen Überlegungen in großer philosophischer Nähe zu Nietzsche aufhält, ist unverkennbar. „‚Ich finde, daß alles gut ist'" (Camus 2002: MS, S. 159), – mit diesem Satz vollzieht der sich auflehnende Mensch zugleich die Affirmation seines Schicksals und die Identifikation mit seinem Los. Er hat, mit Nietzsche gesprochen, gelernt, „das Nothwendige an den Dingen als das Schöne [zu] sehen" und ist auf diese Weise „Einer von Denen" geworden, „welche die Dinge schön machen" (Nietzsche, FW 276, KSA 3, S. 521). Sisyphos ist Camus' vollkommene absurde Gestalt, der absurde Held und der Held des Absurden, weil er das Absurde nicht nur nicht verleugnet, sondern es in der Aktualisierung von Nietzsches existenzieller Formel *Amor fati* sogar umarmt.

3 Nietzsche in Der Mensch in der Revolte

Während Nietzsche in *Der Mythos des Sisyphos* zwar eminent wichtig ist, eine explizite Beschäftigung mit seiner Philosophie aber unterbleibt, hat Camus ihm in *Der Mensch in der Revolte* ein eigenes Unterkapitel mit dem Titel „Nietzsche und der Nihilismus" gewidmet. Das heißt allerdings nicht, dass Nietzsche nun eine noch wichtigere Rolle spielen würde als im früheren Essay. Eher ist das Gegenteil der Fall. Und das hängt, wie wir sehen werden, nicht zuletzt damit zusammen, dass Camus Nietzsche jetzt kritischer sieht.

Im Buch über die Revolte wird Nietzsche als besonders scharfsichtiger Denker präsentiert, der so klar wie niemand vor ihm erkannt hat, dass der Nihilismus – d. h. grob gesagt der Verlust des Glaubens an die Sinnhaftigkeit des Lebens und eine radikale existenzielle Desorientierung[5] – als Konsequenz des Todes Gottes krisenhaft heraufzieht. Tatsächlich hat Nietzsche den Zusammenhang zwischen der Erosion des christlichen Glaubens und dem Erstarken des Nihilismus in aller Deutlichkeit herausgearbeitet. Vor allem das sogenannte ‚Lenzer Heide-Fragment' (Nietzsche, NL 1887, KSA 12, 5[71]) ist in dieser Hinsicht ein Dokument eindrucksvoller Klarsicht. Es beleuchtet die maßgebliche und ambivalente Rolle Gottes in der ‚Sinngeschichte' des Abendlandes. Einstmals war Gott, wie Nietz-

5 Vgl. zu Nietzsches Verständnis des Nihilismus Brock 2015. Dort bemühe ich mich zu zeigen, dass der Nihilismus im Grunde zwei Gestalten hat: eine historische, nämlich die Selbstaufhebung der höchsten Werte des Abendlandes, und eine anthropologische, d. i. die in der menschlichen Selbstbezüglichkeit angelegte Möglichkeit, sich selbst radikal zu verneinen.

sche ausführt, ein mächtiges Bollwerk gegen den Nihilismus, der den am Leben leidenden Menschen, der sich den Launen des Schicksals hoffnungslos ausgeliefert sah, leicht befallen konnte. Mit der Einführung Gottes ließ sich das Leiden jedoch ertragen, weil es erstens als Folge menschlicher Hybris erklärt und eben dadurch in einen Sinnzusammenhang gebracht wurde, und weil zweitens die ebenfalls durch Gott begründete Würde des Menschen trotz der Korruption der menschlichen Natur durch die Sünde gewahrt blieb. Laut Nietzsche hat Gott jedoch mittlerweile ausgedient. Denn in einem Zeitalter des technischen und medizinischen Fortschritts, in einer Zeit wachsenden politischen Mitspracherechts, wird Gott als Tröster des Menschen, als Sinnstifter des Lebens und Erklärer des Leidens sowie als Garant des absoluten Wertes des Menschen nicht mehr benötigt. Mehr noch: Seine Existenz ist unglaubwürdig geworden, sodass er nicht nur antiquiert scheint, sondern sich in ein regelrechtes Ärgernis des vermeintlich aufgeklärten Verstandes verwandelt hat. Als Gegenmittel gegen den Nihilismus taugt die „christliche Moral-Hypothese" (Nietzsche, NL 1887, KSA 12, 5[71], S. 211) jedenfalls nicht mehr. Im Gegenteil ist es, meint Nietzsche, höchste Zeit, sich von einer Weltinterpretation zu lösen, die man eigentlich guten (intellektuellen) Gewissens nicht mehr unterschreiben kann. Andernfalls verwandelt sich das ehemalige Antidot in ein Gift, das die Krankheit des Nihilismus nicht mehr bekämpft, sondern erst recht befeuert. Gott war also, wie gesagt, dereinst

> das große Gegenmittel gegen den [...] Nihilismus. Aber unter den Kräften, die die Moral großzog, war die Wahrhaftigkeit: diese wendet sich endlich gegen die Moral, entdeckt ihre Teleologie, ihre interessirte Betrachtung – und jetzt wirkt die Einsicht in diese lange eingefleischte Verlogenheit, die man verzweifelt, von sich abzuthun, gerade als Stimulans. Zum Nihilismus. (Nietzsche, NL 1886, KSA 12, 5[71], S. 211)

Camus bewundert nun aber nicht nur Nietzsches Klarsicht,[6] sondern er schätzt auch seinen Mut, insofern er dem Nihilismus nicht ausweicht, obwohl er genau um den gefährlichen Rückkoppelungseffekt weiß, mit dem es philosophische Abenteurer zwangsläufig irgendwann einmal zu tun bekommen: „Und wenn du lange in einen Abgrund blickst, blickt der Abgrund auch in dich hinein." (Nietzsche, JGB 146, KSA 5, S. 98) Zu Nietzsches Mut zum Abgrund gehört auch und wesentlich, die Konsequenzen des Todes Gottes zu Ende zu denken, zu begreifen, dass eine „[d]es göttlichen Willens beraubt[e] [...] Welt gleicherweise ihrer Einheit und ihrer Finalität beraubt" (Camus 2016: HR, S. 97) ist. Es kommt darauf an, den Zusammenhang zu sehen, der besteht zwischen „der Unfähigkeit zu glauben", die Nietzsche bei sich und anderen diagnostiziert, und dem „Schwund

6 „In ihm tritt zum ersten Mal der Nihilismus ins Bewusstsein." (Camus 2016: HR, S. 95)

der elementaren Grundlage jedes Glaubens, des Vertrauens zum Leben“ (Camus 2016: HR, S. 96). Nietzsche ist allerdings kein bloßer Schwarzseher. Und als ein solcher würde ihn Camus auch nicht interessieren. Wohl aber knüpft Camus in seinem Essay an Nietzsche als „den Kliniker“ des Nihilismus an, der „nie anders gedacht [hat] als im Hinblick auf den Untergang, nicht um ihn zu verherrlichen, denn er erriet das schmutzige und berechnende Gesicht, das er schließlich annehmen würde, sondern um ihn zu vermeiden und ihn umzuwandeln in Wiederaufstieg“ (Camus 2016: HR, S. 95). Der Wiederaufstieg ist jedoch, davon ist Nietzsche überzeugt, nicht möglich ohne einen vorausgehenden Abstieg – und zwar in die Tiefen des Nihilismus. Aus diesem Grund verschiebt man das Problem des Nihilismus und verdrängt die vom Sinnverlust ausgehende Gefahr nur, wenn man versucht, mit Hilfe verschiedener Gottessurrogate, das Absurde von sich fernzuhalten, um dem destruktiven Einfluss, den die Entdeckung der Sinnleere am Grund aller Dinge und Prozesse auf das eigene Leben nehmen könnte, auszuweichen. Als maßgeblicher Vertreter einer metaphysischen Revolte[7], geht Nietzsche den Nihilismus dagegen geraden Weges an. Er treibt ihn „zur letzten Konsequenz“ und vollzieht, gleichsam als antichristlicher Descartes, „die methodische Verneinung, die Vernichtung dessen, was den Nihilismus vor sich selbst verbirgt, der Idole, die den Tod Gottes verschleiern“ (Camus 2016: HR, S. 96). Die Schleier, die den Tod Gottes so umwehen, dass sie den Blick in den Abgrund dieses „ungeheure[n] Ereignis[es]“ (Nietzsche, FW 125, KSA 3, S. 481) verhindern, müssen gelüftet werden. Andernfalls bleibt man im passiven Nihilismus gefangen, der vor allem in Form einer moralisch durchtränkten Erzählung die geistige Atmosphäre bildet, in der die Menschen leben. Es ist die moralische Geschichte von Schuld und Erlösung, von Strafe und Belohnung, die den narrativen Mittelpunkt des Christentums bildet. Camus stellt heraus, dass sie einen fatalen moralischen Zirkel ins Leben ruft, denn sie endet, wie sie beginnt, in der Verstrickung von Belohnung und Strafe: „Für das Christentum setzen Belohnung und Strafe eine Geschichte voraus. Aber mit unausweichlicher Logik bedeutet die ganze Geschichte am Ende nur Belohnung und Strafe [...]“ (Camus 2016: HR, S. 101). In aller Kürze ist das die Geschichte: Der Mensch sündigt. Er wird daraufhin bestraft, d. h. er wird des Paradieses, in dem er bis dato glücklich lebte, verwiesen. Nun lebt er nicht mehr im Garten Eden, sondern in der Welt. Das Weltleben ist mit Mühsal und Leid verbunden. Eines Tages jedoch, wenn das Reich Gottes anbricht, wird der Mensch erlöst oder möglicherweise auch für seine Sünden im Leben bestraft

7 Als Vertreter einer metaphysischen Revolte erkennt Nietzsche, dass es, nachdem „der Thron Gottes einmal umgestürzt [ist], [...] nun an ihm ist, jene Gerechtigkeit, jene Ordnung und Einheit, die er vergeblich auf seiner Lebensstufe gesucht hat, mit eigenen Händen zu erschaffen und damit die Absetzung Gottes zu rechtfertigen“ (Camus 2016: HR, S. 45–46).

werden. Das bedeutet freilich – und das ist für Nietzsche und Camus die philosophische Krux des Ganzen –, dass im Rahmen dieser Geschichte das Leben in der Welt gleichsam zu einer Strafexpedition wird. Die Gegenwart wird als problematischer, im Grunde unhaltbarer Zustand begriffen, den es zu überwinden gilt. Sie wird zugunsten einer zu einem Erlösungsgeschehen verklärten Zukunft abgewertet, die nebenbei auch noch Ressentiments und Rachegelüste teils befriedigt und teils befeuert, wenn man an das Jüngste Gericht und daran denkt, dass das Reich Gottes einem gewissen Prozentsatz der Menschen verschlossen bleiben könnte. Der „wahre Sinn" des Lebens, der für Camus und Nietzsche in der Gegenwart, im irdischen Leben Hier und Jetzt liegt, wird dabei zugunsten eines imaginierten Sinns geleugnet – und darin liegt (nach Nietzsche und Camus[8]) der nihilistische Kern des Christentums beschlossen:

> Das Christentum glaubt gegen den Nihilismus anzukämpfen, weil es der Welt eine Richtung weist, während es selbst nihilistisch ist, insofern es dem Leben einen imaginären Sinn gibt und die Entdeckung seines wahren Sinns verhindert. (Camus 2016: HR, S. 100)

Nun könnte man meinen, mit dem Tod Gottes verschwinde auch diese Erzählung aus der Welt. Dem ist aber nicht so; jedenfalls bleibt sie in ihren wesentlichen Momenten (Lohn, Bestrafung, Abwertung der Gegenwart zugunsten einer verklärten Zukunft) lebendig, denn Gott lebt auch nach seinem Tod weiter – als Gespenst:

> Überall wo verehrt, bewundert, beglückt, gefürchtet, gehofft, geahnt wird, steckt noch der Gott, den wir todt gesagt haben – er schleicht sich allerwegen herum und will nur nicht erkannt und bei Namen genannt sein. [...] – er lebt f o r t unter der seltsamen und neuen Bedingung, d a ß m a n n i c h t m e h r a n i h n g l a u b t. Aber ein Gespenst ist er geworden! Freilich! (Nietzsche, NL 1881, KSA 9, 14[14], S. 626)

Gott lebt, wie Camus feststellt, für Nietzsche vor allem in Form des Glaubens an die Zielhaftigkeit der Geschichte phantomhaft fort. Solange dieser etwa auch den Sozialismus prägende Glaube fortbesteht, bleibt der Nihilismus wirkmächtig:

> Der Sozialismus ist nur ein degeneriertes Christentum. Er hält in der Tat diesen Glauben an die Finalität der Geschichte aufrecht, der das Leben und die Natur[9] verrät, der ideale Ziele

---

8 Inwieweit sich das Christentum diesen Vorwurf wirklich gefallen lassen muss, steht auf einem anderen Blatt. Es gibt eine Reihe spannender, bedenkenswerter Auseinandersetzungen von Seiten der Theologie mit Nietzsche. Für das Problem des Nihilismus ist Romano Guardinis Nietzsche-Lektüre besonders anregend und ergiebig (vgl. dazu Voigt 2017).

9 Die Natur wurde mit der Einführung der moralischen Kategorien von Strafe und Belohnung in Geschichte transformiert (vgl. Camus 2016: HR, S. 100) bzw. zugunsten der Geschichte abgewertet.

den realen unterschiebt und dazu beiträgt, den Willen und die Vorstellungskraft zu entnerven. Der Sozialismus ist nihilistisch in dem nunmehr genauen Sinn, den Nietzsche dem Wort gegeben hat. Nihilist ist nicht derjenige, der an nichts glaubt, sondern derjenige, der nicht glaubt, an das, was ist. (Camus 2016: HR, S. 100 – 101)

Der von Nietzsche angestrebten Umwandlung des Nihilismus in einen Wiederaufstieg (vgl. Camus 2016: HR, S. 95) steht insbesondere das mit dem Glauben an die Finalität der Geschichte verwobene Wertesystem der herrschenden Moral im Weg. Sie ist gleichsam nihilistisch tiefenimprägniert und muss daher überwunden werden – daher Nietzsches erbarmungsloser Angriff auf die Moral und alle Ideale, die den Blick auf den Wert der Natur, des Lebens und den Augenblick verstellen. „Der Nihilismus", führt Camus aus, „ist das Endprodukt unserer sogenannten höchsten Werte" – und diese Werte garantieren zugleich das Fortleben des Nihilismus, der sich im Wertesystem in seiner passiven Form weitgehend unentdeckt und dementsprechend unbehelligt eingerichtet hat. Hier luxuriert er wie die Made im Speck. Der von Nietzsche mit *Menschliches, Allzumenschliches* ins philosophische Leben gerufene freie Geist wird die nihilistischen Werte jedoch – das ist Nietzsche Vision in den Worten Camus' – „zerstören", er wird „die Illusionen aufdecken, auf denen sie beruhen, das Feilschen, das ihnen zugrunde liegt, und das Verbrechen, das sie begehen, indem sie die klarsichtige Vernunft an ihrer Aufgabe hindern: den passiven Nihilismus umzuwandeln in einen aktiven" (Camus 2016: HR, S. 101). Diese Umwandlung ist jedoch eine gefährliche Zuspitzung. Der Mensch wird dadurch radikal auf sich selbst zurückgeworfen. Er muss, in nie gekanntem Maß, Verantwortung für sich selbst übernehmen: „In dieser von Gott und moralischen Idolen befreiten Welt ist der Mensch nun allein und ohne Herrn. [...] Ihm allein kommt es zu, die Ordnung und das Gesetz zu finden." (Camus 2016: HR, S. 101) Das ist eine unerhörte Aufgabe, die Nietzsche zugleich euphorisch stimmt und erschauern lässt. Er erwartet gewaltige, auch zerstörerische Umwälzungen. Als freiem Geist ist ihm angesichts dieser Zukunftsaussicht jedoch nicht bange, sondern er sieht vor seinem geistigen Auge vor allem einen Horizont ungeahnter Freiheit auftauchen – und das stimmt ihn heiter:

Das grösste neuere Ereigniss, – dass „Gott todt ist", dass der Glaube an den christlichen Gott unglaubwürdig geworden ist – beginnt bereits seine ersten Schatten über Europa zu werfen. Für die Wenigen wenigstens [...] deren A r g w o h n in den Augen stark und fein genug für dies Schauspiel ist, scheint eben irgend eine Sonne untergegangen, irgend ein altes tiefes Vertrauen in Zweifel umgedreht: ihnen muss unsre alte Welt täglich abendlicher, misstrauischer, fremder, „älter" scheinen. [...] – und was Alles, nachdem dieser Glaube untergraben ist, nunmehr einfallen muss, weil es auf ihm gebaut [...] war: zum Beispiel unsre ganze europäische Moral. Diese lange Fülle und Folge von Abbruch, Zerstörung, Untergang, Umsturz, die nun bevorsteht: wer erriethe heute schon genug davon, um den Lehrer und Vorausverkünder dieser ungeheuren Logik von Schrecken abgeben zu müssen, [...] woran liegt es doch,

> dass selbst wir [die freien Geister, E.B.] ohne rechte Theilnahme für diese Verdüsterung, vor Allem ohne Sorge und Furcht für uns ihrem Heraufkommen entgegensehn? [...] seine Folgen für uns sind [...] nicht traurig und verdüsternd, vielmehr wie eine neue [...] Art von Licht, Glück, Erleichterung, Erheiterung, Ermuthigung, Morgenröthe... [...] – endlich erscheint uns der Horizont wieder frei [...]. (Nietzsche, FW 343, KSA 3, S. 573–574)

Nietzsche fährt in diesem Aphorismus schwere rhetorische Geschütze auf. Mit großer Wortgewalt beschwört er eindrückliche (Sprach-)Bilder herauf, wie als wollte er sich selbst von seiner unerhörten Heiterkeit angesichts all der zu erwartenden Schrecken überzeugen. Er scheint sich selbst Mut zuzusprechen für die bevorstehende Aufgabe, über den Nihilismus zur Affirmation des Lebens zu gelangen. Auf die Befreiung der Welt von Gott soll nicht das Chaos folgen, denn, so Camus, „[a]uch das Chaos ist eine Knechtschaft. Freiheit gibt es nur in einer Welt, wo das Mögliche zu gleicher Zeit bestimmt wird wie das Unmögliche. Ohne Gesetz keine Freiheit." (Camus 2016: HR, S. 102) Der Tod Gottes zwingt den Menschen zur Selbstgesetzgebung, wenn er nicht „den Weg ins Dunkle" antreten will, wo seine neu gewonnene Freiheit sich als „die entsetzliche Freiheit des Blinden" erweist. Aus diesem Grund, so Camus weiter, fordert Nietzsche die Verwandlung des Todes Gotte in „‚einen fortwährenden Sieg über uns selbst'"; und er fasst zusammen: „Mit anderen Worten mündet mit Nietzsche die Revolte in die Askese ein." (Camus 2016: HR, S. 102)

Die Situation nach dem Tod Gottes ist also äußerst komplex; und sie ist heikel. Schwierig ist sie auch für Nietzsche den Philosophen. Denn er steht gleich doppelt zwischen den Stühlen der Freiheit und des Chaos sowie der Askese und der Unfreiheit. Als vorausschauender Denker blickt er der zweifachen Gefahr eines heraufziehenden Chaos auf der einen Seite und eines Rückfalls des Menschen in die geistige Selbstversklavung durch Flucht in wiederum metaphysische Selbst- und Weltdeutungen auf der anderen Seite entgegen. In dieser intrikaten Lage setzt Nietzsche auf die bedingungslose Affirmation der Situation: „Man muss das Unannehmbare annehmen und sich ans Unhaltbare halten." (Camus 2016: HR, S. 104) Der Zusammenbruch der teleologischen Weltdeutung macht den Blick frei für die Unschuld des Werdens. Eine neue Weltinterpretation wird möglich, in der in einem menschlichen Willensakt Freiheit und Notwendigkeit konvergieren. Für Camus entwickelt Nietzsche in diesem Zusammenhang seine tiefsten Gedanken. Wie wir gleich sehen werden, erregen gerade sie aber auch Camus Widerstand:

> Sobald man erkennt, dass die Welt kein Ziel verfolgt, schlägt Nietzsche vor, ihre Unschuld anzuerkennen sowie die Tatsache, dass unser Urteil nicht zuständig ist für sie, da man sie auf keine Absicht hin verurteilen kann, und infolgedessen alle Werturteile zu ersetzen durch ein einziges Ja, eine unumschränkte und begeisterte Zustimmung zu dieser Welt. So wird aus tiefer Verzweiflung die unendliche Lust entspringen, aus der blinden Knechtschaft die

> Freiheit ohne Gnade. [...] Der tiefste Gedanke Nietzsches ist, dass die Notwendigkeit der Erscheinungen, wenn sie absolut und ohne Risse ist, keine Art von Zwang in sich schließt. Die totale Zustimmung zu einer totalen Notwendigkeit, das ist seine paradoxe Definition der Freiheit. (Camus 2016: HR, S. 104)

An dieser Stelle wird Camus' geistiger Widerstand gegenüber Nietzsche wach. Es regt sich in ihm ein Widerstreben, das im Kern moralisch und politisch ist. Von hier aus ist auch verständlich, warum sich Camus im *Mythos des Sisyphos* mit kritischen Tönen gegenüber Nietzsche zurückhält. Der *Sisyphos* argumentiert dezidiert existenziell. Es handelt sich um einen existenzphilosophischen Essay, in dem Moral und Politik keine tragenden Rollen spielen. Das ist in *Der Mensch in der Revolte* anders. Zwar ist der Geist der Revolte bereits im *Mythos des Sisyphos* federführend. So ist das Glück, das Sisyphos in der Verachtung der Götter genießt, das Glück eines Revoltierenden. Der revoltierende Sisyphos probt jedoch den *existenziellen* und nicht den *politischen* Aufstand. Sisyphos sagt nein zu den Göttern und ja zu sich selbst. Er tut das in aller Einsamkeit, während er allein einen Berg hinabsteigt. Das Problem, auf das er reagiert, ist das des scheinbar verlockenden Selbstmordes angesichts des Absurden bzw. in Anbetracht einer Existenz, die im Zeichen des Absurden zu führen ist. In *Der Mensch in der Revolte* ist das *setting* ein anderes. Camus versucht sich hier an einer Zeitdiagnose. Er konstatiert, in einem „Zeitalter der Ideologien" (Camus 2016: HR, S. 15) zu leben. Das Erschreckende eines solchen Zeitalters ist, dass „das Verbrechen anfängt, seine Gründe in der Vernunft zu suchen" und auf diese Weise ausgerechnet in der Philosophie eine Verbündete findet, die ihm „ein unwiderlegbares Alibi" (Camus 2016: HR, S. 13) liefert. In einem solchen Zeitalter rückt das Problem des Selbstmordes in den Hintergrund, jetzt „muss man sich mit dem Mord auseinandersetzen" (Camus 2016: HR, S. 15). Das ist also der Rahmen, innerhalb dessen Camus in *Der Mensch in der Revolte* über Nietzsches Philosophie nachdenkt. Das heißt: Nietzsches Philosophie wird hier nicht, oder jedenfalls nicht in erster Linie, daraufhin abgeklopft, inwieweit sie Gründe für eine Bejahung des Lebens liefern kann, sondern sie wird kritisch hinterfragt, ob und inwiefern sie sich vor den Karren von Ideologien spannen lässt, die nach Vernunftgründen suchen, um den Mord zu rechtfertigen. Camus wechselt mithin die Perspektive auf Nietzsches Philosophie, und es ist überaus interessant, dass dieser Wechsel ein und dasselbe Konzept in ein ganz anderes Licht taucht: War der *Amor fati* eben noch (im *Sisyphos*) der emanzipatorische Gedanke *par excellence*, hat er sich plötzlich (in *Der Mensch in der Revolte*) im schlimmsten Fall in ein Instrument zur Rechtfertigung des Mordes verwandelt. Denn: „Alles bejahen setzt voraus, dass man den Mord bejaht." (Camus 2016: HR, S. 109) Nietzsches Philosophie ist gewiss keine Verherrlichung des Mordes. Auch liefert sie keinen philosophischen Persilschein für Nazismus, Rassismus und Antisemitismus. Camus weiß das, und es ist ihm wichtig, Nietzsche gegen diese Vorwürfe zu verteidigen: „Versichern wir zuerst, dass es uns immer unmöglich sein wird, Nietzsche mit Rosenberg zu verwechseln. Wir müssen die Anwälte Nietzsches sein." (Camus 2016: HR, S. 109) Gleichwohl lässt sich nicht leugnen, „wie dieser Philosoph zum Kronzeugen von Rassismus, Selektion und dem Recht des Stärkeren berufen ward" (Meyer 2015, S. 127). Camus ist offenbar der Auffassung, dass Nietzsche einen solchen Missbrauch seiner Philosophie hätte antizipieren können. Er wirft ihm vor, blind gegenüber der möglichen und vorauszusehenden Vereinnahmung seiner Philosophie durch die Ideologen und Tyrannen zu sein: „Die Philosophie säkularisiert die Ideale. Aber es kommen die Tyrannen und säkularisieren die Philosophien, die ihnen dazu das Recht geben." (Camus 2016: HR, S. 112)[10] Nietzsche hätte ihnen, so darf man Camus wohl verstehen, dieses Recht nicht geben dürfen.

4 Eine offene Frage zum Schluss: Finden wir bei Camus eine soziale Form des *Amor fati?*

Genau wie bei Nietzsche wird auch in Camus' *Mythos des Sisyphos* der Gedanke der ewigen Wiederkunft in der „einsamste[n] Einsamkeit" (Nietzsche, FW 341, KSA 3, S. 570) gewälzt. In Werken wie *Der Mensch in der Revolte* und *Die Pest* geht es Camus aber nicht nur um den einzelnen Menschen in der Welt, sondern auch und gerade um Solidarität und Freundschaft. Mit großem Einfühlungsvermögen notiert er, dass Nietzsche „Freiheit mit Einsamkeit verwechselt [hatte] in der Art eines stolzen Geistes." (Camus 2016: HR, S. 108) Vielleicht kommt bei Nietzsche darum die soziale Dimension des *Amor fati* zu kurz und damit auch dessen moralische oder soziale Möglichkeiten. Warum nicht einen geteilten Augenblick vorstellen, der zur absoluten Daseinsaffirmation einlädt? Als einen solchen Augenblick lese ich die Szene in *Die Pest*, in der Rieux und Tarrou gemeinsam im Meer schwimmen:

10 In diesem Kontext geht es natürlich nicht nur um Nietzsches *Amor fati*, sondern auch um sein schwer fassbares und viel diskutiertes Konzept des Übermenschen (vgl. Camus 2016: HR, S. 113) sowie den ebenso umstrittenen Willen zur Macht. Zu letzterem bemerkt Camus: „Der Nietzscheismus, die Theorie des Willens zur individuellen Macht, war dazu verurteilt, sich einzufügen in einen Willen zur totalen Macht." (Camus 2016: HR, S. 111).

> Rieux sprang als erster. [...] Ein schweres Klatschen sagte ihm, dass Tarrou gesprungen war. Rieux drehte sich auf den Rücken und lag reglos angesichts des umgekippten Himmels mit dem Mond und den Sternen. Dann vernahm er ein immer deutlicher in der Stille der Einsamkeit und der Nacht seltsam klares Klatschen im Wasser. Tarrou näherte sich, er hörte bald sein Atmen. Rieux drehte sich wieder um, schloss zu seinem Freund auf und schwamm im gleichen Rhythmus. [...] Einige Minuten schwammen sie im gleichen Takt und mit der gleichen Kraft, einsam, fern von der Welt, endlich von der Stadt und der Pest befreit. [...] Wieder angezogen brachen sie auf, ohne ein Wort gesprochen zu haben. Aber ihre Herzen fühlten gleich, und die Erinnerung an diese Nacht war wohltuend für sie. (Camus 2011: P, S. 292–293)

In dieser Szene offenbart sich nicht nur das Soziale als ein Kraftfeld der Lebensbejahung, sondern auch die Tiefe, die potenziell in der Gewöhnlichkeit liegt (hier des abendlichen Schwimmens im freien Meer, das jenseits der Pest für die Bewohner und Bewohnerinnen Orans das Gewöhnlichste von der Welt ist). Auch das Gewöhnliche, das Camus im *Sisyphos* vor allem noch mit dem Absurden und dem Daseinsekel verbindet, kann vor dem Hintergrund des Unheimlichen und Schrecklichen (hier der Pest), die das Gewöhnliche stets bedrohen, als liebens- und lebenswert erfasst werden (vgl. Brock 2018 und Brock 2019). Camus' Erweiterung des *Amor fati* um die Dimension des Sozialen macht ihn zum wahren nietzscheanischen Philosophen im onfrayschen Sinne.

Bibliographie

Beauvoir, Simone de (2007): *Soll man de Sade verbrennen? Drei Essays zur Moral des Existenzialismus.* Reinbek b. Hamburg.

Brock, Eike (2015): *Nietzsche und der Nihilismus.* Berlin, München, Boston.

Brock, Eike (2018): „Der Kampf um das Gewöhnliche". In: Günter Gödde/Jörg Zirfas (Hrsg.): *Kritische Lebenskunst. Analysen – Orientierung – Strategien.* Stuttgart, S. 347–354.

Brock, Eike (2019): „Vom Horror der Endlichkeit. Versuch über ein Zentralmotiv des Horrorgenres". In: Eike Brock/Thorsten Lerchner (Hrsg.): *Denken des Horrors, Horror des Denkens. Unheimliches, Erschreckendes und Monströses aus philosophischer Perspektive.* Würzburg, S. 13–49.

Camus, Albert (2002): *Der Mythos des Sisyphos.* Reinbek b. Hamburg.

Camus, Albert (2003): „Das Meer". In: Albert Camus: *Ein Lesebuch mit Bildern*, ausgewählt von Barbara Hoffmeister. Reinbek b. Hamburg, S. 135–144.

Camus, Albert (2011): *Die Pest.* Reinbek b. Hamburg.

Camus, Albert (2016): *Der Mensch in der Revolte.* Reinbek b. Hamburg.

Heidegger, Martin (2006): *Sein und Zeit.* Tübingen.

Meyer, Martin (2015): *Albert Camus. Die Freiheit leben.* Stuttgart.

Onfray, Michel (2015): *Im Namen der Freiheit. Leben und Philosophie des Albert Camus.* München.

Sartre, Jean Paul (2006): *Das Sein und das Nichts. Versuch einer phänomenologischen Ontologie.* Reinbek b. Hamburg.
Voigt, Albrecht (2017): *Wirkliche Göttlichkeit oder göttliche Wirklichkeit. Die Herausforderungen der Gegensatzproblematik in Romano Guardinis latentem Gespräch mit Friedrich Nietzsche.* Dresden.

Philipp Schwab

Überwindungen des Nihilismus? Kierkegaard – Nietzsche – Camus

Abstract: Overcoming Nihilism? Kierkegaard – Nietzsche – Camus. The aim of the paper is to unfold a parallel movement in Camus's readings of Kierkegaard and Nietzsche, respectively, in *The Myth of Sisyphus* and *The Rebel*. In doing so, the article intends to show that both readings are essentially two-sided: on the one hand, Camus celebrates and affirms both Kierkegaard and Nietzsche for initially having discovered the absurd; on the other hand, he sharply criticizes both, again in a similar fashion, for ultimately having undone and destroyed the tension of the absurd. In Camus's reading, Kierkegaard dissolves the absurdity of the finite by transposing it into the essence of a *transcendent* and otherworldly god. Nietzsche, in return, annuls the absurd and rebellion by means of an unconditional and undifferentiated affirmation and apotheosis of the *immanent* world. The paper concludes by discussing critical perspectives on Camus's interpretations.

Einleitung

In der deutschsprachigen Existenzphilosophie ist die wesentliche Zusammengehörigkeit von Kierkegaard und Nietzsche ein fester Topos. Prominent und in bemerkenswerter Kontinuität formuliert ist dieses Motiv insbesondere bei Karl Jaspers, beginnend mit der *Psychologie der Weltanschauungen* von 1919. Exemplarisch fasst Jaspers die Zusammengehörigkeit beider Denker etwa in dem kurzen Text *Kierkegaard* von 1951 in das folgende, suggestive Bild:

> Beim Gang in die neue Welt sehen wir Kierkegaard und Nietzsche wie Sturmvögel vor einer Wetterkatastrophe: Sie zeigen die Unruhe, die Hast, – dann die Kraft und Klarheit eines augenblicklichen hohen Fluges, und wieder etwas wie Kreisen und Taumeln und Absturz.
>
> Sie selber wissen sich als Seezeichen; an ihnen ist Orientierung möglich – aber indem man sich in Distanz zu ihnen hält. Ihnen zu folgen, wird von ihnen selbst verwehrt.
>
> Sie wollen wirken, aber in der Weise, daß sie den anderen zu sich selbst bringen, alles Entscheidende von ihm erwarten, es ihm nicht geben. (Jaspers 1968, S. 309)

Im Vorblick auf die folgende Untersuchung seien zwei charakteristische Züge aus dem Passus herausgehoben. Zum einen geschieht der Ausweis einer Duplizität und Zusammengehörigkeit beider Denker bei Jaspers offenkundig nicht in neutraler historischer oder historiographischer Absicht, sondern in einem eminent

https://doi.org/10.1515/9783110760163-008

geschichtlichen Sinne. Kierkegaard und Nietzsche erscheinen bei Jaspers als die zentralen Merkzeichen eines Umbruchs und einer Zeitenwende – und zugleich als Vorboten einer Krise, als Seismographen einer *kommenden* Erschütterung, die offenbar erst in der Jaspers gegenwärtigen Zeit zum vollen Austrag gerät. In der ersten seiner Vorlesungen über *Vernunft und Existenz* von 1935 mit dem Untertitel „Die geschichtliche Bedeutung Kierkegaards und Nietzsches" hat Jaspers diese Zeitenwende näher benannt: Es sei „in der Wirklichkeit des abendländischen Menschen [...] in aller Stille etwas Ungeheures geschehen: ein Zerfall aller Autoritäten, die radikale Enttäuschung eines übermütigen Vertrauens zur Vernunft, eine Auflösung der Bindungen, die alles, schlechthin alles möglich zu machen scheint." (Jaspers 2018, S. 8) Und die Bedeutung von Kierkegaard und Nietzsche bestehe gerade darin, dieses Umbruchsgeschehen als erste klar und hellsichtig analysiert zu haben – eben damit seien sie, wie Jaspers schreibt, die „eigentlich großen Denker ihres Zeitalters" und zugleich, als Vorboten, „die nicht mehr ignorierbaren Philosophen unserer Zeit" (Jaspers 2018, S. 8). Zum anderen zeichnet sich bei aller Emphase auf die Bedeutung Kierkegaards und Nietzsches aber in der eingangs zitierten Passage auch ab, dass Jaspers mehr in den Fragen und Analysen beider Denker Orientierungspotenzial sieht als in ihren Antworten – sei es doch, wie es im Zitat heißt, „verwehrt", ihnen einfach zu „folgen". Das bedeutet offenbar: Kierkegaard und Nietzsche stellen und beschreiben sozusagen das *Problem* der „neuen Welt" (Jaspers 1968, S. 309), ohne aber sogleich auch die Lösung zu formulieren. Die Lösungen sind vielmehr – so die Perspektive von Jaspers – der Gegenwart als je noch zu findende aufgegeben.

Diese Annahme einer wesentlichen Zusammengehörigkeit Kierkegaards und Nietzsches – gerade im Sinne einer geschichtlichen Vergewisserung über die ‚geistige Situation der Zeit'[1] – ist beileibe keine isolierte Einzelmeinung, vielmehr zieht sie sich als Paradigma quer durch die deutschsprachige Existenz- und Existenzialphilosophie. So sind auch in Karl Löwiths wirkmächtiger Studie *Von Hegel zu Nietzsche. Der revolutionäre Bruch im Denken des neunzehnten Jahrhunderts* (1941) Kierkegaard und Nietzsche zentrale Protagonisten eben jener titelgebenden Wende. Noch deutlicher hat Löwith schon 1933 beide in einer separaten Abhandlung nebeneinandergestellt, unter dem Titel *Kierkegaard und Nietzsche oder philosophische und theologische Überwindung des Nihilismus.* Kierkegaard und Nietzsche erscheinen also wiederum so, dass ihnen die kritische Diagnose einer Erschütterung der eigenen Gegenwart gemeinsam ist – während sich ihre

1 Auch im entsprechend betitelten Werk findet sich die Zusammenstellung beider, vgl. bes. Jaspers 1998, S. 12–13.

Antworten auf diese Krise aber wesentlich unterscheiden.[2] Schließlich hat sich auch Martin Heidegger wiederholt zur Duplizität von Kierkegaard und Nietzsche geäußert, obschon deutlich wandlungsreicher. Anfang der 1930er Jahre benennt er beide etwa noch als die „großen Mahner und Wegbereiter im 19. Jahrhundert" (Heidegger 2001, S. 13).[3] Schon 1936 allerdings – und wohl gerade gegen Jaspers und Löwith – heißt es dann, die „jetzt üblich gewordene Zusammennennung von Kierkegaard und Nietzsche" sei zwar „in mancher Hinsicht berechtigt, aber im Grunde philosophisch unwahr und irreführend" (Heidegger 1988, S. 43), und wenig später spricht Heidegger gar von einer „geschichtlich unmögliche[n] Verkoppelung von Nietzsche und Kierkegaard" (Heidegger 1997, S. 479). Noch in dem Bedürfnis der kritischen Abstoßung bestätigt Heidegger so die Prominenz der Duplizität von Kierkegaard und Nietzsche.

Vor diesem in groben Zügen umrissenen Hintergrund lässt sich das Anliegen des folgenden Versuchs zur Trias ‚Kierkegaard – Nietzsche – Camus' einordnen. Richtet man nämlich von Jaspers, Löwith und Heidegger aus den Blick auf den französischen Existenzialismus, so zeigt sich zunächst doch ein deutlich abweichendes Bild. Gewiss sind auch hier Kierkegaard und Nietzsche als Impulsgeber präsent.[4] Sie sind aber einerseits als Referenzfiguren doch nicht derart prominent wie im deutschsprachigen Raum, was allein schon die textuelle Basis der Auseinandersetzung zeigt – vor allem aber sind sie andererseits nicht durchgängiger Bezugspunkt als in ihrer Gemeinsamkeit zusammengehöriges ‚Doppel'.

Entgegen diesem ersten Eindruck soll im Folgenden die These entwickelt werden, dass sich gerade bei Albert Camus doch eine prägnante Doppelbestimmung von Kierkegaard und Nietzsche findet, die dem soeben skizzierten Paradigma der deutschsprachigen Existenzphilosophie mindestens in einem Grundzug verwandt ist. Zwar behandelt Camus Kierkegaard und Nietzsche nicht direkt gemeinsam und nicht unmittelbar nebeneinander, aber doch, wie gezeigt werden soll, in einer bemerkenswerten *Parallelität*.[5] Dabei fasst Camus, so die These,

---

2 Im Auftakt der Abhandlung führt Löwith übrigens die titelgebenden Begriffe der diesem Band zu Grunde liegenden Tagung – „‚Leben' und ‚Existenz'" – als die „charakteristischen Grundbegriffe gegenwärtiger Philosophie" gerade auf Nietzsche und Kierkegaard zurück (Löwith 1987, S. 53).

3 Heidegger bemerkt dabei übrigens parallel zur zitierten Äußerung von Jaspers: „Wir müssen sie hören, aber nicht hörig werden" (Heidegger 2001, S. 11).

4 Vgl. zu Kierkegaards Bedeutung für den Existenzialismus im Überblick Stewart 2009 und im Einzelnen die Beiträge in Stewart 2011, zu Nietzsches Bedeutung im Überblick Stephan 2020, S. 352–364 und die Beiträge des vorliegenden Bandes.

5 In der Forschung werden Camus' Bezugnahmen auf Kierkegaard und Nietzsche bisweilen gemeinsam behandelt, aber dabei doch mehr ergänzend nebeneinandergestellt als eingehend in ihrem Verhältnis ausgetragen. Vgl. z. B. Golomb 1995, S. 127 und Pieper 2014.

näher besehen beide Denker wesentlich *zweiseitig:* Einerseits benennt Camus gerade Kierkegaard und Nietzsche affirmativ als zentrale ‚Entdecker' und Analysten der Grundsituation des Menschen der Gegenwart, und zwar im Sinne einer wesentlich *absurden* respektive *nihilistischen* Grundsituation. Andererseits aber wendet sich Camus kritisch sowohl in seiner Interpretation Kierkegaards als auch in der Nietzsches gegen die ‚Antworten' beider auf diese von ihnen freigelegte Grundsituation. Camus weist nämlich bei beiden jeweils prominente Optionen des ‚Umgangs' mit dem Nihilismus respektive seiner ‚Bewältigung' letztlich zurück: Sowohl Kierkegaards ‚Sprung' als auch Nietzsches ‚absolute Affirmation' erscheinen Camus als *Entschärfungen* des Nihilismus, ja geradezu als Flucht vor dem Absurden, als ein ‚Ausweichen' (*évasion*) – und damit als ‚Überwindungen' des Nihilismus in dem Sinne, dass sie diesen aufheben, umgehen und ungeschehen machen. Die Pointe der Auslegung wird sich dann darin zeigen, dass nach Camus beide, wenn auch auf unterschiedliche Weise, die Grundspannung des Absurden gerade durch seine *Hypostase* und *Apotheose* vernichten.

Diese Parallelität der Interpretation von Kierkegaard und Nietzsche bei Camus ist in ihrer Zweiseitigkeit von Affirmation und Kritik im Folgenden näher und auch differenzierter zu entwickeln. Leitend ist dabei nicht vornehmlich das Anliegen, den französischen Existenzialismus der – ja gerade von Camus scharf kritisierten – deutschsprachigen Existenzphilosophie anzunähern. Wohl aber ist es eines der Ziele des Beitrags, am exemplarischen Fall von Camus den geschichtlichen Horizont und Bezugsraum auszuloten, in dem ein Denken konkreter Existenz aufbricht, nämlich im Angesicht des Nihilismus und mit der Frage seiner Bewältigung – ein Horizont, der sich eben wesentlich im Rückbezug auf Kierkegaard und Nietzsche erschließt.[6] Terminologisch sei dabei noch angemerkt, dass die Bestimmung, die hier titelgebend ins Zentrum des triangulären Verhältnisses ‚Kierkegaard – Nietzsche – Camus' gerückt wird, die des Nihilismus, ein breites Spektrum an Bedeutungsmöglichkeiten eröffnet. Bereits bei Nietzsche selbst ist sie, wie alle seine sogenannten ‚großen Begriffe', notorisch vielschichtig und schillernd.[7] Um die drei verhandelten Positionen aufeinander beziehen zu können, wird die operative Bestimmung des Nihilismus zunächst bewusst in einer gewissen methodischen Unschärfe verstanden und folgt dabei einer Minimaldefinition: ‚Nihilismus' bezeichnet im Allgemeinen die Grunderfahrung einer Sinnleere oder Sinnlosigkeit des Daseins – und zwar als eine solche, die zugleich

---

6 Vgl. zu Kierkegaard und Nietzsche als ‚Initialfiguren' der Existenzphilosophie im Ganzen nun bes. Victor 2021a, zum beiderseitigen Einfluss auf Camus v. a. Victor 2021a, S. 270 – 272.

7 Vgl. dazu Schwab 2011 und bes. auch Sommer 2006, Brock 2016, van Tongeren 2018.

krisenhaft als Sinn*verlust* erscheint, und damit je schon eine geschichtliche Konnotation hat.

Die angezeigten Perspektiven sollen im Folgenden in drei Schritten konkretisierend eingeholt werden: Der erste Schritt widmet sich Camus' Kierkegaard-Interpretation, wie sie sich im *Mythos des Sisyphos* findet (I). Der zweite Schritt soll vor diesem Hintergrund eine Parallelaktion in Camus' Nietzsche-Interpretation in *Der Mensch in der Revolte* aufweisen (II). Schließlich führt der dritte Schritt die Fäden zusammen und diskutiert dabei ausblickhaft auch kritische Perspektiven auf Camus' Auslegungen von Kierkegaard und Nietzsche (III).

I Das Absurde und der Sprung – Camus über Kierkegaard

Im *Mythos des Sisyphos* findet sich tatsächlich eine einzige Partie, in der auch Camus, wenngleich ohne ausführliche Erläuterung, Kierkegaard und Nietzsche unmittelbar nebeneinander stellt – und zwar nicht zufällig gleich am Eingang zu der Galerie jener Philosophen, die „von einer Kritik des Rationalismus aus das absurde Klima erkannt haben“ (Camus 2020a: MS, S. 45/1983a, S. 158).[8] Beide erscheinen dabei geradezu als Entdecker und Initiatoren des absurden Denkens: „Seit Zarathustras großem Ruf ‚von Ohngefähr [...]‘“ und seit Kierkegaards Beschreibung der *Krankheit zum Tode* als Verzweiflung nämlich, so Camus, „lösten die bezeichnenden und quälenden Themen des absurden Denkens einander ab.“ (Camus 2020a: MS, S. 35/1983a, S. 150 – 151)[9] Während aber Nietzsche im folgenden Durchgang durch das Panorama der Denker des Absurden auffällig abwesend ist,[10] kommt Kierkegaard darin eine, wenn nicht die zentrale Rolle zu.

Dabei lassen sich in dem Bild, das Camus im *Mythos des Sisyphos* von Kierkegaard zeichnet,[11] zunächst zwei Seiten recht klar unterscheiden. Kierkegaard wird in der Reihe von Camus' Referenzautoren zwar nicht als erster behandelt, er ist aber der historisch frühste und erhält einen besonderen Akzent. Kurz und klar heißt es bei Camus: „Von allen vielleicht der anziehendste, Kierkegaard, tut we-

8 Alle Übersetzungen Camus' und Kierkegaards sind geringfügig modifiziert.

9 Camus' Bezugsstellen sind hier bei Nietzsche Za III Sonnen-Aufgang, KSA 4, S. 209 und bei Kierkegaard KT, S. 13/SKS 11, S. 133.

10 Ausdrücklich erscheint Nietzsche hier nur einmal am Rande und durch die Brille Schestows, vgl. Camus 2020a: MS, S. 38/1983a, S. 152.

11 Vgl. zu Camus' Verhältnis zu Kierkegaard im Überblick Stewart 2009, S. 441 – 443 und Stan 2011 sowie zukünftig die ausführliche Untersuchung von Abel 2020, vorab auch bereits Abel 2018 und Abel 2019, zur Diskussion der im Folgenden behandelten Punkte auch Pieper 2014, S. 87 – 93.

nigstens für einen Teil seiner Existenz mehr, als das Absurde nur zu entdecken, er lebt es." (Camus 2020a: MS, S. 38/1983a, S. 152–153) In Kierkegaards Denken und Ringen, das jeden „Trost", jede „Moral" und „alle beruhigenden Grundsätze" ablehne – geradezu in der „verzweifelten Freude eines freiwillig Gekreuzigten" – zeige sich „der absurde Geist selbst im Kampf mit einer Wirklichkeit, die stärker ist als er" (Camus 2020a: MS, S. 38/1983a, S. 153). Nähme man diese Passagen allein für sich, so erschiene Kierkegaard als derjenige, der sich kompromisslos und wohl am radikalsten der Absurdität des Daseins stellt, ihr gerade nicht ausweicht oder sie entschärft. Dies ist aber eben nur die eine, affirmative Seite in Camus' Darstellung – und ihr steht sogleich die andere, kritische entgegen. Diese ist bei Camus, im Passus zum philosophischen Selbstmord, ebenfalls kurz und klar formuliert: „Auch Kierkegaard macht den Sprung." (Camus 2020a: MS, S. 50/1983a, S. 162) Und das heißt, für Camus: Wie „ausnahmslos alle" Vertreter der „existenziellen Philosophien" (Camus 2020a: MS, S. 45/1983a, S. 158), so gehe auch Kierkegaard zwar vom Absurden aus, dies aber nur, um ihm in letzter Geste doch auszuweichen – und zwar in der Bewegung einer *Transzendierung*, einer Übersteigung des Absurden ins Religiöse.[12]

Um diese beiden Seiten von Camus' Kierkegaard-Interpretation – die Entdeckung des Absurden in seiner Schärfe, und die *Ent*schärfung des Absurden durch den Sprung – noch etwas genauer in den Blick zu bekommen, ist zunächst auf eine terminologische Eigentümlichkeit in der Auslegung Camus' hinzuweisen:[13] Es finden sich nämlich bei Kierkegaard in der Tat Partien, die jener Grunderfahrung sehr nahekommen, welche Camus in eigener Sache ‚das Absurde' nennt – nur dass diese Erfahrung bei Kierkegaard gerade nicht mit dem Begriff des Absurden bezeichnet wird. Camus selbst zitiert etwa Stellen aus *Furcht und Zittern* und der *Krankheit zum Tode*, die in der Bestimmung nicht des Absurden, sondern der *Verzweiflung* zentrieren (vgl. Camus 2020a: MS, S. 35, 54/1983a, S. 150–151, 164–165). Beschreibungen einer solchen Grunderfahrung der Sinnlosigkeit stehen bei Kierkegaard insbesondere im Kontext der Figurationen *ästhetischer* Existenz,[14] wie sie etwa im ersten Teil von *Entweder/Oder* den Ausgangspunkt

---

12 Diese Bewegung formuliert Camus vorab global für alle verhandelten Positionen der Existenzphilosophie, vgl. Camus 2020a: MS, S. 45/1983a, S. 158.

13 Vgl. zu der im Folgenden diskutierten Begriffsverschiebung des Absurden zwischen Camus und Kierkegaard Stewart 2009, S. 442 und Berthold 2013, S. 138.

14 Diese augenscheinliche Nähe bedeutet noch keineswegs, Camus gänzlich mit dem ‚ästhetischen Stadium' identifizieren zu müssen, wie dies wohl zuerst von Richter formuliert wurde (vgl. Richter 1959, S. 121). In verwandter Richtung, aber ohne den Rekurs auf die Stadientheorie, hat Golomb die einflussreiche These von Camus als „inverted Kierkegaard" formuliert (Golomb 1995, S. 127). Für einen erhellenden ‚Dialog' zwischen dem Ästhetiker aus *Entweder/Oder* und Camus

seiner Existenzanalysen bezeichnen.[15] Eine besonders sprechende Partie dieser Art findet sich den Briefen des ‚jungen Menschen' aus der Schrift *Die Wiederholung:*

> Mein Leben ist bis zum Äußersten gebracht; ich ekele mich des Daseins, das geschmacklos ist ohne Salz und Sinn. Wäre ich auch hungriger als Pierrot, so möchte ich doch nicht die Erklärung fressen, die die Menschen anbieten. Man steckt den Finger in die Erde, um zu riechen, in welchem Lande man ist, ich stecke den Finger ins Dasein – es riecht nach nichts. Wo bin ich? Was will das besagen: die Welt? Was bedeutet dieses Wort? Wer hat mich in das Ganze hineinbetrogen, und lässt mich nun dastehen? Wer bin ich? Wie bin ich in die Welt hineingekommen, warum bin ich nicht gefragt worden, warum nicht mit Bräuchen und Regeln bekannt gemacht worden [...]? Wie bin ich Interessent geworden, in der großen Entreprise, die man die Wirklichkeit nennt? (Kierkegaard: W, S. 70 – 71/SKS 4, S. 68)

Das entspricht in der Tat recht genau Camus' Definition des Absurden, nach der dieses nämlich nicht einfach dem Dasein als solchem zukommt, sondern entsteht in einer *Kollision* zwischen dem Sinnbedürfnis des Menschen einerseits – und der Welt andererseits, die einen solchen Sinn gerade nicht gewährt.[16] Man kann durchaus auch die bei Kierkegaard beschriebene Daseinserfahrung eine ‚nihilistische' nennen, eben in ihrem prononcierten Fokus auf das *Nichts* – aber jedenfalls heißt sie bei Kierkegaard selbst an keiner Stelle eine ‚absurde'.

Hingegen verwendet Kierkegaard selbst den Ausdruck ‚das Absurde' in einem qualifizierten Sinne *umgekehrt* genau für dasjenige, was in Camus' Deutung das Absurde gerade entschärft, nämlich für den *Sprung* – oftmals in der Formel eines Glauben „in Kraft des Absurden".[17] Der Glaube nämlich ist dasjenige, was jenseits und im Abbruch aller menschlichen und verständigen Berechnungen und Erwägungen seinen Ort hat, wie dies besonders Kierkegaards Pseudonym Johannes de Silentio am Isaak-Opfer Abrahams in *Furcht und Zittern* ausführt: Das Absurde

---

vgl. Thurnherr 2004. Für einen Überblick über die Forschungspositionen vgl. insgesamt Abel 2019, S. 12 – 15.

15 Vgl. zu dieser These Schwab 2014, S. 95 – 96, vgl. in *Entweder/Oder* vor allem die „Wechselwirtschaft", wo der Ästhetiker u. a. von dem „Nichts" spricht, das „sich durch das Dasein schlingt" (Kierkegaard: EO1, S. 310/SKS 2, S. 280).

16 Vgl. bes. Camus 2020a: MS, S. 33/1983a, S. 149: „Ich sagte vorschnell, dass die Welt absurd ist. Diese Welt an sich ist nicht vernünftig, das ist alles, was man von ihr sagen kann. Absurd aber ist der Zusammenstoß (*confrontation*) dieses Irrationalen und dieses heftigen Verlangens nach Klarheit, das im tiefsten Inneren des Menschen resoniert." Bei Kierkegaard wird eine solche Kollision etwa auch dort greifbar, wo der Ethiker aus dem zweiten Teil von *Entweder/Oder* den Ästhetiker so beschreibt, dass dieser sich „lieber als einen Gläubiger in der Welt betrachten wolle, der nicht bezahlt worden sei, als die Forderung streichen" (Kierkegaard: EO2, S. 218/SKS 3, S. 198).

17 Vgl. z. B. in der *Wiederholung* Kierkegaard: W, S. 57/SKS 4, S. 55, in *Furcht und Zittern* Kierkegaard: FZ, S. 34, 36/SKS 4, S. 131 – 132.

des Opfers, der Glaube wider alle „menschliche Berechnung" hat seinen tiefsten Punkt darin, dass „bei Gott alles möglich ist" (Kierkegaard: FZ, S. 34, 47/SKS 4, S. 131, 141).

Von hier aus lassen sich die beiden Seiten von Camus' Kierkegaard-Interpretation subtiler fassen als zuvor. Keineswegs nämlich handelt es sich bei der Begriffsverschiebung in Camus' Auslegung um eine simple und bloße Inversion, näher besehen stellt sich vielmehr das Verhältnis beider Seiten komplexer dar als eine einfache Antithese: Kierkegaards religiöser Sprung ist nicht das *einfache Andere* des Absurden, durch das dieses bloß vernichtet würde, er ist vielmehr seine *Transposition.* Gemäß Camus' Auslegung *versetzt* nämlich Kierkegaard die diesseitige Erfahrung des Absurden aus der Endlichkeit *in* den jenseitigen Glauben – und lässt das solchermaßen verewigte Absurde gerade als Rettung vor dem Absurden der Endlichkeit erscheinen. Camus beschreibt diese Transposition in mehreren Schlaufen, etwa in den folgenden Worten:

> Gerade das, was ihn [Kierkegaard, P.S.] am Sinn und an der Tiefe dieses Lebens verzweifeln ließ, schenkt ihm jetzt seine Wahrheit und seine Klarheit. [...] Er macht das Absurde zum Kriterium des Jenseits, während es nur ein Bodensatz diesseitiger Erfahrung ist. [...] Zwischen dem Irrationalen der Welt und der revoltierenden Sehnsucht des Absurden [*la nostalgie révoltée de l'absurde*] hält er nicht das Gleichgewicht. Er respektiert nicht das Verhältnis zwischen beiden, das im eigentlichen Sinne das Gefühl der Absurdität ausmacht. In der Gewissheit, dem Irrationalen nicht entgehen zu können, will er sich wenigstens vor dieser verzweifelten Sehnsucht retten, die ihm unfruchtbar und belanglos erscheint. Aber wenn er auch in diesem Punkt seiner Beurteilung recht haben mag, verhält es sich nicht so mit seiner Verneinung. Wenn er seinen Schrei der Revolte durch eine besinnungslose Zustimmung ersetzt, dann muss er schließlich das Absurde, das ihn bisher erleuchtete, ignorieren und die einzige Gewissheit, die er nun hat, das Irrationale, vergöttlichen. (Camus 2020a: MS, S. 50–51/1983a, S. 162–163)

Das heißt: Kierkegaards Sprung entschärft nach Camus das Absurde gerade dort, wo er es selbst namhaft macht und auf die Spitze stellt, und zwar eben dadurch, dass er es vergöttlicht. Dies vollzieht sich gemäß Camus' Interpretation näher so, dass Kierkegaard die konstitutive Spannung und Kollision des Absurden auflöst und dabei dem nun vergöttlichten Absurden die *Versöhnung* zuschreibt – im *credo quia absurdum est.*[18] Diese restlos affirmative Bewegung ist es, die für Camus Kierkegaards Ausweichen gegenüber dem Absurden ausmacht, seine *Entschärfung durch Transposition* – und zwar, was für das Folgende im Blick zu behalten

---

18 Im *Mythos des Sisyphos* findet sich diese Formel nicht; Kierkegaard verwendet sie zwar nicht an zentraler Stelle, aber doch mehrmals in seinem Werk, meist mit Verweis auf Tertullian. Vgl. z. B. NB10, S. 39/SKS 18, S. 39–40.

ist, als Bewegung einer „besinnungslose[n] Zustimmung (*adhésion forcenée*)" (Camus 2020a: MS, S. 50 – 51/1983a, S. 162 – 163).

Im Übergang zum Folgenden sei bemerkt, dass diese kritische Analyse Camus' in einem weiten Sinne durchaus eine *nietzscheanische* genannt werden kann, näher eine psychologisch-symptomatologische. Strukturell lässt sich nämlich die Bewegung, die Camus am religiösen Existenzialisten herauspräpariert, folgendermaßen fassen: Aus dem Leiden an der Sinnlosigkeit des Daseins einerseits und aus der Unmöglichkeit, ohne ein ‚wozu?' leben zu können andererseits, setzt der Philosoph sich ein sinnstiftendes Ideal *jenseits* dieser Welt entgegen[19] – wenn auch die Pointe hier eben darin besteht, dass diese ‚wahre Welt' nicht als Idee und Vernunft erscheint, sondern geradezu als die Apotheose des Absurden, die aber gleichwohl die Endlichkeit depotenziert.

II Der Nihilismus und die Affirmation – Camus über Nietzsche

Nietzsche hat zwar im *Mythos des Sisyphos* auf den eröffnenden Seiten gleich zweifach seinen Auftritt (vgl. Camus 2020a: MS, S. 15, 19/1983a, S. 135, 138) und ist zweifellos ein wesentlicher Bezugspunkt Camus' in diesem Text – eine ins Einzelne gehende Besprechung, die mit der Kierkegaards vergleichbar wäre, findet sich hier aber nicht. Gleichsam nachgeholt wird die entsprechende Analyse Nietzsches in Camus' *Der Mensch in der Revolte*, im Essay „Die metaphysische Revolte" und näher in dem Abschnitt, der bezeichnenderweise den Titel trägt „Die absolute Bejahung".[20]

Auch hier lassen sich, so die leitende These, zwei Seiten von Camus' Auslegung unterscheiden, und zwar in struktureller Parallelität zu seiner früheren Interpretation Kierkegaards. In Camus' nun deutlich breiterer, schlaufenhaft verfahrender Darstellung Nietzsches sind diese beiden Seiten allerdings weniger kompakt und scharf abgehoben, vielmehr komplexer ineinander verschlungen.

---

19 Vgl. bei Nietzsche etwa, in unterschiedlichen Registern und Perspektivierungen, die Bewegungen in Za I Hinterweltlern, KSA 4, S. 34 – 38; GM III 28, KSA 5, S. 411 – 412 und GD Vernunft, KSA 6, S. 74 – 79.

20 Vgl. aus der Forschung zu Camus' affirmativem Bezug auf Nietzsche bes. Victor 2021b, zur kritischen Seite der Rezeption bes. Pieper 1991, zur Abwägung beider Seiten, auch im Kontext des Gesamtansatzes von *Der Mensch in der Revolte*, Stephan 2020, 358 – 364, zur Diskussion auch Pieper 2014, S. 93 – 98.

Eine anzeigende Figuration wird aber bereits im Eingangspassus des ganzen Abschnitts sichtbar:

> Von dem Augenblick an, da der Mensch Gott einem moralischen Urteil unterwirft, tötet er ihn in sich selbst. Aber was ist dann die Grundlage der Moral? Man leugnet Gott im Namen der Gerechtigkeit, aber kann man die Idee der Gerechtigkeit verstehen ohne die Idee Gottes? Stehen wir dann nicht im Absurden? Es ist eben das Absurde, das Nietzsche geradewegs angeht. Um es besser überwinden zu können, treibt er es auf die Spitze: Die Moral ist Gottes letztes Gesicht, das man zerstören muss vor dem Neuaufbau. Gott ist dann nicht mehr und gewährleistet nicht mehr unser Sein; der Mensch muss sich entschließen zu handeln, um zu sein. (Camus 2020b: HR, S. 91/1983b, S. 88)

Hier zeigt sich zunächst wieder das bekannte Motiv des Absurden,[21] insofern steht Camus' Ansatz in einem dem *Mythos des Sisyphos* mindestens noch verwandten Feld. Wenigstens formal deutlich wird dabei auch eine Doppelseitigkeit, in der Nietzsche in den Blick genommen wird: In der ersten, affirmativen Perspektive erscheint Nietzsche – mit deutlichen Anspielungen auf die *Fröhliche Wissenschaft* 125 zum ‚Tod Gottes' – nun seinerseits als der paradigmatische Denker, der in einer entgötterten Welt mit dem Absurden ernst macht und es konsequent ergreift. Die komplementäre, zweite und kritische Perspektive richtet sich dann offenbar darauf, dass Nietzsche das Absurde auf die Spitze treibe, *um* es zu überwinden, nämlich um einen ‚Neuaufbau' zu entwerfen – und in diesem überwindenden Neuaufbau muss sich das Motiv einer ‚Entschärfung' des Absurden aufweisen lassen. Das zeigt im Grundansatz deutliche Parallelen zu Camus' These, Kierkegaard entdecke das Absurde allein, um ihm letztlich doch überwindend-entschärfend auszuweichen. Keineswegs klar ist aber vorab, worin im Falle Nietzsches das kritisierte Element genau liegen soll. Dass der Mensch – wie es im Passus abschließend heißt – im Angesicht des Gottestods auf sich selbst und sein Handeln zurückgeworfen wird, bedeutet ja kaum *per se* ein Ausweichen vor dem Absurden.

Demgemäß erscheint auch Camus' Zugriff auf Nietzsche in *Der Mensch in der Revolte* insgesamt zunächst durchaus affirmativ. Nietzsche tritt hier im Doppel nicht mit Kierkegaard, sondern mit Stirners *Einzigem* auf, und erhält dabei recht

---

21 Bei Nietzsche selbst wird der Begriff des ‚Absurden' eher selten in signifikanter Bedeutung verwendet, vgl. aber z. B. eine zentrale Partie aus der *Geburt der Tragödie*, direkt auf die ‚Weisheit des Silen' bezogen, in der Nietzsche vom „Entsetzliche[n] oder Absurde[n] des Seins" spricht (GT 7, KSA 1, S. 57), etwas ausführlicher entwickelt in der Vorfassung in *Die dionysische Weltanschauung* (DW 3, KSA 1, S. 566 – 568, 570). Vgl. dazu auch Pieper 2014, S. 93. Die Formel *credo quia absurdum est* findet sich mehrfach in Nietzsches Texten, markant und affirmativ etwa in der Vorrede zur *Morgenröthe* (vgl. M Vorrede 3 – 4, KSA 3, S. 15).

klar den Vorzug: „Schon Stirner wollte, nach Gott selbst, jede Vorstellung von Gott im Menschen austreiben. Aber im Gegensatz zu Nietzsche ist sein Nihilismus zufriedengestellt. Stirner lacht in der Sackgasse, Nietzsche rennt gegen die Mauern an." (Camus 2020b: HR, S. 91/1983b, S. 88) Wiederum erscheint Nietzsche als derjenige, der sich konsequent der Spannung des Absurden aussetzt und zugleich die permanente ‚Revolte' vollzieht.

In entsprechender Weise wird auch an mehreren Stellen der folgenden Erläuterung deutlich, dass Camus sich gegen verkürzende, nämlich einseitig auf Nietzsches ‚neue Werte' abstellende Interpretationen gerade verwahrt. So heißt es zum Einstieg in die vertiefte Auseinandersetzung: „Mit Nietzsche *scheint* der Nihilismus prophetisch zu werden." (Camus 2020b: HR, S. 95/1983b, S. 91; Herv. P.S.) Dabei verweist das ‚Prophetische' offenbar auf genau diejenige Seite Nietzsches, die zuvor kritisch als ‚Überwindung' des Absurden und ‚Neuaufbau' perspektiviert wurde. Gegen den Schein eines prophetischen Nihilismus aber wendet Camus sogleich ein, man könne Nietzsche „nichts anderes entnehmen als die niedrige, gemeine Grausamkeit, die er mit aller Kraft hasste, solange man nicht an die erste Stelle in seinem Werk, weit vor den Propheten, den *Kliniker* stellt" (Camus 2020b: HR, S. 95/1983b, S. 91; Herv. P.S.). Und das heißt offenbar: eben den *Analysten* und *Diagnostiker* des Nihilismus und des Absurden, ergänzt doch Camus, Nietzsche habe „den Nihilismus erkannt und ihn analysiert wie einen klinischen Fall." (Camus 2020b: HR, S. 95–96/1983b, S. 92) In diesem Sinne führt Camus Nietzsche als denjenigen vor, der „den Nihilismus in seine letzte Konsequenz treibt", dabei „die ganze Last des Nihilismus und der Revolte auf sich nimmt" und keinen einfachen Ausweg zulässt (Camus 2020b: HR, S. 96–97/1983b, S. 92–93). Selbst noch in der Diskussion von Nietzsches Rezeption durch den Nationalsozialismus schreibt Camus bemerkenswerterweise: „[D]ie Ungerechtigkeit, die Nietzsche angetan wurde, werden wir nie gutmachen können" – und gerade gegen die Verzerrungen dieser Rezeption, ihre Verkehrung ins „Gegenteil" und ihre „Parade von Lügen" gelte: „Wir müssen die Anwälte Nietzsches sein." (Camus 2020b: HR, S. 108–109/1983, S. 101–102) Gleichwohl möchte Camus Nietzsche in dieser Frage aber keineswegs vollständig aus der „Verantwortung" entlassen (Camus 2020b: HR, S. 111/1983b, S. 103). Vorzuwerfen sei ihm in der Tat, die Verzerrungen seines Denkens seinerseits mindestens *möglich* gemacht zu haben (vgl. Camus 2020b: HR, S. 109/1983b, S. 102).

Das führt zur Grundfrage: Worin liegt dann, angesichts dieses zunächst durchaus affirmativen Porträts, auf der tiefsten Ebene die von Camus kritisierte Seite Nietzsches? Kurz gesagt, und aus Camus' schlaufenartiger Darstellung pointierend herausgehoben: Sie liegt eben im titelgebenden Gedanken einer absoluten *Affirmation*, eines restlosen *Jasagens* zu allem, was ist – und zwar sofern diese Affirmation nach Camus die Revolte und den Widerstand letztlich aufhebt.

Würdigt Camus Nietzsche mithin als scharfen *Analysten* des Nihilismus im Angesicht des Todes Gottes, so kritisiert er aber sodann die *Konsequenzen*, die Nietzsche aus dem Gottestod zieht. Indem Nietzsche insbesondere darauf abziele, die Welt ‚nach' dem Tod Gottes aus den Schuldbestimmungen der Moral zu befreien, und indem er zugleich jede jenseitige oder finale Zweckhaftigkeit der Welt leugne, schlage die Revolte in eine pauschale und differenzlose Affirmation um. Dies liest sich bei Camus etwa wie folgt:

> Sobald man erkennt, dass die Welt kein Ziel verfolgt, schlägt Nietzsche vor, ihre Unschuld anzuerkennen und zu bejahen, dass sie keinem Urteil unterliegt, da man sie auf keine Absicht hin verurteilen kann, und infolgedessen alle Werturteile zu ersetzen durch ein einziges Ja, eine unumschränkte und begeisterte Zustimmung zu dieser Welt. (Camus 2020b: HR, S. 104/1983b, S. 98)

In den kritisch perspektivierten Motiven der Bejahung zieht Camus offenkundig mehrere ‚große Worte' aus Nietzsches Spätwerk zusammen, ohne sie durchweg und im Einzelnen auszuweisen, so die „Unschuld des Werdens", die „Ewige Wiederkunft" und insbesondere das „*amor fati*" (Camus 2020b: HR, S. 104 – 105/1983b, S. 98 – 99).[22] Mit der absoluten Affirmation einer in Unschuld werdenden Welt geht nämlich nach Camus die „Anerkennung des Schicksals", ja letztlich die „Vergöttlichung des Schicksals" einher (Camus 2020b: HR, S. 104/1983b, S. 99). Eben in dieser Apotheose aber benehme sich Nietzsche selbst die Möglichkeit der Revolte. Was nämlich in der absoluten Affirmation zu Grunde gehe, sei das Individuum, und mit ihm der revoltierende Gegenhalt: „Die Revolte, in der der Mensch sein eigenes Sein forderte, verschwindet hinter der völligen Unterwerfung des Individuums unter das Werden." (Camus 2020b: HR, S. 105/1983b, S. 99) Und so *verrät* Nietzsche in den Augen von Camus letztlich geradezu jene Revolte, die er zuerst auf sich genommen hatte:

---

22 Camus gibt dabei an, sich auf Nietzsches „letzte Philosophie" seit „1880" zu beziehen und – was freilich nicht dasselbe ist – einen „Kommentar zum *Willen zur Macht*" zu geben (Camus 2020b: HR, S. 96 Anm./1983b, S. 92 Anm.). Gewiss finden sich bei Nietzsche selbst und namentlich im Nachlass Partien, die einige der von Camus' vereinigten Motive zusammenziehen. Vgl. z. B. NL 1888, KSA 13, 16[32], S. 492: „Eine solche Experimental-Philosophie, wie ich sie lebe, nimmt versuchsweise selbst die Möglichkeiten des grundsätzlichen Nihilismus vorweg: ohne daß damit gesagt wäre, daß sie bei einem Nein, bei einer Negation, bei einem Willen zum Nein stehen bliebe. Sie will vielmehr bis zum Umgekehrten hindurch – bis zu einem dionysischen Jasagen zur Welt, wie sie ist, ohne Abzug, Ausnahme und Auswahl – sie will den ewigen Kreislauf, – dieselben Dinge, dieselbe Logik und Unlogik der Knoten. Höchster Zustand, den ein Philosoph erreichen kann: dionysisch zum Dasein stehn —: meine Formel dafür ist amor fati…".

> [D]ie metaphysische Revolte war in ihrer ersten Bewegung nur Protest gegen die Lüge und das Verbrechen der Existenz. Das nietzscheanische Ja, uneingedenk des ursprünglichen Neins, leugnet die Revolte selbst zu gleicher Zeit, da es die Moral leugnet, welche die Welt, wie sie ist, ablehnt. (Camus 2020b: HR, S. 110/1983b, S. 103)

Oder in der prägnantesten Formulierung gefasst, die sich in der Nietzsche-Darstellung findet: Die Kritik Camus' gilt jener „rasenden Zustimmung (*adhésion forcenée*), die Nietzsches Werk krönt" (Camus 2020b: HR, S. 109/1983b, S. 102) – eine Bejahung, die die Revolte und den Widerstand aufhebt, welche der ‚Kliniker' Nietzsche zuallererst aufgewiesen hatte. *Adhésion forcenée*, eine ‚rasende' oder auch ‚besinnungslose Zustimmung', die zugleich die Spannung der Revolte respektive des Absurden entschärft – diese schon im *Mythos des Sisyphos* verwendete Formulierung bindet Camus' Nietzsche-Kritik offenkundig zusammen mit der Kierkegaards.

III Schlussbemerkungen – Kritik und Methode

Abschließend und formelhaft lässt sich die im *Mythos des Sisyphos* und *Der Mensch in der Revolte* aufgewiesene parallele Doppelbewegung von Camus' Lektüren Kierkegaards und Nietzsches folgendermaßen zusammenfassen: Einerseits *affirmiert* Camus die konsequente Entdeckung des Absurden bei Kierkegaard wie auch der Revolte im Nihilismus Nietzsches – andererseits aber *kritisiert* er bei beiden die Entschärfung ihrer ursprünglichen Einsicht. Diese Entschärfung besteht bei beiden wiederum parallel, wenn auch unter *umgekehrten* Vorzeichen, in einer restlos *bejahenden* Hypostase und Apotheose, die zugleich die Spannung des Absurden und der Revolte vernichtet: Kierkegaard vergöttlicht das Absurde, indem er es ins Jenseits der *Transzendenz* transponiert, damit aber zugleich die Spannung des Absurden im Endlichen auflöst. Nietzsches Vergöttlichung hingegen wendet sich genau umgekehrt in die *Immanenz* der Welt; indem er aber aus der Revolte gegen Gott und die Moral die Konsequenz zieht, das Dasein dieser Welt sei differenzlos zu bejahen, vernichtet er zugleich seinerseits die Widerständigkeit der Revolte. Dabei ist zu beachten, dass in beiden Fällen nicht etwa das Absurde als solches vollständig getilgt oder ‚aufgehoben' wird, indem es beispielsweise in ein rationales Weltbild überführt und aufgelöst würde. Die Pointe in Camus' doppelter Interpretation besteht vielmehr darin, dass in beiden Fällen die *Kollision* und damit auch der ‚*Skandal*' des Absurden abgespannt wird, und zwar durch seine apotheotische *Rechtfertigung:* Im Falle Kierkegaards erhält das Absurde seine Rechtfertigung dadurch, dass es dem jenseitigen Gott selbst zugeschrieben wird – bei Nietzsche hingegen dadurch, dass auch die ‚schrecklichen' Seiten des diesseitigen Daseins als unterschiedslos und unhintergehbar dem ‚Leben' und ‚Schicksal' zugehörig bejaht werden müssen.

Trotz der keineswegs marginalen Differenzen bleibt aber die Grundbewegung eine parallele: In beiden Fällen gilt Camus' Zustimmung einer wesentlichen Entdeckung, nämlich der Spannung des Absurden und der Revolte im Nihilismus, die als *Aufgabe* des modernen Menschen erscheint – die Kritik aber gilt dem Versuch einer *Überwindung* des je Entdeckten, die gewissermaßen in der Bewältigung des Nihilismus ‚zu weit geht', indem sie ihn allzu schnell wieder ungeschehen zu machen droht.

Im Sinne eines Ausblicks sei noch angezeigt, dass sich gegen die skizzierten Interpretationen Camus' gewiss Einwände formulieren lassen. In *inhaltlicher* Hinsicht wäre vor allem rückzufragen, ob denn in Kierkegaards Sprung und Nietzsches Affirmation tatsächlich jede Spannung und Kollision derart restlos aufgehoben ist, wie es Camus' Kritik nahelegt. Dagegen ließe sich in beiden Fällen die *Permanenz* eines Gegenhalts und Widerstands ins Feld führen: Bei Kierkegaard ist der Glaube keineswegs die ein für alle Mal im Sprung erreichte ‚Beruhigung' der Existenz, sondern beständiger Austrag des persönlichen *Gegensatz*verhältnisses zu Gott in ‚Furcht und Zittern'.[23] Parallel kann auch Nietzsches Affirmation so gelesen werden, dass sie nicht als blanke, indifferente und pauschale Bejahung figuriert, sondern als eine solche, die beständig einen Gegenpol und Widerstand voraussetzt und nötig hat; eine Bejahung also, die gleichfalls nicht ein für alle Mal postuliert, sondern als beständige Aufgabe und Forderung begriffen wird[24] – und damit im starken Sinne als eine *tragische* Affirmation zu verstehen ist.[25]

Insbesondere ließe sich sodann in *methodischer* Hinsicht einwenden, dass Camus dem spezifisch *experimentellen* Charakter, der gerade das Denken Kierkegaards und Nietzsches auszeichnet und beide auch verbindet, nicht hinlängliche Aufmerksamkeit schenkt[26] – und dass er deshalb Motive wie den Sprung Kierkegaards oder die Bejahung Nietzsches allzu umstandslos als eindeutige philosophische Positionen und Lehren identifiziert. Mit dieser Rückfrage ist bei Kierkegaard auf das konsequent ausgeübte Verfahren der indirekten Mitteilung verwiesen,[27] bei Nietzsche etwa auf die Bestimmungen der Perspektivität, der Experimentalphilosophie und der ständig sich auf sich zurückbeugenden Inter-

23 Vgl. dazu gegen Camus' Interpretation bes. Stewart 2009, S. 442 und Berthold 2013, S. 148.
24 Vgl. in diesem Sinne als Kritik an Camus bes. Pieper 2014, S. 94–95.
25 Vgl. zu dieser These Schwab 2011.
26 Vgl. zu dieser Parallele zwischen Kierkegaard und Nietzsche Schwab 2015 und Victor 2021a.
27 Vgl. zur indirekten Mitteilung als Methode Kierkegaards Schwab 2012.

pretation.[28] Kierkegaards Pseudonyme etwa sind nicht einfach verschwindende Masken, ‚hinter' denen sich die blanke ‚Meinung' des Verfassers verbirgt, sondern Ausdrucksgestalten für konkurrierende existenzielle Optionen, die *ohne* einfaches Resultat den Leserinnen und Lesern zur Aneignung überantwortet werden. Ebenso sind Nietzsches späte sogenannte ‚Lehren' nicht geronnene und einsinnige Theoreme, sondern in sich vielgestaltige Komplexe, die von Nietzsche selbst beständig und bis zum Schluss der gegenkreuzenden Befragung unterzogen werden.

Diese Einwände haben gewiss ihre Anhaltspunkte in Camus' Interpretationen, ihnen lässt sich gleichwohl wiederum von Camus her entgegnen, und zwar in dreifachem Sinn: Zum ersten macht Camus selbst mehrfach und bestimmt die Grenze und die konsequente Einseitigkeit seiner Interpretationen deutlich, so insbesondere im *Mythos des Sisyphos*, wo er gar vom bewussten „Eigensinn (*entêtement*)" seiner selektiven Auslegung spricht (Camus 2020a: MS, S. 50 Anm./ 1983a, S. 162 Anm.).[29] Korrespondierend heißt es etwa in *Der Mensch in der Revolte*, zu Camus' Zugriff auf die Geschichte der metaphysischen Revolte insgesamt: „Diese Hypothese ist nicht die einzig mögliche; sie ist im Übrigen weit davon entfernt, alles zu erhellen." (Camus 2020b: HR, S. 23/1983b, S. 33) Camus visiert also in beiden Werken – und darin durchaus Vorgehensweisen von Kierkegaard und Nietzsche selbst verwandt – die besprochenen Denker allein in einem spezifischen Experimentalaufbau und unter nachdrücklich begrenzter Perspektive an.

Zum zweiten erscheint aber gerade die von Camus durchgeführte, gleichsam ‚rücksichtslose' Befragung Kierkegaards und Nietzsches philosophisch produktiv, und dies gerade in ihrem provokatorischen Potenzial: Sie stößt dazu an, Kierkegaards und besonders Nietzsches Denken kritisch daraufhin ‚abzuklopfen', ob und inwieweit ihre *Antworten*, und seien diese auch noch so sehr gebrochen-experimentell, der Schärfe ihrer *Ausgangsfragen* entsprechen – und wo sie möglicherweise doch auch hinter diese zurückfallen.

Zum dritten schließlich ist Camus doch auch selbst auf die Fragen der *Methode* bei Kierkegaard wie bei Nietzsche aufmerksam und reflektiert auch in dieser Hinsicht die Grenze der eigenen Auslegung. Im *Mythos des Sisyphos* etwa leitet

28 Vgl. die Ausführungen in Schwab 2015 und (im Blick auf die aphoristische Form) Schwab 2019.
29 Camus bemerkt hier gegen mögliche Einwände: „Aber ich untersuche nicht die Philosophie Kierkegaards, oder Schestows, oder, weiter unten, Husserls (dazu bedürfte es eines anderen Zusammenhangs und einer anderen Geisteshaltung), ich entnehme ihnen ein Thema und untersuche, ob seine Folgerungen mit den bereits festgelegten Regeln übereinstimmen" (Camus 2020a: MS, S. 50 Anm./1983a, S. 162 Anm.).

Camus gerade die oben zitierte kritische Partie zum Sprung mit einer methodischen Einschränkung ein:

> Gewiss, es ist schwer, bei einem so schwer greifbaren Autor klare Aussagen zu umreißen. Aber trotz anscheinend gegensätzlicher Schriften und über allen Pseudonymen, über allem Spiel und Lächeln taucht überall in seinem Werk so etwas wie die Ahnung (und zugleich die Befürchtung) einer Wahrheit auf, die schließlich in den letzten Werken herausbricht: Auch Kierkegaard macht den Sprung. (Camus 2020a: MS, S. 50/1983a, S. 162)

Entsprechend benennt Camus schon im ersten Zugriff auf Kierkegaard als Entdecker des Absurden diesen als einen „Don Juan der Erkenntnis" – übrigens eine nietzschesche Formulierung[30] –, der „die Pseudonyme und Widersprüche vervielfacht" (Camus 2020a: MS, S. 38/1983a, S. 153).

Beinahe noch nachdrücklicher sind die methodischen Verweise in der Nietzsche-Lektüre in *Der Mensch in der Revolte*. Gleich im Eingangspassus zum ‚Kliniker' schärft Camus als Warnung gegen jede vorschnelle Lektüre Nietzsches ein: „Der provisorische, methodische, mit einem Wort strategische Charakter seines Denkens kann nicht bezweifelt werden." (Camus 2020b: HR, S. 95/1983b, S. 91) Schließlich vereint Camus, im Passus zur nationalsozialistischen Interpretation, eine Bemerkung zur Methode mit einer Kritik an Nietzsche, die gleichfalls mindestens bedenkenswert erscheint: „Sobald man den methodischen Aspekt von Nietzsches Denken vernachlässigt (und es ist nicht sicher, dass *er selbst* sich immer daran gehalten hat), kennt seine revoltierende Logik keine Grenzen mehr." (Camus 2020b: HR, S. 109/1983b, S. 102; Herv. P.S.)

Kurzum: Nur scheinbar paradox, erweist sich Camus' Lektüre Kierkegaards und Nietzsches trotz und gerade in ihrer eingestanden kritischen Fokussierung als eine prinzipiell offene. In dieser Weise gelesen eröffnet sich ein weites Feld auch jenseits von Camus' expliziten Bezugnahmen auf beide Autoren. So wäre insbesondere zu fragen, ob nicht Camus gerade in seinen affirmativen Figuren des Absurden – etwa Don Juan, dem Schauspieler, dem Eroberer[31] – sich von Textschichten bei Kierkegaard und Nietzsche hat inspirieren lassen, die Alternativen zu jenen Konzepten bieten, die er als ‚Hauptlehren' kritisch diskutiert. Und umgekehrt ließe sich untersuchen, ob zentrale Figuren von Camus nicht doch sogar als transformierende Aneignungen und Übersetzungen eben jener zurückgewiesenen ‚Lehren' zu lesen sind – so etwa, nicht zuletzt, der doch deutlich an

30 Die Wendung findet sich in der *Morgenröthe*, vgl. M 327, KSA 3, S. 232.

31 Dies die Figuren im Kapitel „Der absurde Mensch", vgl. Camus 2020a: MS, S. 81–109/1983a, S. 185–205. Zum Vergleich von Camus' Don Juan mit dem Kierkegaards, obgleich sich eine Lektüre des entsprechenden Essays aus *Entweder/Oder* nicht nachweisen lässt, vgl. Stan 2011, S. 81–84.

Kierkegaards ‚Wiederholung' und Nietzsches ‚Ewige Wiederkehr' gemahnende Mythos des Sisyphos selbst.

Bibliographie

Abel, Johannes (2018): „Albert Camus' Reading of Søren Kierkegaard's *Fear and Trembling*". In: *Journal of Camus Studies*, S. 7–20.

Abel, Johannes (2019): „Camus's Kierkegaardian Conception of a Good Life". In: *Journal of Camus Studies*, S. 11–22.

Abel, Johannes (2020): *‚...wenn die Nacht das Denken überkommt.' Eine Untersuchung zur Idee des gelingenden Lebens bei Kierkegaard und Camus*. Univ. Diss. Freiburg.

Berthold, Daniel (2013): „Kierkegaard and Camus, Either/Or?". In: *International Journal for Philosophy of Religion* 73, S. 137–150.

Brock, Eike (2015): *Nietzsche und der Nihilismus*. Berlin, Boston.

Camus, Albert (1983a): „Le Mythe de Sisyphe". In: *Œuvres complètes*. Bd. 1. Paris, S. 118–252.

Camus, Albert (1983b): „L'Homme révolté". In: *Œuvres complètes*. Bd. 3. Paris, S. 11–339.

Camus, Albert (2020a): *Der Mythos des Sisyphos*. 26. Aufl. Hamburg.

Camus, Albert (2020b): *Der Mensch in der Revolte*. 33. Aufl. Hamburg.

Golomb, Jacob (1995): *In Search of Authenticity*. London, New York.

Heidegger, Martin (1988): „Schelling: Vom Wesen der Menschlichen Freiheit (1809)". In: Ingrid Schüßler (Hrsg.): *Gesamtausgabe*. Bd. 42. Frankfurt a. M.

Heidegger, Martin (1997): „Nietzsche. Zweiter Band". In: Brigitte Schillbach (Hrsg.): *Gesamtausgabe*. Bd. 6/2. Frankfurt a. M.

Heidegger, Martin (2001): „Die Grundfrage der Philosophie". In: Hartmut Tietjen (Hrsg.): *Gesamtausgabe*. Bd. 36/37. Frankfurt a. M., S. 1–80.

Jaspers, Karl (1968): „Kierkegaard". In: Hans Saner (Hrsg.): *Aneignung und Polemik. Gesammelte Reden und Aufsätze zur Geschichte der Philosophie*. München, S. 296–311.

Jaspers, Karl (1998): *Die geistige Situation der Zeit*. Berlin, New York.

Jaspers, Karl (2018): „Vernunft und Existenz". In: Dominic Kaegi (Hrsg.): *Gesamtausgabe*. Bd. I/8. Basel, S. 1–98.

Kierkegaard, Sören (EO1): „Entweder / Oder. Erster Teil". In: *Gesammelte Werke und Tagebücher*. 1. Abteilung. Simmerath 2004.

Kierkegaard, Sören (EO2): „Entweder / Oder. Zweiter Teil". In: *Gesammelte Werke und Tagebücher*. 2. und 3. Abteilung. Simmerath 2004, S. 1–377.

Kierkegaard, Sören (FZ): „Furcht und Zittern". In: *Gesammelte Werke und Tagebücher*. 4. Abteilung. Simmerath 2004.

Kierkegaard, Sören (KT): „Die Krankheit zum Tode". In: *Gesammelte Werke und Tagebücher*. 24. und 25. Abteilung. Simmerath 2004, S. 1–134.

Kierkegaard, Sören (NB 10): „Journal NB10". In: *Deutsche Søren-Kierkegaard-Edition*. Bd. 6. Niels Jørgen Cappelørn (Hrsg.). Berlin, Boston 2015.

Kierkegaard, Søren (SKS): *Søren Kierkegaards Skrifter*. Niels Jørgen Cappelørn (Hrsg.). Kopenhagen 1997–2012.

Kierkegaard, Sören (W): „Die Wiederholung". In: *Gesammelte Werke und Tagebücher*. 5. und 6. Abteilung. Simmerath 2004, S. 1–97.

Löwith, Karl (1987): „Kierkegaard und Nietzsche oder philosophische und theologische Überwindung des Nihilismus". In: *Sämtliche Schriften*. Bd. 6. Stuttgart, S. 53–74.
Pieper, Annemarie (1991): „Nihilismus und Revolte. Camus' Nietzschekritik". In: *Zeitschrift für philosophische Forschung* 45, S. 171–185.
Pieper, Annemarie (2014): „Absurdes Dasein. Albert Camus auf der Spur von Kierkegaard und Nietzsche". In: Rudolf Langthaler/Michael Hofer (Hrsg.): *Existenzphilosophie. Anspruch und Kritik einer Denkform*. Wien, S. 84–98.
Richter, Liselotte (1959): „Camus und die Philosophen in ihrer Aussage über das Absurde". In: Albert Camus: *Der Mythos von Sisyphos. Ein Versuch über das Absurde*. Hamburg, S. 113–141.
Schwab, Philipp (2011): „Die tragische Überwindung des Nihilismus. Nietzsches ‚Philosophie des Tragischen' von der *Geburt der Tragödie* bis zum Spätwerk". In: Lore Hühn/Philipp Schwab (Hrsg.): *Die Philosophie des Tragischen. Schopenhauer – Schelling – Nietzsche*. Berlin, Boston, S. 575–621.
Schwab, Philipp (2012): *Der Rückstoß der Methode. Kierkegaard und die indirekte Mitteilung*. Berlin, Boston.
Schwab, Philipp (2014): „Ethik und Ethikkritik. Philosophie der Existenz bei Kierkegaard und Nietzsche". In: Hans Feger/Manuela Hackel (Hrsg.): *Existenzphilosophie und Ethik*. Berlin, Boston, S. 89–111.
Schwab, Philipp (2015): „Critique of ‚the System' and Experimental Philosophy – Nietzsche and Kierkegaard". In: Katia Hay/Leonardo R. dos Santos (Hrsg.): *Nietzsche, German Idealism and Its Critics*. Berlin, Boston, S. 223–245.
Schwab, Philipp (2019): „‚Rück- und vorsichtig lesen'. Nietzsches ‚aphoristische' Denkform". In: Ralph Häfner/Sebastian Kaufmann/Andreas Urs Sommer (Hrsg.): *Nietzsches Literaturen*. Berlin, Boston, S. 49–82.
Sommer, Andreas Urs (2006): „Nihilism and Skepticism in Nietzsche". In: Keith Ansell Pearson (Hrsg.): *A Companion to Nietzsche*. Oxford, S. 250–269.
Stan, Leo (2011): „Albert Camus. Walled within God". In: Jon Stewart (Hrsg.): *Kierkegaard and Existentialism*. Burlington, Farnham, S. 63–94.
Stephan, Paul (2020): *Links-Nietzscheanismus*, Bd. 2: *Aneignungen Nietzsches*. Stuttgart.
Stewart, Jon (2009): „France: Kierkegaard as a Forerunner of Existentialism and Poststructuralism". In: Jon Stewart (Hrsg.): *Kierkegaard's International Reception. Northern and Western Europe*. Burlington, Farnham, S. 421–474.
Stewart, Jon (Hrsg.) (2011): *Kierkegaard and Existentialism*. Burlington, Farnham.
Thurnherr, Urs (2004): „Konsequentes Schreiben. Albert Camus' ‚Mythos des Sisyphos' im Lichte der Diapsalmata aus Søren Kierkegaards Entweder-Oder". In: Cornelia Blasberg/Franz-Josef Deiters (Hrsg.): *Denken/Schreiben (in) der Krise – Existentialismus und Literatur*. St. Ingbert, S. 255–274.
van Tongeren, Paul (2018): *Friedrich Nietzsche and European Nihilism*. Newcastle upon Tyne.
Victor, Oliver (2021a): *Kierkegaard und Nietzsche: Initialfiguren und Hauptmotive der Existenzphilosophie*. Berlin, Boston.
Victor, Oliver (2021b): „‚Ich verdanke Nietzsche einen Teil dessen, was ich bin'. Zur Nietzsche-Rezeption von Albert Camus". In: Dennis Sölch/Oliver Victor (Hrsg.): *Geschichte und Gegenwart der Existenzphilosophie*. Basel, S. 227–262.

Emanuel Seitz

Der Wille zur Macht ist die letzte Karte: Leben und Skepsis bei Cioran und Nietzsche

Abstract: The Will to Power is the Last Card: Life and Scepticism with Cioran and Nietzsche. As this article shows, Emil Cioran is an existentialist sceptic. What keeps people from committing suicide is not the perpetuation of the absurd, but a lasting doubt that renders them incapable of action. He opposes Nietzsche's will to power with a will to nothing. This nothingness, which exists in the world, enables man to distance himself from his own life and weakens the will. Nietzsche's philosophy of becoming, however, aims at strengthening the will, although it is clear that this strengthening can only be achieved by delusional inventions of being. The strong will of a truly free spirit overcomes the desperate indecision of an existentialist sceptic and decides either for a clear-cut yes to life or for a clear-cut yes to nothingness, but not for something in between. He chooses powerful freedom or free death.

Einleitung

Paris 1939: Ein rumänischer Stipendiat des *Institut français* in Bukarest berichtet über die Fortschritte an seiner Promotion, einem Werk, an dem er seit mehreren Jahren arbeiten soll. Die Arbeit scheint voranzugehen, sie ändert sich. Das Böse und die Sünde bei Nietzsche und Kierkegaard waren das ursprüngliche Thema, jetzt ginge es mehr – heißt es in diesem Bericht – um den Konflikt zwischen Leben und Bewusstsein bei Nietzsche. Der junge Gelehrte legt ein kommentiertes Inhaltsverzeichnis bei, das den Lesern genügend Hoffnung macht, um die gesuchten Gelder weiter zu bewilligen. Es lautet, paraphrasiert (Cioran 2015, S. 143):

> Kapitel 1): *Der Konflikt zwischen Leben und Bewusstsein.* Nietzsche behandele diesen Konflikt nicht als ein metaphysisches Problem, nicht als einen Konflikt zwischen Geist und Materie, sondern als ein Problem der Psychologie. Das Geistige und Geistliche verlieren ihre Autonomie, das Leben tritt an die Stelle des Absoluten und die Erfahrung des Individuums hat den Vorrang vor dem Geistigen.
>
> Kapitel 2): *Die Werte des Lebens und die Werte des Geistes.* Obwohl sie aus dem Leben heraus entstehen, wenden sich die Werte des Geistes am Ende gegen das Leben. Diese Dialektik soll beschrieben werden anhand der Werke von Simmel, Bergson, Nietzsche und Kierkegaard.

https://doi.org/10.1515/9783110760163-009

Kapitel 3): *Das Problem der Dekadenz – Das Böse in der Geschichte.* Der Geist wirkt bei Nietzsche subversiv und verringert den Anteil des Irrationalen am Leben. Aus einem solchen Mangel an Wahnsinn entsteht eine lebensarme und lebensfeindliche Dekadenz.

Kapitel 4): *Auf den Trümmern der Normen.* Nietzsches Vitalismus ist mit der Erschaffung einer Moral unvereinbar. Jenseits einer autonomen Vernunft lässt sich eine Ethik nicht begründen.

Die Angestellten des *Institut français* hatten guten Grund, den Plänen des jungen Mannes Vertrauen zu schenken. Obwohl nicht einmal dreißig Jahre alt, hatte der Philosoph bereits vier Monographien auf Rumänisch veröffentlicht: *Auf den Gipfeln der Verzweiflung* (1933), *Das Buch der Täuschungen* (1936), *Die Verklärung Rumäniens* (1936) und *Von Tränen und von Heiligen* (1937). Doch das Vertrauen wurde enttäuscht. Emil Cioran tat – nichts. Der Plan blieb ein tönendes Wort, ein leeres Versprechen, eine bloße Versuchung.

Das Geld sicherte die Existenz eines städtischen Asketen, der eine Dachstube im *Quartier latin* zum Ort seiner modernen Einsiedelei wählte. Ein Einsiedler, auch ein listiger wie Cioran, ist ein Meister des Nichtstuns. Er muss sich bemühen, noch unbrauchbarer und untätiger zu sein als ein Clochard unter den Brücken der Seine. Seine Übungen trainieren sich das Können und die Möglichkeiten ab, die ihn nützlich werden lassen könnten. Er wird frei von den Zwängen und den Ehren der geschäftigen Welt.

Ein wirklich freier Philosoph übt in seinen Askesen genau *einen* Gedanken ein, und dieser eine Gedanke macht das Ganze seines Denkens aus (vgl. Heidegger 1996, S. 427). Er ist das ‚Woher', von wo aus das ‚Wie' seines Vollzugs einfach und geschlossen wird. Es verhält sich nicht anders als bei einem geübten Kartenspieler. Er muss zunächst das Ganze auf seiner Hand kennen, bevor er sein Blatt in der günstigsten Reihenfolge bis zur letzten Karte ausspielen kann.

Das Leben und seine Aufgabe

Von wo aus beginnen also Nietzsche und Cioran ihr Denken? Mit dem Leben. Beide nehmen den Konflikt zwischen Leben und Bewusstsein als ein ‚psychologisches' Problem. Doch was heißt hier Leben? Und was heißt hier Psyche? Die Psychologie hat bei Nietzsche und Cioran nichts mit der empirischen Erforschung der Gefühlslagen und ihrer Rückführung auf vermeintliche oder erfundene Ursachen zu tun, wie es heute in der wissenschaftlichen Psychologie geschieht. Was Cioran und Nietzsche mit der Seele meinen, ist antik gedacht und bedeutet die *Entelechie* eines Leibes, die gesamte Wirklichkeit eines Lebewesens (vgl. Arist. de an. 412a 20 – 22). Eine solche Lehre untersucht das Tätigsein eines Menschen im

Ganzen, die Person insgesamt mit ihrem Bewusstsein, ihrem Glauben, ihren Gewohnheiten und ihren Lebensweisen. Eine solche Seelenkunde ist eine Charakterlehre und ein Teil der Rhetorik, einer dialektischen und philosophischen Rhetorik im Dienst des Wahren und Wahrscheinlichen. Sie prüft die Glaubwürdigkeit des Sprechers und fragt, *wer* dieser Mensch ist, der hier das Wort ergreift. Kann er bezeugen, was er fordert, weil er ist, was er will?

Eine philosophische Seelenkunde prüft *ad hominem*. In *Schopenhauer als Erzieher* schreibt Nietzsche: „Ich mache mir aus einem Philosophen gerade so viel als er im Stande ist ein Beispiel zu geben.“ (Nietzsche, SE, KSA 1, S. 350) Dieses Beispiel meint nichts Illustratives und Beiherspielendes, sondern den Beweis für die eigene Lehre in der Person des Philosophen. Der Charakter legt ein Zeugnis ab über das eigene Wesen und macht die Gedanken glaubwürdig oder unwahrscheinlich, wahrhaft und redlich oder listig und verlogen. In Nietzsches Übersetzung der aristotelischen Rhetorik hat der Charakter sogar den „ersten und wesentlichen Einfluss auf den Glauben“ (Nietzsche, Aristoteles Rhetorik I, KGA II/4, S. 541; Arist. rhet. 1056a: κυριωτάτην ἔχει πίστιν τὸ ἦθος). Eine derartige philosophische Psychologie prüft die Art des Menschen und untersucht, ob diese Menschenart und dieses Menschenleben beispielhaft für die gesamte Menschheit sein können.

Auch Nietzsche muss in diesem Sinn vor der Menschheit ein Zeugnis ablegen. Er muss von sich beweisen, dass er ein Vorbild ist. Als er in *Ecce homo* sich selbst sein Leben erzählt, tut er es in der Absicht zu sagen, *wer* er sei. Seine Identität ist nicht in Thüringen, in Basel oder im Piemont zu finden; er ist derjenige, der „mit der schwersten Forderung an die Menschheit“ herantritt (Nietzsche, EH, KSA 6, S. 257). Diese Forderung ist die Umwertung aller Werte. Bislang hat man die Realität um ihren Wert, ihren Sinn und ihre Wahrhaftigkeit gebracht, indem man eine Welt der Ideale erfand. Solche hinterweltlichen Erfindungen mit ihren fernen Löhnen sind aber gerade ein Entzug und eine Flucht vor dem Dasein, eine Feigheit der Philosophen, Theologen und Moralisten, die nicht genügend Kraft und Härte besitzen, um sich der eigentlichen Wahrheit, ihrem Sinn und Wert zu stellen. Sie berauben die Menschen „um die einzige Welt, die es gibt“ (Nietzsche, EH, KSA 6, S. 374), um die einzige *wirkliche* Realität, die Realität *dieser* Erde. Ohne eine solche Realität gibt es kein Ziel, keine Vernunft und keine Aufgabe mehr im Leben, nur einem verlogenen Hinter-Sinn, ein Jenseits, das in Wirklichkeit Nichts ist.

Der Sinn der Erde ist der Übermensch, ein Menschentypus höchster Wohlgeratenheit, ein höchstes Beispiel für die Menschheit (Nietzsche, EH, KSA 6, S. 300; Z, KSA 4, S. 14). Zarathustra ist dessen Fürsprecher und Nietzsche lediglich der Verfasser des Zarathustra, ein Sekretär der Worte desjenigen, der den Sinn der Erde erkannt hat und ihn der Menschheit verkündet. Wer also Nietzsche und was der Sinn seines Lebens ist, kann klar gesagt werden: Er ist jener Schreiber, der ein

Ohr hatte für die höchste Aufgabe der Menschheit und den Sinn der Erde uns überlieferte.

Wer dagegen ist Emil Cioran? Wofür ist er Zeuge und Beispiel? Auch er ist nicht der listig gescheiterte Promovend, der Nachtspaziergänger in Paris oder ein rumänischer Emigrant, der Camus und Sartre kannte. In den *Gipfeln der Verzweiflung* heißt es: „Ich bin ein sinnleerer Mensch und bedauere es keineswegs, vom Sinn entbunden zu sein" (Cioran 2019a, S. 118). Die Sinnleere ist keine Sinnlosigkeit. Die Sinnleere ist des Sinns beraubt, wo es einmal einen Sinn gegeben hat. Nur dort, wo es einmal ein Bindendes gab, kann ein Mensch von diesem einst Bindenden entbunden worden sein. Sinnlosigkeit würde dort herrschen, wo es gar keinen Sinn geben kann.

So wie Zarathustra der Verkünder des Sinns der Erde ist, so ist Cioran der Verkünder der Sinnleere. Diese Leere kommt nicht aufgrund eines persönlichen Meinens des rumänischen Eigenbrötlers zustande, sondern beruht auf einer grundsätzlichen Bestimmung des Wesens des Menschen. Der Mensch ist, heißt es ebenfalls in den *Gipfeln der Verzweiflung*, ein „vom Leben abgefallenes Tier" (Cioran 2019a, S. 148). Den Tieren bedeutet das Leben noch alles, weil sie ganz wörtlich Lebe-Wesen sind, also Wesen, deren Sein nur im Lebendigsein geschieht. Allein dem Menschen ist das Leben nichts Absolutes und nichts Wesentliches mehr. Für den Menschen ist das Leben nur noch eine Wahlmöglichkeit. Er kann von seinem Leben Abstand nehmen und sich von ihm verabschieden – und selbst wenn er sich *für* das Leben entscheidet, bleibt es für ihn ein Fragezeichen, etwas Zweifelhaftes, denn – so Cioran an derselben Stelle – „das Leben hat nicht nur keinen Sinn, sondern kann auch gar keinen haben" (Cioran 2019a, S. 148). Was weder einen Sinn noch keinen Sinn hat, ist unmöglich sinnhaft. Die Sinnleere des Lebens folgt aus seiner faktischen Sinnlosigkeit. Die Einsicht, dass ein Sinn der Erde unmöglich ist, hat dem Menschen die Möglichkeit eines Sinnes für sein Leben geraubt.

Diese Erkenntnis macht Cioran zu einem Skeptiker. Er verneint die Möglichkeit, ohne Zweifel einen Sinn der Erde und des Lebens zu bestimmen. Seine Schreibübungen sind die Exerzitien eines Hohepriesters der Agnostik, der sich um Steigerungen in der Kraft des Negativen bemüht. Eine echte Virtuosität dieser Dekonstruktionen *avant la lettre* wäre erreicht, wenn alle Glaubenssätze zerfallen sind und nicht einmal das Nichts wertvoller als das Sein erscheint.

Nietzsches Philosophie der Umwertung aller Werte ist von einer solchen Skepsis weit entfernt. Er fordert in Konjunktiven des Befehls klare Übungsziele für die Menschheit und eine vollständige Umwendung der Seele und des Charakters. Erst wenn die Menschen die ontotheologischen Begründungen der Werte hinter sich gelassen haben, können sie eine eine Wohlgeratenheit erreichen, die fast schon übermenschlich ist. Die alten ontotheologischen Werte erweisen sich in dieser Umwendung der Seele als falsche Werte, die der Wahrheit der Welt nicht gerecht werden. Die Welt gibt es in Wirklichkeit nämlich nur als ein fortwährendes Werden und als ein Spiel der Kräfte, das nichts ewig Bleibendes hervorbringt. In einem solchen Spiel muss sich jede Kraft ständig erneuern, neu versuchen und durch eine Probe herausfinden, was von dem Vorhandenen sie steigert und für sie einen Wert hat. Ein Gutes für alle – die Idee des Platonismus – kann es in einer solchen Welt nicht geben. Es gibt kein Seiendes, nur ein Noch-nicht-Seiendes gibt es für alle. Es ist die Aufgabe zur Seelenstärke und Seelenhöhe.[1]

Das Leben und der Wille zur Macht

Das Denken Ciorans und Nietzsches ergänzen einander und liefern grundlegende Einsichten in die Existenzbedingungen der Menschen: Nur dort, wo ein Mensch vom Leben abfallen kann, ist es möglich, dass er mit einer falschen Forderung an sich selbst herantritt und ins Unglück stürzt; und nur dort, wo ein Wesen seine Werte selbst erschaffen muss, kann es über die eigenen Werte zweifeln und verzweifeln.

Doch selbst wenn wir diese Stimmigkeit von Schaffen und Zweifel, von Abfall und Aufgabe des Lebens anerkennen – woher kommt sie? Wie können wir sagen, dass diese Denkweisen ‚sich ergänzen'? Etwas kann sich nur ergänzen, wenn es ein Ganzes gibt, worauf die beiden Teile zurückgeführt werden, von denen es heißt, dass sie sich ergänzen würden. Dieses Gemeinsame benennt Cioran in der *Lehre vom Zerfall:* Der Mensch ist, sagt er dort, das „unbestimmteste aller Wesen." (Cioran 1979a, S. 34) Auf dem Grund dieser Unbestimmtheit kann der Mensch zweifeln und muss sich entscheiden. Wäre der Mensch in seinem Wesen festgelegt und in sich bestimmt, hätte es gar keinen Zweck und wäre völlig unvernünftig, wenn er sich seine Werte selbst erschaffen würde. Denn die Werte wären schon da und müssten nur noch erkannt werden.

---

1 Die Postmoderne macht einen großen Fehler, wenn sie meint, ihre generelle Skepsis vor der Wahrheit sei mit Nietzsche zu vergleichen. Deleuze schreibt: „Nietzsche kritisiert nicht die falschen Anmaßungen der Wahrheit, vielmehr die Wahrheit selber und als Ideal" (Deleuze 2013, S. 104). Das Gegenteil ist der Fall: Der Wille zur Wahrheit und der Wert der Wahrheit sind Nietzsches Problem, und zwar genau dann, wenn sie nicht eine „Beständigung des Werdens in die Anwesenheit" (vgl. Heidegger 1996, S. 584) bedeuten. Das Werden als das Wahre (und ‚Ganze' und Eine) fordert einen umgekehrten Platonismus und eine dynamische (leibnizianische) Perspektivierung, aber weder das Nichtsein von Wahrheit noch eine bloß subjektive Kontingenz des Wahren.

Die Bestimmung des Menschen als das unbestimmteste aller Wesen gibt den entscheidenden Grund für die Möglichkeit des Zweifels und des Schaffens an und ist selbst nicht unbegründet. Nietzsche stellt sich diesem Problem in seinem Spätwerk und entdeckt hierbei die Phänomenologie des Nihilismus. Nihilismus bedeutet, dass die obersten Werte *sich* entwerten, sich selbst entwerten, weil das Bewusstsein zur Einsicht kommt, „daß mit dem Werden nichts erzielt werden soll und daß unter allem Werden keine große Einheit waltet" (Nietzsche, NL 1887, KSA 13, 11[99]).

Dieses Zitat stammt aus den Vorstudien (Nr. 351) zu dem geplanten und nicht zu Ende gebrachten Hauptwerk Nietzsches, das den Titel tragen sollte: Der Wille zur Macht (als „Schluß-Resümé" wohl des ersten Buches, vgl. Nietzsche, NL 1888, KSA 13, 12[1]). Man erlaube mir an dieser Stelle eine philologische Zwischenbemerkung: Das Buch zählt seit der Neuauflage der Werke unter Giorgio Colli und Mazzino Montinari zu den vermeintlich erfundenen Büchern Nietzsches. Die beiden Philologen haben mit dieser Edition ein Entnazifizierungsverfahren gegen Nietzsche eingeleitet, an dessen Ende er als kulturpolitisch harmloser „Demokrat und Pazifist" erscheinen soll (Montinari 1991, S. 73). Der Kern dieses Verfahrens ist das Unverfügbar-Machen des *Willens zur Macht*. Nach wie vor ist aber Nietzsches Nachlass aus dem Umkreis dieses Buches inhaltlich – gerade in der Behandlung des Nihilismus – von höchstem philosophischem Interesse. Man wünscht sich eine bessere, *thematisch geleitete* Nachlass-Sammlung als jene von Heinrich Köselitz und Elisabeth Förster-Nietzsche (Gast/Förster-Nietzsche 1996), die sich zuverlässig auf Nietzsches eigene Pläne und Entwürfe des *Willens zur Macht* stützt (Nietzsche, NL 1886, KSA 12, S. 7–8; NL 1888, KSA 13, 12 [1]). Der *Antichrist* leistet inhaltlich erkennbar viel weniger als das alte Buchprojekt, selbst wenn Nietzsche diesen Fluch auf das Christentum in seiner eigenen Korrespondenz als ‚Umwertung aller Werte' bezeichnet (vgl. Sommer 2013, S. 1–8). Faktisch ist das Hauptwerk Nietzsches nicht dieser Nachlass, entgegen der Meinung Heideggers, sondern das „höchste Buch, das es giebt" (Nietzsche, EH, KSA 6, S. 259), das ist: der Zarathustra, worin Skepsis und Nihilismus überwunden werden.

Ein Nihilismus gründet sich auf eine verbesserte Einsicht in die Wahrheit allen Geschehens: Weil die Welt wesentlich ein Werden ist, gibt es sie nur grundlos, ohne Zweck und ohne Ziel. In den Fragmenten zum *Willen zur Macht* schreibt Nietzsche in diesem Sinn:

> Es ist eine bloße Erfahrungssache, daß die Veränderung nicht aufhört: an sich haben wir nicht den geringsten Grund zu verstehen, daß auf eine Ver<änderung> eine andere folgen müsse. Im Gegentheil: ein erreichter Zustand schiene [...] nicht sich erhalten <zu> wollen... [...] Gerade an allem Lebendigen ist am deutlichsten zu zeigen, daß es alles thut, um nicht sich zu erhalten, sondern um mehr zu werden. (Nietzsche, NL 1888, KSA 13, 14[121])

Dieser Wille, mehr zu werden und über sich hinaus zu wachsen, ist der Wille zur Macht, und dieser Wille bedeutet Leben. Er ist die treibende Kraft und das dynamische Prinzip, das das Werden der Welt am Laufen hält.

Werden bedeutet – wie Nietzsche in einem anderen Fragment schreibt – „Erfinden, Wollen, Selbstverneinen, Sich-selbst-überwinden: kein Subjekt, sondern ein Thun, Setzen, schöpferisch" (Nietzsche, NL 1886, KSA 12, 7[54]) Wer sich also selbst seine Werte erfindet, setzt sich ein Sein, auch wenn das Seiende selbst nichtig ist. Die Welt des Seins erlügt sich ein Beständiges im Werden, aber gerade dieser Wille zur Täuschung ist der höchste Wille zur Macht und damit auch der höchste Wille zum Leben. Der Wille zur Macht ist der Wille zum Sein.

Nietzsches Irrtum

Diese Apotheose des Willens gehört für Cioran im Weltbuch der Philosophie zu dem Kapitel der gut gemeinten Überschätzungen. „Nietzsche irrte", schreibt er im *Buch der Täuschungen*, „als er, von der Offenbarung des Lebens hingerissen, im Willen zur Macht das grundlegende Problem und die wesentliche Modalität des Seins bloßlegte. Der Mensch will im Angesicht des Lebens wissen, ob er ihm seine letzte Einwilligung erteilt." (Cioran 2019b, S. 125)

Schauen wir uns die Stelle genau an: Nietzsche irrte sich, heißt es, als er hingerissen war. Wer hingerissen ist, gibt sich einer Begeisterung hin, einer inneren Spannung, einer Ekstase im Dasein. Er wird ein Jünger des Dionysos, wie sich Nietzsche einmal selbst nannte, und steht in seinem Rausch und seinem besinnungslosen Taumeln einem Satyr näher als einem Heiligen (vgl. Nietzsche, EH, KSA 6, S. 258). Der Berauschte fühlt und empfindet, doch er weiß nicht, was er tut.

Cioran erlebt diese Augenblicke des Außer-sich-stehens in der Musik, der Mystik und der Erotik (Cioran 2019b, S. 137). In diesen Momenten, in denen das Selbst vergessen wird und sich der Mensch zur reinen Energie verwandelt, kann das Absolute in der Welt erfahren werden. Die Ekstase ist Gegenwart, die nicht enden will, und die „ewige Gegenwart ist *Dasein*" (Cioran 2019a, S. 119). Der Geist verlischt in der Ekstase und mit ihm auch jeder Zweifel.

Der frühe Cioran stand der Philosophie des dionysischen Taumelns noch sehr nahe, doch ab dem Zeitpunkt, an dem er französisch schreibt, verschiebt sich die Grundlage seines Denkens. In den *Syllogismen der Bitterkeit* von 1952 heißt es: „Wir haben mit Nietzsche an den ewigen Bestand der Rauschzustände geglaubt; dank der Reife unseres Zynismus sind wir weiter gegangen als er. Die Idee des Übermenschen erscheint uns nur noch als ein Hirngespinst." (Cioran 2016, S. 26 – 27) Der frühe Cioran pflegte die lebendige Selbstverneinung im Rausch, der späte Cioran treibt die rationale Lebensverachtung des Zynismus aufs Äußerste (vgl. Cioran 2019c, S. 13).

Diese Verschiebung geht auf einen Fortschritt in der Übung des Denkens zurück und nicht auf eine Korrektur einer alten Erkenntnis durch eine neue und aktuellere. Auch der späte Cioran war nämlich nie ein reiner Skeptiker, sondern immer – wie er selbst in *Gevierteilt* schreibt – ein „Verehrer des Zweifels, ein Zweifler in Trance, ein Fanatiker ohne Credo" (Cioran 2017, S. 77). Cioran entwickelte sich von einem zweifelnden Ekstatiker zu einem ekstatischen Zweifler.

Die wesentlichen Modalitäten der menschlichen Empfindung des Daseins sind genau diese beiden Weisen des Erlebens: die ekstatische Begeisterung erfährt das Grundproblem des Lebens positiv, die zweifelnde Verzweiflung äußert sich in einer „negativen Form der Begeisterung" (Cioran 2019b, S. 97). Eine Begeisterung entsteht, wo es einen Glauben gibt. Die großen Enthusiasten, die großen Ja-Sager in der Welt – selbst wenn es nur das Ja zum Übermenschen ist – müssen neue Ideen und Idole errichten, zu denen sie Ja sagen können. Diese Ideen und Idole müssen Täuschungen sein, weil sie ein kommendes Sein postulieren, das aber vergehen wird, weil die Welt ein Werden ist. *Also sprach Zarathustra* kann in diesem Sinn selbst nichts anderes sein als ein *Buch der Täuschungen*, weil ontologisch in einer Welt des Werdens das Schaffen gleichbedeutend ist mit Täuschung.

Cioran sucht in der Verzweiflung eine wahrhaftere Form des Lebens als die positive, sich selbst aber belügende Begeisterung Nietzsches. Der „Trommler der Trügnisse" aus Naumburg (Cioran 2019b, S. 163) schlägt in Ciorans Ohren sein Fell für das Unwesentliche am Dasein, für das Vergängliche und Sinnlose. Erst ein existenzieller Zweifel bringt die wahre und entscheidende Frage ans Licht: Soll das Tier, das vom Leben abfallen kann, dem Leben seine Einwilligung erteilen oder nicht?

Die Herkunft des Irrtums

Der Wille zur Macht, der Wille, mehr zu werden, über sich hinauszugehen und seinem Selbst das Äußere zu unterwerfen, soll bei Cioran kein wesentliches Problem des Menschen sein. Er begründet diese Aussage mit folgendem Satz: Der Mensch „kann stark sein, selbst wenn er darbt" (Cioran 2019b, S. 125). Der darbende Mensch ist ein ohnmächtiger Mensch. Wer ohnmächtig ist, hat keinen Willen zur Macht oder kann seinen Willen zur Macht nicht verwirklichen. Wenn dieser Ohnmächtige dann noch ein Darbender ist, leidet dieser Mensch ohne Wirksamkeit, hat Hunger ohne Anlass und fühlt eine Not ohne Notwendigkeit. Die

Darbenden gehören zu den Verzweifelten, zu den Fanatikern ohne Credo, zu den Helden und Enthusiasten eines tragsamen Geistes.

Das Ertragen-Können soll stärker sein als das Schaffen-Können, die passive Kraft stärker als die aktive Kraft. Das Stärkere liegt *per definitionem* näher am Wesen: Wo etwas in seiner stärksten Form anwesend ist, ohne Beraubungen und Behinderungen, ohne Wucherungen und unterentwickelte Potentiale, ohne den Verfall eines Könnens aufgrund von Alter, Siechtum oder Nachlässigkeit, da zeigt sich das Wesentliche in seinem Können ohne die umstandsmäßigen Beschränkungen.

Die Frage, welche Art zu leben die stärkste ist, führt also zur Frage nach dem Wesen des Lebens und des Menschen. Nietzsche bestimmt den Willen zur Macht als ein Verhalten, das jedes Leben zeigt und für jedes Leben charakteristisch ist. Dieser Wille ist das innere Gegenstück zu dem, was sich in der Welt als Kraft verwirklicht. Präzise gesagt, ist es kein Wille im Sinn einer geistigen Entscheidung, keine Urteilskraft, sondern ein inneres Verlangen, das bei allem Lebendigen entsteht. Dieses Verlangen will dem Werden den Charakter des Seins aufprägen – das ist die Hauptbedeutung des Willens zur Macht (Nietzsche, NL 1886, KSA 12, 7[54]).

Wo Macht im Spiel ist, soll das Werden verschwinden und eine bleibende Form entstehen; doch diese Form kann eigentlich nicht bleibend sein, wenn Nietzsches Interpretation allen Geschehens, allen Welt-Geschehens, wahr ist. Denn wenn die Welt in Wahrheit ein Werden ist, bedeutet dieser Wille zum Sein eine Selbstbelügnis und Selbstbetrügnis. Diese Täuschung durch das Sein leugnet Nietzsche auch nicht, entscheidend ist aber, dass der seinsprägende Wille zur Macht das dynamische Prinzip des Werdens und nicht das Prinzip des statischen Seins bedeutet, denn das jeweils Wirkende will nicht bloß das Sein überhaupt, sondern ein Mehr-Sein. Erst mit diesem Willen zum Mehr-Sein gleicht sich das Individuum in seinem Willen zur Macht wieder an die Wahrheit allen Geschehens an und wird in seinen Erfindungen und Machtergreifungen selbst wieder ein Werdender, der sich und andere ständig überwinden muss

Dass die stärkste Existenz auch die mächtigste sein soll, verweist auf die Metaphysik des Thomas Hobbes. Dessen mechanistischer Begriff von Kraft, von *power*, ersetzt die antike Lehre von der δύναμις, die ein Können oder eine Möglichkeit zu wirken bezeichnet, durch eine Lehre, in der jedes Können bloß Macht bedeutet. Aufgrund einer inneren Disposition ist der Träger eines Könnens zu wirken befähigt, ohne aber wirken zu müssen. Die äußere Verwirklichung kann hinter den eigenen Möglichkeiten zurückbleiben. Der hobbesianische Begriff von Macht kennt dagegen eine solche disponible Stärke nicht und ersetzt sie durch eine Art von Energie, die sich in verschiedenen Formen äußert. Die Macht oder *power* ist dann in ihrer Möglichkeit immer nur so groß, wie sie sich tatsächlich im

Äußeren verwirklicht. Wo von Macht die Rede ist, fallen die mögliche Wirksamkeit und die tatsächliche Wirksamkeit in Eins (Hobbes 1839, Kap. X).

Zum Wesen eines Könnens gehört es, dass es Dinge fertigbringt (Heidegger 1981, S. 100 – 101). Wer die Kraft hat, ein Gewicht zu heben, dem fällt es nur durch Zufall aus der Hand. Und wer Karten spielen kann, der bringt auch eine Partie zu Ende. Gleichzeitig lässt sich aber eine Fertigkeit auch durch Übung scheinbar unendlich steigern, solange äußere Umstände dieses Mehr-Wollen nicht verhindern. Genauso wie eine Kraft in ihrem Wirken fertig wird, ist sie in ihrer Entstehung immer unfertig: Je mehr die Kraft geübt wird, desto mehr Kraft entsteht. Weil eine übungsfähige Kraft in ihrem Wesen ein lebendiges Werden ist, nutzt sie sich im Gebrauch nicht ab und wird nicht konsumiert.

Diese Paradoxie der Kraft in ihrem inneren Wirken erkannte Nietzsche und nahm sie für das Wesentliche allen Lebens. Ob die Erfindung eines Seins wahr ist oder nicht, spielt keine Rolle, solange diese Erfindungen die Kraft ins Unendliche zu steigern fähig sind. Nietzsche erkannte den Nutzen auch der Lügen für eine Vertikalspannung, die das Werden im Leben verstetigt.

Der Preis einer wohltätigen Lüge war für Emil Cioran zu hoch. Im *Buch der Täuschungen* heißt es: „Die erste Frage angesichts des Lebens fällt mit einem Aufruf zur Aufrichtigkeit zusammen. Ob wir hernach Macht anstreben oder nicht, ist unerheblich. Die Menschheit dürstet nach Macht, um die letzte Karte des Lebens auszuspielen" (Cioran 2019b, S. 125). Der Wille zur Macht soll also nicht das Primäre sein, nicht das Wesentliche des Lebens, sondern bloß dessen letzte Karte. Die letzte Karte... Dieser Ausdruck lässt alle Fragen offen: Ist diese letzte Karte der Trumpf, den der Lebende in seinem Lebensspiel aufhebt, um das Höchste zu stechen und das Leben zu gewinnen? Oder ist diese letzte Karte eine Lusche, die man nur noch abwerfen muss, wenn die wichtigsten Stiche bereits gefallen sind?

Um diese Frage beantworten zu können, müssen wir uns einem weiteren Grundsatz der Philosophie Ciorans zuwenden: „Fortschritt im Wesen ist Rückschritt im Leben", heißt es am Ende des *Buches der Täuschungen* (Cioran 2019b, S. 189). Weil der Mensch wesentlich ein Tier ist, das vom Leben abfallen kann, erkauft er sich das Wesentlich-Werden gleichzeitig mit einem Verlust an Lebendigkeit. Der Aufruf zur Aufrichtigkeit, dieser Gewissensruf, den Cioran die primäre Frage des Lebens genannt hat, ist ein Ruf zum eigentlichen Selbstsein. Doch dieses Selbstsein kann nur erreicht werden, wenn man auf das Leben verzichtet. „Dass der Wille zur Macht die letzte vom Leben ausgespielte Karte ist", schreibt er an späterer Stelle, „beweist der höchste Zugang zur Macht bei jenen, die nichts mehr zu verlieren haben, oder jenen, denen das Leben nichts geboten hat. Jesus: der schwächste Mensch ist der stärkste gewesen. Seelenstärke gibt es allein im biologischen Mangel." (Cioran 2019b, S. 134) Wer den Willen zur Macht ausspielt, verliert das Leben und wird wesentlich. Erst die Sezession von der biologischen

Natur des Lebe-Wesens bringt den Menschen zur Eigentlichkeit seines Daseins. Im Kern ist der Mensch Abfall. Im Abfall zeigt er Verzweiflung, im Machtwillen Begeisterung.

Nichts werden

Dass im Abfall die höchste Stärke liegen soll, gründet in der nihilistischen Interpretation allen Geschehens, die Cioran durch seine gesteigerte Skepsis gewonnen hat. Nietzsche glaubte noch, im Erfinden den Nihilismus überwinden zu können, der das absolute Werden kennzeichnet, weil es sinnlos ist. Diesen Glauben hat Cioran verloren. Der Sinn der Erde kann nicht darin bestehen, einer Existenz, die gar nicht sinnhaft sein kann, einen erlogenen Sinn zu verleihen.

Nietzsche kann dem Werden noch ein *amor fati* abgewinnen, eine Liebe zur eigenen Erfindung und zum eigenen Schicksal. Für Cioran wird der Selbstmord zur Obsession eines Menschen, der weder sterben noch leben kann. So wie Nietzsche die subjektivistische Metaphysik vollendet hat (Heidegger 1996, S. 425–432), vollendet Cioran den Nihilismus in seiner subjektivistischen Form. Er bringt für das Werden auch die Todeskarte ins Spiel und bestimmt das Werden als eine „Überwindung des Nichts, das nur dort erscheinen kann, wo die Zeitlichkeit dem Dasein wesentlich wird. Denn die Zeitlichkeit führt ein Moment des Nichts ins Werden ein" (Cioran 2019a, S. 119).

Da unser Dasein wesentlich zeitlich verfasst und wir Menschen ein Sein zum Tode sind, muss auch unserem Werden ein solches Moment des Nichts wesentlich zugehören. Dieses Nichts, schreibt Cioran in der *Lehre vom Zerfall*, „ist das Dasein selbst; es ist das Ganze […], es vermag zwar nicht dem Leben einen Sinn zu geben, bewirkt aber, dass es fortdauert als das, was es ist: ein Zustand des Nicht-Selbstmordes" (Cioran 1979a, S. 27). Mit einer rechten Haltung zum Nichts im Dasein können wir das Sterben lernen, ohne zu sterben. Unvermittelt verwandelt sich die Sinnlosigkeit der Welt aus einem Problem der Erkenntnis zu einem Problem des Lebens und der Existenz.

Existenzfragen: Der Selbstmord und die Revolte

Cioran entdeckt hier die existenzialistische Dimension des heideggerianischen Denkens. Die Leitfrage des Existenzialismus in seiner subjektivistischen Form, zu der sich Cioran bekannte (vgl. Große 2014, S. 109), fällt zusammen mit dem ersten Satz aus Camus' *Mythos des Sisyphos:* „Il n'y a qu'un problème philosophique vraiment sérieux: c'est le suicide" (Camus 1942: MS, S. 15) – es gibt nur ein wirklich ernst zu nehmendes Problem der Philosophie: den Selbstmord. In dieser Frage scheiterte Heidegger. Zwar bestimmte er das Daseins als ein Sein zum Tode und als etwas, das in das Nichts gehalten ist (vgl. Heidegger 1976, S. 115), er bestimmte sogar die Welt als das „Nichts, das sich zeitigt" (Heidegger 2007, S. 272) – aber alle diese Überlegungen bemühen sich nur um eine Erkenntnis. Heidegger stellt nicht die Frage nach dem Selbstmord, er verzweifelt nicht an der faktischen Sinnlosigkeit des Daseins, er hat nur Angst und Langeweile. Dem Leben hat er immer schon seine Einwilligung gegeben. Diese neue Ontologie, die auf dem Nichts beruht, taugt als *ars moriendi* genauso wenig wie ihre ontotheologischen Vorgängerinnnen. Heideggers Lehre von der entschlossenen Sorge um ein eigentliches Dasein gibt kein Beispiel für das Nichts, von dem er redet.

Cioran macht in seiner Philosophie die *ars moriendi* in einem Dasein ohne letzten Grund zu seiner Aufgabe und Lebensform. Das wirkliche Rätsel ist nicht der Akt des Tötens, nicht die Frage, ob ein Dolchstoß, ein Pistolenschuss oder der vorbeifahrende Schnellzug das beste Mittel sind, sich seiner selbst zu entledigen – das eigentliche Problem ist die Frage, ob es Gründe oder keine Gründe gibt, um auf der Welt zu bleiben; und wenn es Gründe gibt, ob diese Gründe gut oder schlecht sind, ob sie die Kraft haben, überzeugen zu können, oder ob sie nicht eher in Zweifel gezogen werden müssen, weil sie Ausflüchte, Besänftigungen oder irrtümliche Wahnvorstellungen sind. Ist das Leben wert, gelebt zu werden? Oder müssen wir nicht eher einen Selbstmörder loben, weil er eine wahre Erkenntnis von der Welt hat, weil er ihre Grundlosigkeit und Wertlosigkeit einsieht und weil er diese Erkenntnis nicht nur als ein intellektuelles Spiel betreibt, sondern sie ernst nimmt und zur Richtschnur seines Handelns macht? Ist der Mensch nicht das einzige selbstmordende Tier – und dieser Selbstmord der Akt seiner höchsten Freiheit? Ein Nihilist, der bereits in den Abgrund des Grundlosen gefallen ist, findet auf solche Fragen keine befriedigende Antwort mehr, die ihn vom Selbstmord abhalten könnte. Er taucht in die ätzende Säure des Absurden, die nur die wenigsten Menschen aushalten können, ohne sich in Nichts aufzulösen. Selbst Camus glaubte noch, einen paradoxen Ausweg gefunden zu haben: Sobald sich der Mensch gegen das Sinnlose auflehnt, würde er angeblich aus dieser Auflehnung, aus dieser Anstrengung und Aufsässigkeit, seinen eigenen Wert erschaffen können. „Cette révolte donne son prix à la vie" (Camus 1942: MS, S. 78), ruft Camus mit dem revolutionären Pathos derer, die noch nicht alt und einsichtig geworden sind. Eine Revolte, die nichts will, außer nicht nichts zu wollen, kennt nur den Preis des Willens und nicht Wert des Gewollten. Der Mensch ist nicht etwas, nur weil er sich empört – das ist ungereimt. Und erst recht liegt der Empörung nicht an sich eine Allgemeinheit zugrunde (vgl. Camus 2020: HR, S. 25–41), sondern leider nur allzu häufig das ganz persönlich Interesse, der Wille zur Macht.

Gegen diese dogmatische Verirrung des Existenzialismus stellt Cioran in *Die verfehlte Schöpfung* süffisant die Frage: „Woher kommt es, dass im Leben wie in der Literatur die Revolte – auch die reine Revolte – etwas Falsches an sich hat, während die Resignation, und entspräche sie auch der Entschlusslosigkeit, immer den Eindruck von etwas Wahrem gibt?“ (Cioran 1979b, S. 95–96) Ein wahrhafter und redlicher Denker verweigert sich nicht dem sinnlosen Nichts und leistet keinen Widerstand gegen den Nihilismus, wenn er ihn einmal als Wahrheit allen Geschehens erkannt hat. Er bäumt sich nicht auf, er wird gefällt. Allein die Unentschiedenheit der Skepsis kann einen Nihilisten vor dem Selbstmord retten. Sie bringt ihn von jedem Entschluss und jedem Handeln ab. So lautet Ciorans skeptische Antwort auf Nietzsche, Camus und Heidegger.

Der Wert der Skepsis: Wegsterben oder Freitod?

Cioran macht ernst mit der Fähigkeit der Skepsis, jedem Gedanken einen gleichwertigen entgegen zu setzen. Dieser Fanatiker ohne Credo enthält sich jeder Parteinahme. Er zeigt, was Wertfreiheit wirklich bedeutet, wenn man sie ernst meint: die völlige Antriebslosigkeit, die Abgewöhnung von allem Wollen, weil jeder Wille, bloß weil er Wille ist, auf einen Mangel an Isosthenie hinweist (vgl. Sextus Empiricus 1997, I,4,8). Wirkliche Skepsis zeigt sich, wie Cioran in *Absturz in die Zeit* beweist, im „Zustand der reinen Unentschiedenheit“ (Cioran 1980, S. 48). Selbst der Teufel ist ihm zu dogmatisch. Der echte Zweifel, wenn er existenziell wird, trocknet die Menschen aus und macht sie unfruchtbar, untätig und gleichgültig – er ist „unverträglich mit dem Leben“ und erschafft „lebendige Leichname“ (Cioran 1980, S. 55). Solange den Menschen Zweifel plagen, kann er nicht mehr Hand an sich legen. Er bleibt ein Suchender und vom Leben Versuchter. Im Zustand der reinen Unentschiedenheit gibt es weder gute noch keine Gründe für eine Entleibung. Während die Pessimisten nicht mehr zweifeln und sich zu Advokaten des Todes, zu Gläubigen des Nichts, zu Parteigängern des Ablebens entwickeln, ist ein Skeptiker jemand, der „seinen Problemen und Fragestellungen keine aktive Kraft mehr abgewinnen kann“ (Cioran 1986, S. 36–37), der zu seiner eigenen Auflösung hintreibt und sich nach ihr sehnt, aber unfähig ist, sie zu vollziehen, weil ihm die Gründe fehlen. Der Skeptiker erteilt weder dem Leben noch dem Tod seine Einwilligung. Beides ist ihm gleich viel wert und er enthält sich des Urteils. Sein Leben endet in einem langsamen Wegsterben.

Dank dieser existenzialistischen Skepsis konnte sich Emil Cioran nie für den Selbstmord entscheiden und ertrug sein Dasein ohne Sinn und Credo, ebenso wahrhaftig wie verzweifelt.

Für Nietzsche ist der Selbstmord dagegen eine Modalität, wie der Wille zur Macht als letzte Karte in einem sinnfreien Leben wieder ins Spiel kommen kann. Die faktische Sinnlosigkeit des Werdens erlaubt in seiner Lehre eine aufrichtige Art zu sterben ebenso wie eine aufrichtige Art zu leben – nur für das verwelkende und unentschiedene Wegsterben eines Emil Cioran sollte sich niemand entscheiden. Es ist die Skepsis generell, die Nietzsche als Ziel des Lebens ablehnt. In einem hinterlassenen Fragment heißt es: „Skeptiker sein, den Skepticismus bekennen – ein schlechter Weg, seinen Weg zu machen!" (Nietzsche, NL 1887, KSA 13, 11[298]) Schlecht ist dieser Weg, weil mit der Skepsis der Wille entartet (Nietzsche, JGB §208, KSA 5, S. 138). Dieses Schwanken zwischen Ja und Nein, die vermeintliche Isosthenie, das wertfreie Erkennen, die demokratisch-pluralistische Vielwollerei und die noble Enthaltung als skeptische Tugend und Toleranz lähmen den Willen in seiner Einheit, wenn sie so ernst gemeint sind wie bei Cioran. Alles andere aber wäre unernst, uneigentlich und damit unwesentlich. Zwar gibt es auch für Nietzsche eine harte und mannhafte Art der Skepsis (Nietzsche, JGB §209 – 211, KSA 5, S. 140 – 145), doch eine solche Skepsis ist nur der Ausschnitt eines Könnens, das einer befehlenden Natur zukommt. Die Befehlenden können als Befehlende keine Skeptiker sein. Ihnen dient die Skepsis, insofern sie „experimentell" ist (Sommer 2020, S. 6). Das Wort „experimentell" trifft, wenn man unter den Experimenten nicht eine frei flottierende Bastelei oder eine assoziative, unsystematische Gedankenfolge versteht. Versuch und Experiment meinen bei Nietzsche die übende Erfahrung der eigenen Kraft. Im Nachlass heißt es: „In Wahrheit heißt etwas wollen ein Experiment machen, um zu erfahren, was wir können; darüber kann uns allein der Erfolg oder Mißerfolg belehren." (Nietzsche, NL 1880, KSA 9, 3[120])

Der Wille zur Macht übt sich in seinen Experimenten und erprobt seine Kraft, bis er sich als Eines und Ganzes setzen kann. Skepsis in ihrer positiv konnotierten Art bedeutet bei Nietzsche diese vorläufige Suchbewegung, bis die Freiheit des Könnens erreicht ist.

Skepsis als „die freie Seite einer jeden Philosophie" (Hegel 1970, S. 229) bedeutet eine Vorbereitung auf dem Weg zum freien Geist. Die freien Geister aber haben Macht über sich gewonnen und können ein klares Ja oder ein klares Nein sagen zu dem, was sie skeptisch untersucht haben. Entschiedenheit und nicht Enthaltung, Entschlossenheit und nicht Isosthenie, Wertschätzung und nicht Wertfreiheit sind die Kennzeichen eines befreiten Geistes. Gleiches gilt auch für den Tod. Das skeptische Herumlavieren und Wegsterben eines Cioran sind die Anzeichen und Indizien für die Willensschwäche eines unfreien Geistes, der noch keine Macht über sich hat und noch nach Ziel und Ende sucht. Ein wirklich freier Geist lobt mit Zarathustra „den freien Tod, der mir kommt, weil ich will" (Nietzsche, Za, KSA 4, S. 94). Wenn der Wille zum Nichts über den Willen zur

Macht siegt, verwandelt sich der Selbstmord in einen Frei-Tod. Ein solcher Tod ist die freie Seite eines jeden Lebens.

Bibliographie

Aristoteles (1923): *The „Art" of Rhetoric.* Cambridge (Arist. rhet.).
Aristoteles (1936): *On the soul.* Cambridge (Arist. de an.).
Camus, Albert (1942): *Le mythe de Sisyphe.* Paris.
Camus, Albert (2020): *Der Mensch in der Revolte.* Hamburg.
Cioran, Emil (1979a): *Lehre vom Zerfall.* Stuttgart.
Cioran, Emil (1979b): *Die verfehlte Schöpfung.* Frankfurt a. M.
Cioran, Emil (1980): *Absturz in die Zeit.* Stuttgart.
Cioran, Emil (1983): *Dasein als Versuchung.* Stuttgart.
Cioran, Emil (2015): „Rapport sur mon activité universitaire pendant l'année 1938–1939". In: *Les Cahiers de L'Herne: Cioran.* Paris, S. 143–145.
Cioran, Emil (2016): *Syllogismen der Bitterkeit.* Frankfurt a. M.
Cioran, Emil (2017): *Gevierteilt.* Frankfurt a. M.
Cioran, Emil (2019a): *Auf den Gipfeln der Verzweiflung.* Berlin.
Cioran, Emil (2019b): *Das Buch der Täuschungen.* Berlin.
Cioran, Emil (2019c): *Gedankendämmerung.* Berlin.
Deleuze, Gilles (2013): *Nietzsche und die Philosophie.* Hamburg.
Gast, Peter/Förster-Nietzsche, Elisabeth (1996): *Friedrich Nietzsche. Der Wille zur Macht. Versuch einer Umwertung aller Werte.* Stuttgart.
Große, Jürgen (2012): „Cioran, Nietzsche und der Nietzscheanismus". In: *Weimarer Beiträge* 58/2, S. 238–261.
Große, Jürgen (2014): *Erlaubte Zweifel. Cioran und die Philosophie.* Berlin.
Hegel, Georg Wilhelm Friedrich (1970): „Verhältnis des Skeptizismus zur Philosophie. Darstellung seiner verschiedenen Modifikationen und Vergleichung des neuesten mit dem alten". In: Hegel, Georg Wilhelm Friedrich: *Jenaer Schriften* 1801–1807. Frankfurt a. M., S. 213–273.
Heidegger, Martin (1976): „Was ist Metaphysik?". In: Martin Heidegger: *Wegmarken.* Frankfurt a. M., S. 103–123.
Heidegger, Martin (1981): *Aristoteles Metaphysik Θ 1–3. Vom Wesen und der Wirklichkeit der Kraft.* Frankfurt a. M.
Heidegger, Martin (1986): *Sein und Zeit.* Tübingen.
Heidegger, Martin (1996): *Nietzsche I.* Frankfurt a. M.
Heidegger, Martin (2007): *Metaphysische Anfangsgründe der Logik im Ausgang vom Leibniz.* Frankfurt a. M.
Hobbes, Thomas (1839): „The Elements of Philosophy. The First Section Concerning Body". In: William Molesworth (Hrsg.): *The English Works of Thomas Hobbes of Malmesbury.* London.
Montinari, Mazzino (1991): *Nietzsche. Eine Einführung.* Berlin.
Nietzsche, Friedrich: *Aristoteles Rhetorik I*, KGA II/4, S. 529–612.
Sextus Empiricus (1968): *Grundriß der Pyrrhonischen Skepsis.* Frankfurt a. M.
Sextus Empiricus (1997): *Esquisses pyrrhoniennes.* Paris.

Sommer, Andreas Urs (2013): *Kommentar zu Nietzsches* Der Antichrist. Ecce homo. Dionysos-Dithyramben. Nietzsche contra Wagner. *Historischer und kritischer Kommentar zu Friedrich Nietzsches Werken.* Hrsg. v. d. Heidelberger Akademie der Wissenschaften. Bd. 6/2 (NK 6/2). Berlin, Boston.

Sommer, Andreas Urs (2020): „‚Große Geister sind Skeptiker'. Friedrich Nietzsche als Lehrer des Zweifels". In: *der blaue reiter* 45, S. 6–9.

Freigeistigkeit und Individualismus

Paul Stephan

Zwischen Verschmelzungswunsch und Selbstbehauptung. Das Problem der Alterität als Grundproblem der Existenzphilosophie von Kierkegaard bis Camus – und seine Gegenwärtigkeit

Abstract: Between Desire for Fusion and Self-Assertion. The Problem of Alterity as Founding Problem of Existential Philosophy from Kierkegaard to Camus – and Its Topicality. In this paper I describe a fundamental conceptual constellation of existential philosophy: the simultaneous positing of the deep ontological alienation between the self and alterity and the demand for overcoming this alienation. Since attempts to do so are destined to fail, I first maintain that existential philosophy is locked within this aporetical structure. I then demonstrate the extent to which classical thinkers like Kierkegaard and Nietzsche are 'existential philosophers' in this sense and consider how Sartre and Camus accounted for this problematic stance. Finally, I reconstruct Camus's later critique of existentialism. This enables a discussion of whether the problem of existential philosophy is still relevant in our postliberal and post-democratic society.

I Einleitung: Himmel und Hölle

„[L]'enfer, c'est les Autres", so lautet wohl Sartres berühmtester Satz aus seinem Drama *Huis clos* (Sartre 2000: HC, S. 93). Um ihn voll zu verstehen, muss man wohl ergänzen: Sie können natürlich nur die Hölle sein, weil sie auch – zumindest für kurze Augen-Blicke (im geradezu wörtlichen Sinne des Wortes) – der Himmel sein können. Doch diese Augen-Blicke der Seligkeit, in denen der Blick des Anderen keine Bedrohung mehr darstellt, sondern das eigene Dasein bestätigt, vergehen schnell und am Ende bleibt jeder allein. Die Entfremdung nicht nur von den anderen Menschen, sondern auch von der sozialen Welt und der Dingwelt als fundamentale Grundtatsache der menschlichen Existenz zu betrachten, das ist eine der entscheidenden Grundthesen des sartreschen Existenzialismus. Der Mensch ist beim ihm „ein Seil über einem Abgrunde" (Nietzsche, Za Vorrede 4, KSA 4, S. 16), ständig hin- und hergerissen zwischen einem tiefen Verlangen nach Verschmelzung mit seinem Anderen und dem ebenso drängenden Willen, ein

https://doi.org/10.1515/9783110760163-010

selbständiger Einzelner zu sein und zu bleiben. Im Balancieren zwischen beiden Extremen besteht seine Freiheit.

Die Hauptthese dieses Textes lautet, dass diese Weltsicht nicht nur Sartre zu eigen ist, sondern ein typisches, wenn nicht gar definitorisches, Charakteristikum aller Repräsentanten dessen darstellt, was man gemeinhin als ‚Existenzphilosophie' bezeichnet.[1] Zunächst soll das beschriebene Grundmotiv näher bestimmt werden (II.), um dann zu zeigen, inwiefern es sich bei Kierkegaard (III.) und Nietzsche (IV.) als wichtigsten Existenzphilosophen des 19. Jahrhunderts finden lässt. Im Anschluss daran soll dieses Grundmotiv bei Sartre (V.) und dem frühen Camus (VI.) nachvollzogen werden, um dann aufzuzeigen, inwiefern der späte Camus (VII.) eine immanente Kritik der Existenzphilosophie entwickelt, die ihn zu einer Art ‚Post-Existenzialismus' führt. Es ist klar, dass diese Darlegung nur sehr skizzenhaft bleiben muss: Dieser Aufsatz soll vor allem der Klärung der Frage dienen, was ‚Existenzphilosophie' überhaupt heißt und inwiefern es berechtigt ist, diesen Begriff zu verwenden. Dazu ist es unvermeidlich, einige grobe Linien zu zeichnen.

An was sich das ‚Problem der Alterität' freilich heute am dringlichsten zeigt, ist zweifelsohne das Corona-Virus, diese eigenartige Entität zwischen Kleinstlebewesen und organischem Partikel, die uns nicht nur mit der schmerzhaften, ‚höllischen' Seite der Alterität in ungewohntem Maße konfrontiert, sondern auch mit unserer unsere Selbstliebe kränkenden Abhängigkeit von anderen Menschen und sozialen Strukturen. Im abschließenden Abschnitt dieses Aufsatzes (VIII.) soll daher ausgehend von dieser neuen Herausforderung nach der Aktualität der Existenzphilosophie für die Gegenwart gefragt werden.

II Die Grundkonstellation

Unter ‚Alterität' verstehe ich ‚Andersheit' im umfassendsten Sinne: den Anderen, die Andere, das Andere, die Anderen; andere Subjekte, die soziale Mitwelt, die Dingwelt, vielleicht sogar der mir verborgene Teil meiner eigenen Seele ... Alles eben, was andersartig ist.

Man mag sich hier natürlich sofort fragen: Was soll so ein weiter Begriff überhaupt bezeichnen? Ist er nicht viel zu weit gefasst? Wie könnte man ihn sinnvollerweise zum Ausgangspunkt irgendwelcher Überlegungen nehmen? Handelt es sich hier womöglich um ein Paradebeispiel eines abgehobenen

1 Vgl. zum gemeinen Begriff von ‚Existenzphilosophie' etwa die gute Zusammenfassung in Burkard 2008.

Sprachspiels im Sinne des späten Wittgenstein, auf das man sich am besten gar nicht erst einlassen sollte? In der Tat: Im Alltag begegnet uns nicht ‚die Alterität', sondern immer nur konkrete einzelne Alteritäten. Man kann sich kaum vorstellen, dass man in irgendeiner alltäglichen Situation von ‚der Alterität' sprechen würde.

Ich denke, die Kritik, dass hier möglicherweise schon im Ansatz ein Denkfehler vorliegt, nämlich eine unzulässige Verallgemeinerung, ist berechtigt. Auch wenn alle Existenzphilosophen gegen die klassische Metaphysik, den Essentialismus, die Ontologie, die Abstraktion etc. polemisieren, müssen sie sich doch den Vorwurf gefallen lassen, selbst von einer abstrakten und abgehobenen Ausgangsprämisse auszugehen, die ihrerseits höchst ontologisch ist: Nämlich davon, dass es überhaupt sinnvoll ist, von so etwas wie ‚die Alterität' zu sprechen. Der Verdacht liegt nahe, dass wir es auf einem derart hohen Niveau mit der berühmten „Nacht" zu tun haben, „worin [...] alle Kühe schwarz sind", von der Hegel sprach (Hegel 1973, S. 22): Lebensweltliche, konkrete Probleme werden dazu missbraucht, dürre Allgemeinplätze zu begründen. Denn im Alltag, im konkreten Umgang mit den Dingen, stoßen wir ja stets auf zweierlei Arten von Widrigkeiten: Probleme, die sich lösen lassen, und solche, die sich nicht lösen lassen. Der Alltagsverstand lehrt: Man soll die lösbaren Probleme lösen und durch Nachdenken und Übung versuchen, den Bereich der lösbaren Probleme immer mehr zu vergrößern. Über Probleme, die sich auf lange Sicht oder gar aus prinzipiellen Gründen nicht lösen lassen, lohnt es sich eigentlich gar nicht erst nachzugrübeln. Die Kategorie der ‚Alterität schlechthin' ergibt aus diesem Blickwinkel heraus betrachtet eigentlich keinen rechten Sinn.

Die Existenzphilosophen nun nehmen von vornherein eine diesem Pragmatismus geradezu entgegengesetzte Perspektive ein: Sie gehen zunächst davon aus, dass der Mensch in allen soeben genannten Hinsichten (andere Subjekte, Dingwelt, soziale Welt, inerte Seite des Selbst) grundsätzlich und ontologisch entfremdet ist. Dies mag man nun mit einem Achselzucken abtun, denn es stellt ja noch kein *Problem* dar: Es ergibt sich ja aus der Alltagserfahrung, dass es eine grundsätzliche Fremdheit zwischen mir und der ‚Alterität' gibt; das wird, wenigstens auf einer phänomenologischen Ebene, wohl niemand leugnen wollen, selbst wenn man diese Erfahrung (wie bspw. Schopenhauer) als Doxa abtun sollte, die über die tieferliegende Einheit zwischen allen Entitäten hinwegtäuscht.

Zum *Problem* wird die Alterität in der existenzphilosophischen Perspektive nur durch den *Widerspruch*, der dadurch entsteht, dass sich die Subjekte nach einer Kittung der ontologischen ‚Urwunde' sehnen, nach einer Verschmelzung zwischen Ich und Alterität; eine Verschmelzung, die zugleich aber auch gefürchtet wird und deren Realisierung notwendig scheitert. Die ethische Konsequenz der Existenzphilosophen lautet freilich nicht, diese widersprüchliche Konstellation zu verlassen – das hielten sie auch für schlichtweg unmöglich –, sondern sich bewusst in sie hineinzubegeben, sie auszuhalten und aus ihr das Beste zu machen; zwischen diesen beiden Extremen zu leben ohne zu verzweifeln oder aufzugeben, aber auch ohne sich beispielsweise in religiösen Illusionen zu sonnen, die eine Einheit von Ich und Alterität vorgaukeln, wo in Wahrheit gar keine sei.

Der erwähnte pragmatische Skeptiker mag natürlich auch an dieser Stelle mit den Achseln zucken und die Frage aufwerfen, ob hier nicht etwas ganz Banales, Alltägliches in große Worte verpackt und überdramatisiert wird. Natürlich versuchen wir, mit der Welt in einer Art ‚Einheit' zu leben (wenn man überhaupt so sprechen möchte) und daran scheitern wir immer wieder. Doch zugleich gelingt es uns ja auch oft, unsere Wünsche zu realisieren. Der Tod macht uns letztendlich allen einen Strich durch die Rechnung, doch genau über den sollte man sich eben gar nicht so viele Gedanken machen[2]: Unterm Strich kommt die Menschheit ja doch ganz gut voran.

Es ist angesichts dessen, dass diese pragmatische Denkweise ja äußerst naheliegt und auch von vielen vertreten wird, umso bemerkenswerter, dass es trotzdem so viele Menschen gibt, die sie nicht teilen. Oft ist es die Religion, die das Bedürfnis nach einer ‚tieferen' Einheit zwischen Ich und Alterität thematisiert und dem Anspruch nach befriedigt; in der Moderne ist es die Existenzphilosophie, die in für zahllose Menschen evidenter Form über die ‚großen Fragen' des Lebens nachsinnt, nun freilich keine Erlösung mehr verspricht. Der Soziologe Hartmut Rosa spricht etwa, explizit Motive der Existenzphilosophie aufgreifend,[3] von einem „Resonanzverlust", der in modernen Gesellschaften zur Massenerfahrung werde: Man erlebe sich als vereinzeltes Individuum, das einer als kalt, glatt und sinnlos wahrgenommenen Lebenswelt gegenübersteht, jedoch trotzdem nach „Resonanz", also einem lebendigen In-Eins-Sein mit der Alterität, strebt. – Wichtig ist dabei freilich, dass Rosa explizit zwischen „Konsonanz" als Verschmelzung und „Resonanz" unterscheidet: In Resonanz-Beziehungen bleibt eine gewisse Unterschiedenheit zwischen den in Beziehung stehenden Entitäten stets erhalten, sie bewegen sich im Zwischenraum zwischen Dissonanz und Konso-

---

2 Interessanterweise kommt – in expliziter Abgrenzung zu Heidegger – Sartre in *L'être et le néant* zu einer ganz ähnlichen Schlussfolgerung: Der Tod sei gewissermaßen ‚so absurd', dass man ihn in seinen alltäglichen Entwürfen gar nicht zu beachten brauche (Sartre 1976: EN, S. 576–592).

3 In seinem entsprechenden Hauptwerk *Resonanz. Eine Soziologie der Weltbeziehung* (Rosa 2016) werden alle vier hier behandelten Kernautoren mehrmals als Kronzeugen der modernen „Resonanzkatastrophe" angeführt und dabei auch mehrmals in einem Atemzug genannt (etwa Nietzsche und Camus [vgl. S. 448–452], Sartre und Kierkegaard [vgl. S. 190], Sartre und Camus [vgl. S. 324], Kierkegaard und Nietzsche [S. 533–534, 538–540, 566]). Aber auch Peter Sloterdijk – dessen Denken, wie sich später noch zeigen wird, ebenfalls einen wesentlichen existenzphilosophischen Zug aufweist – gehört zu Rosas meistzitierten Theoretikern.

nanz. Dies entspricht der existenzphilosophischen Grundfigur: Weder unterschiedslose Einheit noch völlige Entfremdung seien erstrebenswert, sondern gelungenes Leben bewege sich in der Mitte zwischen beidem.

Außer dem existenzphilosophischen oder dem religiösen Weg gibt es freilich noch andere Möglichkeiten, mit dieser grundlegenden Entfremdungserfahrung umzugehen. Kritisiert wird an den Existenzphilosophen vor allem, dass sie das Problem der Entfremdung ja gerade nicht als konkret-historisches verstehen, das von einer bestimmten kollektiven Seinsweise abhängt, sondern als ontologisches Urübel. Dagegen stehen Philosophien unterschiedlicher Provenienz, die eine gewisse Wahrheit der existenzphilosophischen Grunderfahrung anerkennen, aber zugleich betonen, dass eine Änderung der kollektiven Seinsweise die ‚ontologische Urwunde' heilen könnte.[4] Hierunter fallen kollektivistische Ansätze aller Art, die alle ihre antiindividualistische und mithin auch liberalismuskritische Schlagseite eint: Während den Marxisten beispielsweise die ‚klassenlose Gesellschaft' als Ideal einer solchen versöhnten Menschheit vorschwebt, träumen Vertreter reaktionärer Ansichten von einer Rückkehr zu verlorenen archaischen Ursprüngen.

Einen anderen Gegenentwurf zur Existenzphilosophie könnte man als ‚modernistisch' bezeichnen: Die von den Existenzphilosophen beschriebene Entfremdungserfahrung wird anerkannt, doch es wird zugleich propagiert, die ‚romantische' Sehnsucht nach einem nicht-entfremdeten Leben aufzugeben. Aus einer solchen Haltung heraus entwickelt beispielsweise Max Stirner seinen radikalen Egoismus (vgl. Stirner 2009) und man könnte eine solche auch bei manchen Vertretern der ‚Post-Moderne' ausfindig machen[5]. Diese Attitüde nähert sich dem Pragmatismus an und unterscheidet sich von diesem vielleicht nur durch ihren militanten Anti-Romantizismus, der sie doch mit dem verbindet, was sie eigentlich zu bekämpfen sucht.

III Kierkegaard als Begründer der Existenzphilosophie

Die existenzphilosophische Herangehensweise ist natürlich älter als die Existenzphilosophie im engeren Sinne. Ganz ähnlich wie Sartre formulierte etwa

4 Vgl. hierzu, unter Bezugnahme auf Marx, Rosa 2016, S. 540, dessen Resonanztheorie zu diesen „Philosophien unterschiedlicher Provenienz" gezählt werden dürfte.

5 Wahrscheinlichsten am ehesten unter diese ‚postmoderne Attitüde' ließe sich der späte Foucault subsumieren.

Schopenhauer: „Die Welt ist eben *die Hölle*, und die Menschen sind einerseits die gequälten Seelen und andererseits die Teufel darin." (Schopenhauer 2004, S. 354) Dieser Pessimismus setzt auch bei Schopenhauer das Bewusstsein einer verlorenen Einheit voraus, die sich bei ihm der Möglichkeit nach qua Selbstaufhebung des Willens wiedererlangen ließe. Genau, dass er eine Versöhnung von Ich und Alterität qua Aufgabe des Egoismus für möglich hält, lässt Schopenhauer jedoch als einen eher religiösen als existenzphilosophischen Denker erscheinen: Erst Nietzsche sollte aus Schopenhauers Grundannahmen ein genuin existenzphilosophisches Weltbild formen – genau unter Aufgabe der Aussicht auf eine mögliche Erlösung.[6]

Derjenige, der gemeinhin als Begründer der Existenzphilosophie gilt – und beispielsweise von Sartre und Camus auch als solcher behandelt wird[7] –, ist Søren Kierkegaard. Dementsprechend steht er zwischen Religiosität und Atheismus: Einerseits betont er in seinen Schriften die Notwendigkeit einer Erlösung qua „Sprung in den Glauben"[8], andererseits verficht er die Unmöglichkeit dieser Erlösung und die Vergeblichkeit aller religiösen Bemühungen. Die Existenz ist für ihn grundsätzlich paradox und absurd; der vollkommene Glaube bestünde genau darin, diese Absurdität vollkommen zu akzeptieren, was jedoch nie vollständig gelingt. Sein wohl nicht zufällig nie systematisch ausgearbeitetes, sondern fragmentarisch bleibendes Denken – das Kierkegaard zumal oftmals nicht direkt, sondern hinter der Maske verschiedener Pseudonyme ausspricht –, kreist um diese eine Aporie, die wohl tatsächlich niemand zuvor so deutlich entwickelt hat: Es gibt keine Erlösung und trotzdem soll man an die Möglichkeit der Erlösung glauben.[9]

---

6 Zu Schopenhauer als jemanden, der den modernen Resonanzverlust bereits vor Kierkegaard und Nietzsche beschrieb vgl. auch Rosa 2016, S. 533, 540.

7 In *Le Mythe de Sisyphe* schreibt Camus, Kierkegaard sei unter allen Denkern des Absurden „le plus attachant" (Camus 1962: MS, S. 42), da er das Absurde nicht nur entdeckt, sondern gelebt habe. Camus zitiert ihn in diesem Buch häufiger als Nietzsche. Wie Sartre in *L'être et le néant* (vgl. etwa Sartre 1976: EN, S. 64), bezieht sich Camus dabei besonders stark auf Kierkegaards Konzept der Angst (vgl. Camus 1962: MS, S. 61). In dem seinem zweiten Hauptwerk *Critique de la raison dialectique* vorangestellten *Questions de méthode* (Sartre 1960: CRD, S. 13–111) bezeichnet Sartre Kierkegaard sehr eindeutig als den Begründer und wichtigsten geistigen Vorläufer seines Existenzialismus.

8 Diese Kierkegaard oft zugeschriebene Formulierung findet sich in seinen Schriften an keiner einzigen Stelle (vgl. McKinnon 1993), der für den Eintritt in den Glauben notwendige „qualitative Sprung des Einzelnen" (Kierkegaard 1890a, S. 61) ist jedoch eine der Schlüsselkategorien der für die existenzphilosophischen Debatten besonders wichtigen Schrift *Der Begriff Angst*.

9 So etwa der Grundgedanke von *Der Begriff Angst* (Kierkegaard 1890a), aber auch der *Philosophischen Bissen* (Kierkegaard 1890b) und der *Unwissenschaftlichen philosophischen Nachschrift* (Kierkegaard 2003) zu jenen.

IV Nietzsche als Kierkegaards Bruder im Geiste

Nietzsche wurde nicht zufällig Kierkegaard immer wieder geradezu als ‚geistiger Zwilling' gegenübergestellt.[10] Wie bereits angedeutet, findet sich die kierkegaardsche Grundproblematik auch in seinem Werk. Sie basiert bei ihm auf der Aufnahme und Umdeutung schopenhauerscher Motive, die sich erstmalig in der *Geburt der Tragödie* vollzieht: Das Subjekt wird hier als zerrissen zwischen dem *principium individuationis* dargestellt, für das der Gott Apoll steht, und einer tiefen Einheitserfahrung, die Nietzsche mit dem Gott Dionysos assoziiert. Weder kann das Subjekt diese Einheitserfahrung dauerhaft ertragen – denn sie ist zugleich beglückend wie schmerzhaft –, noch das *principium individuationis*, also den Zustand des alltäglichen pragmatischen Egoismus, ohne ein tiefes Unbehagen erdulden. Allein die Kunst vermag es, beide Grundtriebe für kurze Momente zu versöhnen – eine Versöhnung freilich, die ebenso ephemer wie der Kunstgenuss ist.

Es ist wahr, dass sich Nietzsche in seinen späteren Schriften von dieser existenzphilosophischen Grundfigur distanziert und Gedanken entwickelt, die sich am ehesten als im oben genannten Sinne ‚modernistisch' beschreiben lassen: Er kritisiert selbst ja in seinem der Neuauflage der Schrift von 1886 vorangestellten „Versuch einer Selbstkritik" an seinem Erstlingswerk unter anderem, dass hier noch davon die Rede ist, dass die Kunst die Menschen zu trösten habe. Der spätere Nietzsche scheint demgegenüber eine radikalere Position zu vertreten: Nämlich, dass schon die Haltung der Trostbedürftigkeit als solche ein Irrweg ist; egal, ob sie sich nun christlich, ästhetizistisch oder utilitaristisch zu befriedigen sucht. Zarathustra wird ins Spiel gebracht, um eine Ethik des bedingungslosen *amor fati* zu verkünden, der unbedingten Diesseitigkeit; ein Leben, das keines Trostes mehr bedarf, auch keines künstlerischen (vgl. Nietzsche, GT Versuch 7, KSA 1, S. 21–22).

Man fragt sich freilich bei der Lektüre derartiger Passagen beim späten Nietzsche, ob es sich nicht hier nicht um eine reine Phantasie handelt, die mit der Lebensrealität wenig zu tun hat: Wenn Nietzsche, wie etwa in *Zur Genealogie der Moral* nicht als Ethiker, sondern als Psychologie und Sozialphilosoph spricht, gibt er ja selbst zu, dass eine solche Lebenshaltung nur einer privilegierten sozialen Situation entspringen kann, eine „Herrenmoral" ist. Wer kein „Herr" ist, der kann eine solche Haltung gar nicht einnehmen, wenn er sich nicht völlig selbst verneinen möchte.

---

10 Zwei klassische Autoren, die eine solche Parallelität vertraten, sind etwa Karl Löwith (vgl. Löwith 1987 a u. b) und Karl Jaspers (vgl. Jaspers 1973, S. 7–34).

Neben dieser einseitigen, schrillen und kaum ernstzunehmenden modernistischen Seite – die sich am deutlichsten in Texten seiner letzten Schaffensphase wie dem *Antichrist* und *Ecce homo* ausdrückt –, gibt es aber auch noch eine andere Seite, in der Nietzsche die Schwierigkeit, zu einer Haltung des unbedingten *amor fati*, d. h. einer völligen Bejahung der Alterität, zu finden, mitreflektiert und eine Position vertritt, die ihn eher der Existenzphilosophie annähert.

Sinnbildlich für diese Seite steht das bereits zitierte Motiv des Seiltanzes aus der *Vorrede* in *Also sprach Zarathustra*. Zarathustra fungiert hier als *alter ego* Nietzsches.[11] Der Mensch selbst wird als Seil bezeichnet, „geknüpft zwischen Thier und Übermensch" (Nietzsche, Za Vorrede 4, KSA 4, S. 16). Er zeichnet sich dadurch aus, dass er nicht mehr Tier und noch nicht Übermensch ist; er ist ein Zwischenwesen, eine „Brücke" (ebd.). Während das Tier dasjenige Wesen ist, das zu sehr mit der Alterität verbunden ist, das einfach in seiner Umgebung aufgeht und nicht in einem tiefen Sinne an ihr leidet, dürfte man den Übermenschen wohl als utopische Vision eines mit der Alterität *mit Bewusstsein* versöhnten Menschen verstehen; ein Mensch, der souverän, gleich dem spielenden Kind, von dem Zarathustra an anderer Stelle spricht (vgl. Nietzsche, Za Verwandlungen, KSA 4, S. 29 – 31), mit der Alterität spielt und sie in seinem Sinne umschafft.

Wichtig ist, dass Zarathustra den Übermenschen an dieser Stelle deutlich als *Utopie* konzipiert – nicht als etwas, das man durch das Erlernen einer ‚Lebenskunst' oder einer besonderen Ethik ohne Weiteres werden könnte. Sicher handelt es sich hier um eine diesseitige Vision, doch Zarathustra spricht dezidiert davon, dass man sich für das Ziel der Realisierung des Übermenschen aufopfern, sich selbst überschreiten und sogar untergehen soll. Kritisiert wird dagegen im folgenden Abschnitt der *Vorrede* die Haltung des „letzten Menschen" (vgl. Nietzsche, Za Vorrede 5, KSA 4, S. 18 – 21). Mit diesem scheint Nietzsche den Menschen in der modernen Massengesellschaft karikieren zu wollen: Es handelt sich um einen hedonistischen Typus, der sich am Ende der Geschichte wähnt, und der an der Alterität auch nicht mehr leidet, da er alles, was beschwerlich ist – allzu hoch gesteckte moralische Ziele, anstrengende zwischenmenschliche Beziehungen, politische Hierarchien, Arbeit, die eigenen psychischen Abgründe –, meidet und es verdrängt. Er trägt zwar vielleicht noch Chaos in sich – die Alterität ist noch nicht völlig domestiziert –, doch er tritt zu diesem Chaos in kein wirkliches Verhältnis mehr, lässt sich davon nicht mehr berühren. Seine Lebenswelt ist steril

---

11 Es ist natürlich wichtig, zwischen dem Pseudonym Zarathustra und Nietzsche zu unterscheiden. Doch wie die erwähnte Stelle aus der „Selbstkritik" zur *Geburt der Tragödie* ja hinreichend zeigt, identifiziert sich Nietzsche oft genug selbst mit ihm, so dass eine strikte Trennung zwischen beiden Rollen nicht sinnvoll zu sein scheint: Es sind natürlich Nietzsches Gedanken, die Zarathustra ausspricht und Nietzsche selbst bekennt sich zu ihnen immer wieder.

und leer, sein allein auf diesseitiges Glück ausgerichteter Lebensstil entspricht dem reinsten Nihilismus.

Der „letzte Mensch" entspricht recht genau dem optimistischen Utilitaristen, den Nietzsche schon in der *Geburt der Tragödie* aufs Korn nahm, aber auch dem hemdsärmeligen Pragmatisten, den wir eingangs charakterisierten. Vom modernistischen Typus unterscheidet ihn wohl nur sein völliger Non-Heroismus, der es noch nicht einmal zu einem pathetischen Anti-Heroismus bringt. Ein Spießer, ein „Dutzendmensch", wie ihn etwa auch Kierkegaard immer wieder attackiert (vgl. etwa Kierkegaard 1929, S. 235), der sich nur darum sorgt, sich die feindliche Seite der Alterität möglichst vom Leib zu halten, der bloßen Abwesenheit des Unglücks zuliebe jedoch auf jede tiefere Glückserfahrung verzichtet, die aus der Berührung mit der feindlichen Seite der Alterität auch entstehen könnte – siehe Rosas Resonanz-Theorie.

Auch wenn diese Grundkonstellation also eine andere ist als in der *Geburt der Tragödie*, so folgt sie doch einer ähnlichen Logik: Der Mensch wird als zwischen zwei Extremen befindlich beschrieben – tierische und übermenschliche Verschmelzung mit der Alterität –; während der Typus des „letzten Menschen" dieser Spannung ausweicht und sich in seinem seichten Hedonismus dem Tierischen annähert, plädiert Nietzsche aus dem Munde des Zarathustra für eine heroische Lebenshaltung der Selbstüberschreitung, des Sich-Entwerfens auf ein unerreichbares Ziel. In der mutigen Konfrontation mit den feindseligen Seiten der Alterität soll der heroische Mensch sich selbst behaupten und wachsen – auch wenn er weiß, dass es einen nie tilgbaren Rest ‚Chaos' gibt und geben muss, der sich seinen Bemühungen notwendig entzieht.

V Sartres Ontologie der Existenz

Sartres bleibender philosophischer Verdienst besteht wohl darin, die im 19. Jahrhundert von Denkern wie Kierkegaard und Nietzsche in eher literarisch-fragmentarischer Form artikulierte Weltsicht in *L'être et le néant* in eine systematisch-begriffliche Form übersetzt zu haben. Freilich um den Preis, dass bei ihm aus einer historisch-konkreten Erfahrung endgültig eine ontologische Idee wird. Auf allen drei Ebenen der Alterität – und, wie man ergänzen muss, auch bezogen auf das Verhältnis zum eigenen Selbst – befindet sich der Mensch bei Sartre in einer zutiefst paradoxalen, zerrissenen Konstellation zwischen dem für seine Identität konstitutiven Begehren nach Vereinigung und der Unmöglichkeit, diese zu erreichen. Mit Sartre müssen wir genau diese Widersprüchlichkeit bejahen lernen und das Beste draus machen.

Was bei Nietzsche das Tier ist, könnte man mit Sartre als eine Existenz beschreiben, die bloß ‚en soi' existieren würde – die einfach nur das ist, was sie ist, die keinerlei Überschreitung kennt.[12] Das Ziel des ‚Übermenschen' wäre bei Sartre das Ideal des „en-soi-pour-soi" als Verschmelzung von Bewusstsein und Alterität: Ein gottgleicher Zustand, in dem Bewusstsein und Sein eins wären, aber das Bewusstsein trotzdem Bewusstsein bliebe. Da das Bewusstsein bei Sartre aber wesentlich durch seine Negativität, seinen sich selbst überschreitenden Charakter, gekennzeichnet ist, kann es diesen Zustand niemals geben: Das Bewusstsein bleibt stets eine Entität, die das ist, was sie nicht ist, und das nicht ist, was sie ist. Dieser wesentlich negative Charakter des Bewusstseins ergibt sich freilich genau aus seinem Begehren, das Ideal des „en-soi-pour-soi" zu realisieren.[13]

Der unauthentische Mensch bildet sich beim Sartre von *L'être et le néant* ein, diesen Zustand bereits erreicht zu haben – er gleicht darin Nietzsches „letztem Menschen", der sich mit der Alterität im Einklang dünkt, indem er ihre feindselige Seite verdrängt und unterdrückt. Der authentische Mensch hingegen nimmt diese widersprüchliche Situation bewusst auf sich und weiß darum, dass er dazu verdammt ist, ein Leben lang einem unerreichbaren Ideal hinterherzurennen. Er tut es aber trotzdem, weil er genau darin den tieferen Sinn seiner Freiheit erkennt.

VI Der frühe Camus

Einen ganz ähnlichen Weg beschreitet Camus in *Le Mythe de Sisyphe:* Der Mensch ist auch hier hin- und hergerissen zwischen seinem Begehren nach Sinn und der Absurdität der Welt. Dieses Begehren nach Sinn soll nicht einfach aufgegeben werden – ähnlich wie Sisyphos, der trotzig den Stein nach oben rollt, soll jeder Einzelne lernen, dieses Begehren aufrecht zu erhalten im Angesicht seiner Unstillbarkeit.

Sartre und Camus brachten in ihren Texten das Lebensgefühl zahlloser Menschen zum Ausdruck. Doch man sieht schnell, dass sie in ihnen natürlich eine sehr anspruchsvolle Konstellation als Ideal beschreiben – um ein Bild Nietz-

---

12 Für Sartre haben höhere Tiere übrigens durchaus ein Bewusstsein und sind damit auch frei (vgl. Husson 2005, Fn. 55). (Zur Frage, welchen Status Tiere für Sartre haben, vgl. auch die entsprechenden Hinweise in Damon Borias Beitrag für diesen Band.)

13 Das ist, nebenbei bemerkt, auch der tiefere philosophische Grund, warum für Sartre die Anderen die Hölle sind: Weil das Bewusstsein hofft, durch die Aneignung des Blicks des Anderen diesen Zustand zu erreichen – und er genau deswegen durch den Anderen immer wieder mit der Unmöglichkeit dieses Begehrens konfrontiert wird.

sches[14] aufzugreifen, könnte man sagen, dass sie den Bogen bis zum Äußersten spannen; doch sie zögern, ihn loszulassen und es ist unklar, in welche Richtung er fliegen wird.[15]

VII Camus' Existenzialismus-Kritik

Sartre versucht seine Grundgedanken aus *L'être et le néant* bis zum Ende seines Lebens treu zu bleiben, sie jedoch zugleich mit einer politisch linken Grundhaltung zu verknüpfen und entsprechend mit dem Marxismus in einen fruchtbaren Dialog zu bringen. Wobei man sich des Eindrucks nicht erwehren kann, dass er sich da abmüht, zwei eigentlich unvereinbare Impulse zu synthetisieren: den kierkegaardschen Individualismus einerseits, eine an Hegel und Marx geschulte, und um Freud erweiterte, Sensibilität für die Macht der Alterität andererseits.

Auf die Unvereinbarkeit dieser beiden Impulse wies nicht zuletzt Camus in seinen Schriften nach dem Zweiten Weltkrieg hin, in denen er sich deutlich vom sartreschen Existenzialismus, aber auch demjenigen von *Le Mythe de Sisyphe* von 1942 distanziert. Diese Wende vollzieht sich bereits in dem heute eine überraschende Renaissance erfahrenden Roman *La Peste* von 1947 (Camus 1955: P). Zwar ist auch eine der Grundbotschaften dieses Lehrromans, dass die Pest – die ja sinnbildlich für die negative, dem Menschen abgewandte Seite der Alterität schlechthin stehen soll – niemals auszurotten ist und dennoch ausgerottet werden muss – eine paradoxale Figur, die sehr existenzialistisch anmutet (vgl. Wehe 2020).[16] Doch Camus verlässt die Pfade des Existenzialismus, insofern er betont, dass es gegenüber der Pest letztlich keinerlei Spielraum für Freiheit und Interpretation gibt. Es ist ein Ereignis, dessen Faktizität niemand leugnen kann und demgegenüber im Verlaufe des Romans alle Versuche des Uminterpretierens, des Leugnens, der Flucht, des Aufladens mit metaphysischem Sinn letztendlich ver-

---

14 Vgl. etwa die *Vorrede* von Nietzsche, JGB (KSA 5, S. 13).

15 Gemäß einer von der meinen abweichenden Lesart stellt Sartre in *L'être et le néant* nur dar, wie das *unauthentische* Bewusstsein beschaffen ist. Das Streben nach dem „en-soi-pour-soi" wäre mithin keine Eigenschaft des Bewusstseins schlechthin und würde von dem authentischen Bewusstsein ganz aufgegeben (vgl. Shuttleworth 2021, S. 52). Dies widerspricht freilich nicht nur Sartres eigenen Aussagen in seinem ersten Hauptwerk – das ja beansprucht, eine Beschreibung der menschlichen Existenz *schlechthin* zu sein –, sondern scheint mir der existenzialistischen Ethik auch ihren eigentlichen Witz zu nehmen, der genau darin besteht, das *Paradox* der menschlichen Existenz auf sich zu nehmen.

16 Eine sehr ähnliche Überlegung – und mithin sehr gute Zusammenfassung des existenzialistischen Programms –, findet sich auch am Ende von Simone de Beauvoirs *Pour une morale de l'ambiguïté* (Beauvoir 1947: PMA, S. 228–230), das im selben Jahr erschien.

stummen. Die Pest *ist* und es ist klar, was in ihrem Angesicht zu tun ist. Der sartresche Existenzialismus der Freiheit ist, so muss man den Roman wohl verstehen, eine Haltung für den bürgerlichen Normalzustand – für den Ausnahmezustand ist eine Haltung der Entschlossenheit erforderlich, in der die Freiheit des Einzelnen auf ein Mindestmaß zusammenschrumpft.[17]

In einer Schlüsselpassage des Textes erläutert sein Held, der nüchterne Arzt Rieux, die besagte ethische Grundhaltung und bringt den Gegensatz zwischen dieser Haltung zu derjenigen Sartres – aber auch derjenigen des *Mythe de Sisyphe* deutlich zum Ausdruck, wenn er sagt: „J'étais jeune alors et mon dégoût croyait s'adresser à l'ordre même du monde. Depuis, je suis devenu plus modeste. Simplement, je ne suis toujours pas habitué à voir mourir." (Camus 1955: P, S. 144–145) Durch die Verwendung des Ausdruck „dégoût" spielt der Protagonist auf Sartres Roman *La Nausée* (Sartre 1973) an, in dem Sartre für ‚Ekel' sowohl das titelgebende Wort als auch jenes gebraucht. Mit Ekel ist in dem Roman eine grundlegend ablehnende Haltung der Alterität gegenüber gemeint – eine metaphysische Rebellion gegen die Absurdität der Welt.[18] Vor dem Hintergrund dieses Kampfes wäre es ebenso vertretbar, gegen die Pest zu kämpfen wie sie zu verleugnen[19], und Rieux ein typisches Beispiel für einen unauthentischen Repräsentanten des „esprit de sérieux" (Sartre 1976: EN, S. 74).[20]

Camus nimmt also, mit Rieux als Sprachrohr, dezidiert einen praktischen Standpunkt ein, der eine eindeutige Faktizität und eine klare aus dieser folgenden Moral kennt und gegenüber dem der nihilistische Ekel und die absurde Revolte des Sisyphos verblassen. Dementsprechend setzt sich Camus nun etwa auch deutlich von dem im *Le Mythe de Sisyphe* noch vorbehaltlos bewunderten Nietz-

---

17 Vgl. etwa den Dialog zwischen Rieux und dem auf die bürgerliche Freiheit und das Glück jedes Einzelnen insistierenden Journalisten Rambert, der zu Füßen eines Denkmals der Republik stattfindet. Rambert wirft darin dem Arzt, ganz ähnlich wie Sartre der objektifizierenden Wissenschaft und den Vertretern des Allgemeinwohls, vor, in der Abstraktion zu leben (Camus 1955: P, S. 100–104).

18 Vgl. zu dieser Stelle auch Meyer 2020, S. 291.

19 Vgl. Sartres programmatischen Vortrag *L'Existentialisme est un humanisme*, wo er als Beispiel dafür, dass es keine moralisch richtigen Handlungen *a priori* geben könne, einen jungen Mann heranzieht, der vor der Wahl steht, sich den *Forces Françaises Libres* anzuschließen oder sich um seine Mutter zu kümmern. Beide Entscheidungen seien prinzipiell gleich richtig (vgl. Sartre 1970: EH, S. 39–43). Für den Camus der *Peste* wäre hingegen – wie sich im erwähnten Dialog mit Rambert deutlich zeigt – klar: Nur die erstere Option wäre vertretbar.

20 Es ist davon auszugehen, dass schon der Name ‚Rieux' ein entsprechendes Wortspiel sein soll. Einmal im Roman folgen sogar beide Worte – „sérieux" und der Name des Arztes – unmittelbar aufeinander (vgl. Camus 1955: P, S. 25).

sche[21] ab. 1951 veröffentlicht er einen Essay mit dem Titel *L'Europe de la fidélité*, in dem er für eine Abkehr von den totalitären „philosophies du Nord" (Camus 1951a: RevPro51, S. 15) plädiert, zu denen nun neben denjenigen Hegels und Marx' nun auch Nietzsches zählt. Ein friedliches, demokratisches und solidarisches Europa aus dem Geist der Gewerkschaftsbewegung und des antifaschistischen Kampfes heraus soll sich endgültig von diesen Denkern verabschieden, die nur die Unterwerfung des Einzelnen unter inhumane Ziele propagiert hätten.

Im selben Jahr erschien auch sein zweiter wichtiger Beitrag zur Philosophie, *L'Homme révolté* (Camus 1951b), in dem er vor allem Nietzsches Denkweise einer radikalen Kritik unterzieht. Sein Kernargument – und das dürfte nicht zuletzt auch als ein Argument gegen den in dieser Hinsicht ‚nietzscheanischeren' Sartre gemeint sein: Von einem individualistischen Ausgangspunkt käme man nie und nimmer zu einer solidarischen moralischen und politischen Grundhaltung. Um eine volle Haltung der Solidarität zu entwickeln, müsse man daher mit dem Individualismus und auch dem mit ihm verbundenen Perspektivismus Nietzsches radikal brechen und sich auf den Standpunkt der rebellierenden Unterdrückten stellen.

Mit anderen Worten: Camus plädiert in seinen späteren Schriften für eine Haltung zur Alterität, in der diese klar als dem Individuum gegenüber vorranging vorgestellt wird. Das Individuum soll aufhören, der Alterität gegenüber auf seiner Freiheit zu bestehen, sondern sich ihr hingeben – eine Hingabe, die freilich gerade auch bedeuten kann, politischen Widerstand gegen die bestehende soziale Ordnung zu leisten. Der Einzelne soll eine zugleich solidarische und realistische Grundhaltung ausbilden.

In seinen letzten Schriften spitzt Camus diese Haltung weiter zu einer geradezu christlichen Ethik der Selbstlosigkeit zu. In dem Roman *La Chute* von 1956 erzählt Camus die Geschichte eines Advokaten, der sich zunächst selbstbewusst für die Rechte der Unterdrückten einsetzt, ständig das Wort „liberté" im Mund führt (vgl. Camus 1968: Chu, S. 143) und sich für einen „surhomme" hält (vgl. Camus 1968: Chu, S. 33). Man kann in dieser Figur leicht eine Karikatur auf einen Intellektuellen vom Typus Sartres oder auch Camus' selbst erkennen, der sich darin gefällt, ein Sprachrohr der Unterprivilegierten zu sein – damit jedoch, wie dem Erzähler im Laufe des Romans immer klarer wird, nur seine eigene Eitelkeit befriedigt und eigentlich nicht besser ist als diejenigen, die er anklagt. Der Er-

21 Nietzsche wird in dem Werk nur zustimmend zitiert. Bemerkenswert ist allerdings, dass Camus ihn nicht unbedingt als Philosophen auffasst: „Nietzsche paraît être le seul artiste à avoir tiré les conséquences extrêmes d'une esthétique de l'Absurde, puisque son ultime message réside dans une lucidité stérile et conquérante et une négation obstinée de toute consolation surnaturelle" (Camus 1962: MS, S. 184).

zähler hört in der Folge darum auf, Advokat zu sein und wird Bußrichter, entwickelt eine Haltung der Zerknirschung und Bescheidenheit, die ihn letztlich zum Selbstmord führt: Er bäumt sich also, anders als der absurde Held des *Sisyphe*, nicht mehr gegen das Absurde auf und leistet auch keinerlei politischen Widerstand, sondern kapituliert.

Ähnlich wie Sartre bezeichnet Camus in dem Roman die Welt, historisch konkret gefasst: das Europa nach dem Krieg, „l'Europe des marchands" (Camus 1962: MS, S. 17), der kolonialen Ausbeutung und des Holocaust, als Vorhölle und Hölle. Er bezeichnet die Welt als „malconfort" (Camus 1962: MS, S. 104), womit er einen Kerker meint, in dem man sich nicht bewegen kann und von den anderen angespuckt wird. Der Weg der Erlösung bestehe aber darin, sich bewusst in dieses Gefängnis zu begeben und sich anspucken zu lassen. Gegen Ende des Romans träumt der Protagonist von einer diktatorischen Gesellschaft, in der erst wahre Demokratie herrsche, weil jeder gleichermaßen schuldig sei (vgl. Camus 1968: Chu, S. 147–150). Der Selbstmord am Ende des Buches muss wohl als Vollstreckung des Urteilsspruchs verstanden werden, den der Bußrichter geläutert über sich selbst verhängt.

Freilich ist die Position des Jean-Baptiste Clamence – dessen Name wohl nicht zufällig demjenigen des angeblichen Verfassers der *Philosophischen Bissen*, Johannes Climacus, ähnelt – so exaltiert und extrem, dass es unsinnig wäre, ihn als einfaches Sprachrohr Camus' zu interpretieren. Eher dürfte seine Darstellung eine Art Gedankenexperiment darstellen: Wie sähe die äußerste Antithese des Existenzialismus des *Sisyphe* und der Ethik des Widerstands des *L'Homme révolté* aus? Eine Position, die konsequenterweise in der Aufgabe jedes Willens zur Selbstbehauptung enden muss. Ein Gedankenexperiment, das in der Tat zu denken gibt: Ist es nicht gerade sein Stolz, seine Selbstgerechtigkeit – personifiziert in den großen ‚Advokaten' –, gepaart mit dem unbedingten Willen zur Selbstbehauptung, der Europa in den moralischen Ruin trieb? Und ist die Existenzphilosophie Teil, nicht Lösung des Problems? Das ist eine Frage, die Camus in seinem letzten vollendeten Roman aufwirft, ohne auf sie seinerseits eine befriedigende Antwort zu geben.

VIII Die Aktualität des Existenzialismus in der Post-Demokratie

Wie gesagt stieg *La Peste* im Zuge der aktuellen Corona-Krise zu den meistverkauften Büchern auf.[22] Das überrascht, insofern auch die größten Warner ja wohl kaum der Ansicht sein dürften, dass das Corona-Virus auch nur ansatzweise mit der Gefährlichkeit der empirischen Pest zu vergleichen sei. Es geht hier wohl eher darum, dass der Roman eben eine realistisch-nüchterne und zugleich solidarische Grundhaltung propagiert, die dem aktuellen Lebensgefühl nicht erst seit Corona weitaus besser entspricht als ein nietzscheanischer oder existenzialistischer Individualismus: Wir sehen uns immer wieder mit Krisen konfrontiert, die im Grunde, so jedenfalls die verbreitete Wahrnehmung, nur *eine* wahrhaft menschliche Antwort zulassen und zu deren Lösung auch radikale, den bürgerlichen Normalzustand wenigstens stückweit aushebelnde Notstandsmaßnahmen erforderlich sind. Vielleicht könnte man hier sogar von einem Paradigmenwechsel sprechen, der bereits im Zuge von 9/11 erfolgte, als diese Art Politik zu betreiben erstmals wieder in größerem Maßstab angewandt wurde, während man zuvor ein eher auf Diskussion und Multiperspektivität setzendes Politikmodell präferierte.

Egal, ob ‚Terrorismuskrise', ‚Coronakrise', ‚Klimakrise', ‚Sexismuskrise', ‚Rassismuskrise' oder ‚Wirtschaftskrise': Die Politik erscheint mehr und mehr als eine Abfolge von mit rasanten Ausschlägen in der Erregungskurve verbundenen Ausnahmezuständen, in denen der bürgerliche Normalzustand immer wieder zumindest in Teilen zur Disposition gestellt wird.

Es ist bezeichnend, dass der vielleicht als ‚Neo-Existenzialist' zu bezeichnende Sloterdijk[23] sich nun dafür aussprach, nicht *La Peste*, sondern lieber *Das*

---

22 Vgl. zum Ausmaß des weltweiten Interesses Wehe 2020.

23 In seinem ersten Hauptwerk *Kritik der zynischen Vernunft* spricht Sloterdijk von einer Haltung des „linken Existentialismus" (Sloterdijk 1983, S. 892), die es einzunehmen gälte, sowie von einem „kritische[n] Existentialismus des satirischen Bewußtseins" (Sloterdijk 1983, S. 930), mit dem es das „seriös vorgetragene[]" (Sloterdijk 1983, S. 930) Denken zu konfrontieren gälte. Dementsprechend vertritt Sloterdijk in dem Buch die These von der unaufhebbaren Entfremdung des individualisierten modernen Bewusstseins – und versucht zugleich Wege aufzuzeigen, wie sich diese Entfremdung und Vereinzelung überwinden ließe (vgl. v. a. das Schlusskapitel des Buches; Sloterdijk 1983, S. 928–953). Interessanterweise artikuliert er auch dort schon eine radikale Kritik der modernen Medizin, die seine spätere Kritik an der Corona-Politik vorwegnimmt (vgl. Sloterdijk 1983, S. 489–505). – Vgl. hierzu den Aufsatz *The Limits of the Spheres* (Lucci 2021), in dem der Autor aufzeigt, dass Sloterdijks Denken von derselben Grundkonstellation wie die Existenzphilosophie ausgeht: die gleichzeitige Setzung einer grundsätzlichen Entfremdungserfahrung sowie des Verlangens nach einer Wiederversöhnung.

Decameron von Boccaccio (Boccaccio 2017) zu lesen.[24] Der Witz am *Decameron* besteht freilich darin, dass die Seuche hier gerade *keine* besondere Rolle spielt, sondern genau im Gegenteil die Flucht vor der Seuche im physischen wie auch im psychischen Sinn propagiert wird. Soll doch die ganze Welt zu Grunde gehen – die ‚freien Geister' werden dieser Katastrophe unbekümmert zusehen und sich gegenseitig weiterhin Geschichten erzählen, die die Welt als großes *theatrum mundi* abbilden, in denen ein göttliches Eingreifen in der Regel einen bloßen Priesterbetrug darstellt. Die Absurdität des Spiels der Kräfte der Welt gilt es hinzunehmen und es geschickt zu seinen Gunsten zu manipulieren – nicht in völlig zynischer und amoralischer Weise, aber doch in einer großen Unbekümmertheit, die vom ‚Ernst der Lage' nichts wissen will.

Natürlich ist diese Haltung nicht gerade ‚existenzialistisch' – insbesondere nicht vor dem Hintergrund von Sartres und Beauvoirs Ethik des Engagements –, doch mit der Existenzphilosophie eher zu vereinbaren als der Realismus des späten Camus. Mit Camus müsste man die Haltung der jungen florentinischen Adligen eindeutig als feige und zynische Weltflucht verdammen – Sartre könnte auf der Grundlage seiner eigenen philosophischen Prämissen ein solches Werturteil nicht so ohne Weiteres aussprechen und erst recht nicht Nietzsche, der mit so einer „halkyonischen"[25] Haltung vielmehr eher sympathisieren würde,[26] oder Kierkegaard.

Während Camus' *La Peste* also die Bestsellerlisten stürmt, ist mir jedenfalls nicht zu Ohren gekommen, dass man nun beispielsweise im Zuge der #metoo-Bewegung Beauvoirs Werk neu entdecken[27] oder Sartres *Réflexions sur la question juive* (Sartre 1985) wiederlesen würde als scharfe Analyse einer rechten Ideologie. Selbst Frantz Fanon dürfte heute nur noch wenigen Spezialisten überhaupt bekannt sein.[28]

---

24 So Sloterdijk in seinem umstrittenen Interview für *Le Point* (Sloterdijk 2020).

25 Zum Begriff des Halkyonischen bei Nietzsche vgl. Stephan 2015.

26 Vgl. etwa das Nachlassfragment 1870/71, 7[9] – die einzige Stelle im gesamten Corpus ab 1869 im Übrigen, an der Nietzsche das Hauptwerk Boccaccios überhaupt erwähnt: „Eine Parallele für die Entstehung der olympischen Götterwelt bei dem schrecklichen Hintergrund bietet die Entstehung des *Decamerone* zur *Pestzeit*" (Nietzsche, NL 1870, KSA 7, S. 138).

27 Eine Ausnahme stellt die äußerst bemerkenswerte Studie *On ne naît pas soumise, on le devient* von Manon Garcia dar (Garcia 2018) – in der sie aber unter Rekurs auf Ideen Beauvoirs die Opferhaltung des Mainstream-Feminismus gerade kritisiert: Frauen seien ihr zufolge nicht einfach nur passive Opfer des Patriarchats, sondern reproduzierten es durch ihr submissives Verhalten selbst.

28 Freilich gibt es auch diesbezüglich erfreuliche Ausnahmen wie die Arbeiten der britischen Soziologin Sara Salem, die in Franz Fanon einen hochaktuellen Theoretiker des Zusammenhangs von Kapitalismus und Rassismus erblickt (vgl. Salem 2018 und, speziell auf die Aktualitätsfrage

Vielleicht übertreibe ich, aber ich hoffe, worauf ich hinauswill, ist klar genug: Wir haben es im Augenblick mit einem radikalen Bedeutungsverlust existenzphilosophischer Impulse zu tun bis ins Politische hinein – ein Bedeutungsverlust, der vermutlich etwas mit tiefgreifenderen sozialen und politischen Verschiebungen zu tun hat, vielleicht sogar mit dem endgültigen Ende der liberalen Werteordnung und ihre Ersetzung durch ein alternatives Herrschaftsmodell nach eher chinesischem Vorbild, wie sie beispielsweise auch Sloterdijk dezidiert als langfristige Folge der westlichen Corona-Politik befürchtet.[29] Der heute geforderte Menschentyp ist nicht der nietzscheanische Freigeist, der kierkegaardsche Zweifler oder der sartresche Mensch am Abgrund, der zwischen Angst und Entschlossenheit schwankt (vgl. Sartre 1976: EN, S. 75–77) – er gleicht eher dem nüchternen Arzt Rieux, der einfach tut, was getan werden muss, der ‚resonant'[30] ist, ‚sozial kompetent' und ‚sensibel' für seine Außenwelt.

Dementsprechend ätzen die Herausgeber der *Philosophischen Rundschau* in dem Sonderheft der Zeitschrift zur Corona-Krise, ganz auf der Höhe des Zeitgeists:

> Mit Sloterdijk und Agamben können wir uns zurückversetzen in einen Nachkriegsexistenzialismus, der eine rigorose moralische und künstlerische Selbstentfaltung forderte, sozusagen als heroische Stütze eines Wiederaufbaus der Humanität. Die realen Bemühungen um die Standards der Menschlichkeit und des humanen Überlebens wurden auch damals der Politik überlassen. Deutlich wird bei solchen Wiederholungen nur, dass eine gewisse Nähe zur menschenverachtenden Großen Politik des Geistes der ersten Hälfte des Jahrhunderts auch dann vorhanden bleibt, wo man sich mit dem Wind vom Nationalen der Rechten weg zu einem globalen Sozialismus der Linken hin gedreht hat. (Gessmann/Stekeler-Weithofer 2020, S. 76)

Große Visionen und ein heroischer Individualismus werden lächerlich gemacht, stattdessen ein Pragmatismus vertreten, der sich an das Faktische und an scheinbar selbstevidente moralische Erfordernisse hält. Dieser Realismus des „There is no alternative" – „Man muss doch die Flüchtlinge retten", „Man muss doch die Alten retten", „Man muss doch den Terror bekämpfen" ... – steht freilich sowohl moralisch als auch deskriptiv gesehen auf tönernen Füßen: Er abstrahiert

eingehend, Salem 2017). Auch Carolin Emcke zitiert Fanon, wenn auch in oberflächlicher Manier, in dem dem Rassismus gewidmeten Kapitel ihres Essays *Gegen den Hass* (vgl. Emcke 2016, S. 84–105). – Generell scheint Fanons Name, wenn er überhaupt einmal erwähnt wird, eher als Chiffre zu dienen für etwas ‚Radikales', als dass man sich mit seinen Theorien wirklich auseinandersetzt.

29 Vgl. das bereits erwähnte Interview (Sloterdijk 2020).

30 Der Begriff steht hier bewusst in Anführungszeichen, da dieser Menschentyp natürlich im Sinne eines anspruchsvollen Begriffs von Resonanz, wie ihn Hartmut Rosa ins Spiel brachte, gerade als nicht besonders resonant anzusehen wäre. Dasselbe gilt entsprechend für die anderen Attribute.

davon, dass alle diese Ereignisse ja Auswüchse struktureller Missstände sind, auf die hinzuwiesen als hybrid und veraltet gilt. Dabei ist klar: Solange die Ursachen der Krankheit fortbestehen, kann die bloße Symptombekämpfung ja sogar schädliche Folgen haben. Der moralisierende Blick auf politische Missstände verhindert gerade ihre Auflösung und blockiert die Suche nach nachhaltigen Lösungsansätzen. Man fragt sich zumal: Was soll die Aufgabe von Philosophen und Sozialtheoretikern denn sein, wenn nicht den Erregungskurven des Zeitgeists und dem naiv-moralisierenden Blick auf politische Ereignisse zu widerstehen und die Ereignisse in einem größeren Kontext zu verorten – eben einen „halkyonischen" Blick auf die Dinge einzunehmen?

Es scheint so, dass die Existenzphilosophie vielleicht doch noch nicht ganz überholt ist und es sich noch immer lohnt, sich von der Haltung eines kierkegaardschen, nietzscheschen oder existenzialistischen Heroismus inspirieren zu lassen – gerne erweitert um etwas von der moralischen Haltung des späten Camus. Der Bogen bleibt also gespannt – was die Hoffnung verbürgt, dass wir am Ende vielleicht doch einmal in einer Gesellschaft erwachen könnten, in der die Alterität kein Problem darstellte.

Bibliographie

Beauvoir, Simone de (2017): „Pour une morale de l'ambiguïté". In: Simone de Beauvoir: *Pour une morale de l'ambiguïté suivi de Pyrrhus et Cinéas*. Paris, S. 5–230.

Boccaccio, Giovanni (2017): *Das Decameron*. Ditzingen.

Burkard, Franz-Peter (2008): Eintrag „Existenzphilosophie". In: Peter Prechtl/ Franz-Peter Burkard (Hrsg.): *Metzler Lexikon Philosophie*. Stuttgart, Weimar, S. 174–175.

Camus, Albert (1951a): „L'Europe de la fidélité". In: *La Révolution Prolétarienne* 351, S. 15–16.

Camus, Albert (1951b): *L'Homme révolté*. Paris.

Camus, Albert (1955): *La Peste*. Paris.

Camus, Albert (1962): *Le Mythe de Sisyphe*. Paris.

Camus, Albert (1968): *La Chute*. Paris.

Emcke, Carolin (2016): *Gegen den Hass*. Frankfurt a. M.

Garcia, Manon (2018): *On ne naît pas soumise, on le devient*. Paris.

Gessmann, Martin/Stekeler-Weithofer, Pirmin (2020): „Editorial. Kulturzeitdämmerung?". In: *Philosophische Rundschau* 67, S. 75–85.

Hegel, Georg Wilhelm Friedrich (1973): *Phänomenologie des Geistes*. Frankfurt a. M.

Husson, Laurent (2005): „La nature, entre métaphysique et existence. Sartre et la question de la nature". In: *Le Portique. Revue de philosophie et de sciences humaines* 16 (Onlineausgabe). http://journals.openedition.org/leportique/733, besucht am 18. 02. 2021.

Jaspers, Karl (1973): *Vernunft und Existenz. Fünf Vorlesungen*. München.

Kierkegaard, Søren (1890a): „Der Begriff Angst. Eine simple psychologisch-wegweisende Untersuchung in der Richtung auf das dogmatische Problem der Erbsünde von Vigilius

Haufniensis“. In: *Zur Psychologie der Sünde, der Bekehrung und des Glaubens. Zwei Schriften Sören Kierkegaards*. Leipzig, S. 1–166.
Kierkegaard, Søren (1890b): *Philosophische Bissen oder Ein Bisschen Philosophie von Johannes Climacus*. In: *Zur Psychologie der Sünde, der Bekehrung und des Glaubens. Zwei Schriften Sören Kierkegaards*. Leipzig, S. 167–255.
Kierkegaard, Søren (1929): *Über den Begriff der Ironie. Mit ständiger Rücksicht auf Sokrates*. München, Berlin.
Kierkegaard, Søren (2003): *Abschliessende unwissenschaftliche Nachschrift zu den Philosophischen Brocken*. Simmerath.
Löwith, Karl (1987a): „Kierkegaard und Nietzsche“. In: *Nietzsche. Sämtliche Schriften*. Bd. 6. Stuttgart, S. 75–99.
Löwith, Karl (1987b): „Kierkegaard und Nietzsche oder philosophische und theologische Überwindung des Nihilismus“. In: *Nietzsche. Sämtliche Schriften*. Bd. 6. Stuttgart, S. 53–74.
Lucci, Antonio (2021): „The Limits of the Spheres. Otherness and Solipsism in Peter Sloterdijk's Philosophy“. In: *Angelaki. Journal of the Theoretical Humanities* 26, S. 92–108.
McKinnon, Alastair (1993): „Kierkegaard and ‚The Leap of Faith‘“. In: *Kierkegaardiana* 16, S. 107–125.
Meyer, Michael (2020): „Über Revolten gegen schweigende Marmelade. Bemerkungen zum Zusammenhang von Melancholie und Aggressivität bei Sartre und Camus“. In: Michael Neecke/Rainer Barbey (Hrsg.): *Musik, Melancholie und Tod. Schriftstücke. Beiträge zu Philosophie und Literaturwissenschaft* 3. Berlin, S. 176–207.
Rosa, Hartmut (2016): *Resonanz. Eine Soziologie der Weltbeziehung*. Berlin.
Salem, Sara (2017): „A Revolutionary Lifeline. Teaching Fanon in a Postcolonial World“. In: *Historical Materialism*. https://www.historicalmaterialism.org/blog/revolutionary-lifeline-teaching-fanon-postcolonial-world, besucht am 19.02.2021.
Salem, Sara (2018): „Reading Egypt's Postcolonial State Through Frantz Fanon. Hegemony, Dependency and Development“. In: *Interventions. International Journal of Postcolonial Studies* 20, S. 428–445.
Sartre, Jean-Paul (1960): *Critique de la raison dialectique. Tome I: Théorie des ensembles pratiques*. Paris.
Sartre, Jean-Paul (1970): *L'Existentialisme est un humanisme*. Paris.
Sartre, Jean-Paul (1973): *La Nausée*. Paris.
Sartre, Jean-Paul (1976): *L'être et le néant. Essai d'ontologie phénoménologique*. Paris.
Sartre, Jean-Paul (1985): *Réflexions sur la question juive*. Paris.
Sartre, Jean-Paul (2000): „Huis clos“. In: Jean-Paul Sartre: *Huis clos suivi de Les Mouches*. Paris, S. 9–95.
Schopenhauer, Arthur (2004): *Parerga und Paralipomena. Kleine philosophische Schriften II. Sämtliche Werke*. Bd. V. Darmstadt.
Shuttleworth, Kyle Michael James (2021): *The History and Ethics of Authenticity. Meaning, Freedom and Modernity*. London, New York.
Sloterdijk, Peter (1983): *Kritik der zynischen Vernunft*. Frankfurt a. M.
Sloterdijk, Peter (2020): „Sloterdijk: ‚Le système occidental va se révéler aussi autoritaire que celui de la Chine‘“. In: *Le Point*. https://www.lepoint.fr/politique/sloterdijk-le-systeme-

occidental-va-se-reveler-aussi-autoritaire-que-celui-de-la-chine-18-03-2020-2367624_20.php, besucht am 19. 02. 2021.

Stephan, Paul (2015): „Über die Salzflut gleiten. Zum Begriff des ‚Halkyonischen' bei Nietzsche". In: *Narthex. Heft für radikales Denken* 2, S. 64 – 75.

Stirner, Max (2009): *Der Einzige und sein Eigentum. Ausführlich kommentierte Studienausgabe.* München.

Andrew Inkpin

Mit oder ohne die Anderen? Beauvoir contra Nietzsche über Freiheit

Abstract: With or without the others? Beauvoir contra Nietzsche on freedom. This article focuses on Beauvoir's critique of the „adventurer", an apparently Nietzschean figure she depicts as „inauthentic" in *The Ethics of Ambiguity.* Its aim is to assess whether her criticisms of individualistic freedom amount to a tenable critique of Nietzsche. I start by outlining Beauvoir's conception of the adventurer and his faults, before showing how this figure closely tracks distinctive features of Nietzsche's „free spirit". Finally, I evaluate whether or not Beauvoir's criticisms apply to Nietzsche. Although he can be defended to some extent, I argue that Beauvoir shows how Nietzsche's otherwise potentially attractive concept of the free spirit is more problematic than it might initially appear.

> [L]a liberté humaine est l'ultime, l'unique fin à laquelle l'homme doive se destiner. (Beauvoir 2017: PMA, S. 63)

> Wohl aber dürften wir uns „freizügige Geister" [...] nennen, weil wir den Zug zur Freiheit als stärksten Trieb unseres Geistes fühlen [...]. (Nietzsche, MA II, §211)

In ihrer 1947 erschienenen Abhandlung *Pour une morale de l'ambiguïté* bespricht Simone de Beauvoir kritisch eine Figur, die sie den ‚Abenteurer' nennt und die, obwohl die Verbindung implizit bleibt, sich als stellvertretend für mehrere Motive des Denkens Friedrich Nietzsches erkennen lässt. Auch wenn Beauvoir und Nietzsche in vielen Hinsichten unwahrscheinliche philosophische Gesprächspartner sind, erweist sich diese Kritik als aufschlussreich, weil sie bei einem gemeinsamen Interesse an menschlicher Freiheit ihre Uneinigkeit darüber offenlegt, wie diese Freiheit zu verstehen und zu verwirklichen ist.[1] In diesem Beitrag untersuche ich, inwiefern Beauvoirs Text eine tragfähige Kritik von Nietzsche ergibt,

1 Obwohl Nietzsche als Vorläufer des Existenzialismus gilt, gibt es sehr wenig Literatur zum philosophischen Verhältnis von Beauvoir und Nietzsche. Als Ausnahmen siehe Parker (2014), die Beauvoirs Einstellung zum Begriff des ‚Willens zur Macht' untersucht, Tirrell (1994), die Gemeinsamkeiten in ihrem Identitätsbegriff erkennt, sowie Thorgeirsdottir (2012), die ihr Verhältnis mit Blick auf die Themen von Ehe, Liebe und Geburt thematisiert. Dem vorliegenden Beitrag am nächsten kommt Daigle (2010 – 2011), die ihrer Deutung von Nietzsches ‚Übermenschen' entsprechend mehrere Parallelen zwischen Beauvoirs Ethik und Nietzsche sieht und Beauvoirs Kritik von Nietzsches Individualismus hervorhebt.

https://doi.org/10.1515/9783110760163-011

insbesondere ob individuelle Freiheit auch einer Harmonisierung mit anderen Menschen bedarf. Nach einer Erläuterung von Beauvoirs ‚Abenteurer' sowie ihrer Kritik desselben im ersten Teil verdeutlicht der zweite Teil, wie Nietzsches ‚freier Geist' Beauvoirs Figur zugrunde liegt. Der abschließende dritte Teil erwägt systematisch, inwiefern Beauvoirs Kritik tatsächlich Nietzsche trifft. Dort versuche ich zu zeigen, wie diese Kritik, auch wenn sie nicht in jeder Hinsicht überzeugt, deutlich macht, dass Nietzsches sonst verlockender Begriff des Freigeistes problematischer ist, als er zunächst erscheinen mag.

1 Beauvoirs Kritik des Abenteurers

Beauvoir entwickelt in *Pour une morale de l'ambiguïté* den Grundriss einer existenzialistischen Ethik, bei der die Verwirklichung von Freiheit als einer spezifisch menschlichen Aufgabe im Mittelpunkt steht.[2] Was das für Beauvoir bedeutet, fasst sie in der folgenden Formel zusammen: „Vouloir l'existence, vouloir dévoiler le monde, vouloir les hommes libres, c'est une seule volonté" (Beauvoir 2017: PMA, S. 108). Aber wie ist diese Formel zu verstehen? Auf welche Weise hängen Existenz, Welterschließung und die Freiheit Anderer zusammen?

Der Zusammenhang zwischen Existenz und Welterschließung ergibt sich aus Beauvoirs Anschluss an die Ontologie Jean-Paul Sartres. Unter Freiheit versteht Sartre die Freiheit des Bewusstseins bzw. „Fürsichseins" (*être-pour-soi*), die er zugleich als einen ontologischen Grundzug des Menschen begreift. Demnach impliziert Existenz – als die menschliche Seinsweise – unmittelbar ein Bewusstsein von der Welt (d. h. Welterschließung). Diese Freiheit umfasst zwei hier relevante Aspekte, die Sartre als ‚unmittelbare Strukturen' des Fürsichseins versteht. Einerseits ist der Mensch frei, indem er nicht durch das faktisch Gegebene oder Vergangenes bestimmt wird, d. h. in Sartres Terminologie, indem er als „Faktizität" das Ansichseiende negiert oder „nichtet" (Sartre 2006: EN, S. 118–119). Andererseits ist er als „Transzendenz" frei, indem er sich auf eine Zukunft entwirft: „en étant transcendance [...] [il est] un être qui est originellement pro-jet, c'est-à-dire qui se définit par sa fin" (Sartre 2006: EN, S. 497). Mit diesem Aspekt der „Transzendenz", der für Beauvoirs Ethik zentral wird, verbindet Sartre auch das Schaffen von Werten sowie Möglichkeiten. Kurzum gilt bei Sartre: der Mensch

2 Diese Ethik bildet auch die Grundlage für ihr späteres Hauptwerk *Le deuxième sexe*: „La perspective que nous adoptons, c'est celle de la morale existentialiste." (Beauvoir 1976: DS, S. 33) Ich beschränke mich hier auf *Pour une morale de l'ambiguïté*, weil sich dort Beauvoirs Haltung zu Nietzsches Freiheitsbegriff am besten nachvollziehen lässt.

als bewusstes Wesen existiert als (ist) seine Freiheit im zweifachen Sinne einer Freiheit *von* Faktizität sowie *für* Transzendenz.

Beauvoirs Anschluss an Sartre macht jedoch nicht sofort verständlich, warum unsere Existenz für sie zugleich die Freiheit anderer Menschen fordert. Im Gegenteil. Für Sartre gehört ein Verhältnis zu Anderen *nicht* unmittelbar als ontologische Struktur zum Bewusstsein, sondern bildet eine kontingente, faktische Erweiterung des Fürsichseins (Sartre 2006: EN, S. 321). Weiterhin wird Sartre zufolge unser Verhältnis zu Anderen stets durch Konflikt geprägt, sofern es einen Kampf um die eigene Selbstbehauptung und gegen die Vergegenständlichung durch den Blick des Anderen veranlasst. Gleichwohl teilt Beauvoir als Ausgangspunkt Sartres individualistische Voraussetzung, dass dem Einzelmenschen (dessen Bewusstsein) Freiheit zukommt, und daraus ergibt sich ihre Auffassung der „véritable position du problème" für eine existenzielle Ethik: „Originellement séparés, comment les hommes pourraient-ils se rejoindre?" (Beauvoir 2017: PMA, S. 24)

Man könnte vermuten, eine existentielle Ethik stehe vor einem grundsätzlichen Problem: Wenn Menschen bereits vom Wesen her frei sind, was soll „freie Menschen wollen" bedeuten? Wenn Freiheit (deskriptiv) einen Grundzug des Menschen bildet, wie kann sie uns (normativ) als Ziel aufgegeben sein? Beauvoirs Antwort lautet: Obwohl es „contradictoire" wäre, sich vorsätzlich *nicht frei* zu wollen, kann man aus „Faulheit" (*paresse*) seine Freiheit *nicht wollen* (Beauvoir 2017: PMA, S. 35). Wie Heidegger und Sartre vor ihr unterscheidet sie daher Eigentlichkeit von Uneigentlichkeit. „L'homme authentique" übernimmt die Verantwortung für seine Entscheidungen, statt sie durch objektive oder absolute Werte begründen zu wollen, und organisiert sein Leben unter einem einheitlichen Projekt (Beauvoir 2017: PMA, S. 20, 35–36).

Im Unterschied zu Heidegger und Sartre jedoch bietet Beauvoir eine Typologie mehrerer „attitudes inauthentiques" (Beauvoir 2017: PMA, S. 61). Ausgehend vom „metaphysisch privilegierten" Kind, das seine Freiheit noch nicht kennt, und dem Adoleszenten, der die Verantwortung dieser Freiheit erst aufnimmt, skizziert sie mehrere Entwicklungsstufen auf dem Weg zur „eigentlichen" Freiheitsverwirklichung. Um ihr Verhältnis zu Nietzsche zu verstehen, sind vor allem drei Figuren relevant. Der „homme sérieux" ist ein Konformist, der gesellschaftlich etablierte Werte als Grundlage seiner Identität fraglos übernimmt und dadurch seine Freiheit stets verneint (Beauvoir 2017: PMA, S. 62).[3] Als eine Umkehrung dieser Figur – „le sérieux déçu" (Beauvoir 2017: PMA, S. 68) – versteht Beauvoir den „Nihilisten". Der Nihilist will nichts und leugnet alle bestehenden Werte,

3 Ähnlich beschreibt bereits Sartre (2006, S. 74) „l'esprit de sérieux".

bleibt aber ebenfalls auf die etablierten Werte der „monde sérieux" als die Grundlage seiner ablehnenden Haltung angewiesen (Beauvoir 2017: PMA, S. 69).

Gegenüber diesen beiden Figuren bestimmt Beauvoir die Figur des „Abenteurers" (*aventurier*). Im Unterschied zum Nihilisten einerseits erkennt der Abenteurer Freiheit als Wert an, während andererseits seine „liberté se réalise comme indépendance à l'égard du monde sérieux" (Beauvoir 2017: PMA, S. 75). Beauvoirs Konzeption des Abenteurers hat alsdann zwei Hauptaspekte, von denen der erste in seinem Verhältnis zu Aktivität und Zielen liegt:

> Il se jette avec ardeur dans des entreprises : exploration, conquête, guerre, spéculation, amour, politique, mais il ne s'attache pas à la fin visée ; seulement à sa conquête. Il aime l'action pour l'action, il trouve sa joie à déployer à travers le monde une liberté qui demeure indifférente à son contenu. (Beauvoir 2017: PMA, S. 74)

Dem Abenteurer gilt seine *eigene* Freiheit somit als höchster Wert. Er widmet sich leidenschaftlich der Aktivität, um der Aktivität statt des Zieles willen, weshalb seine Freiheit letztlich auf eine „indépendence abstraite" hinausläuft (Beauvoir 2017: PMA, S. 79).

Für Beauvoir kommt der Abenteurer „très proche [...] d'une attitude authentiquement morale" (Beauvoir 2017: PMA, S. 75). Denn er teilt mit ihrem Existenzialismus nicht nur eine Hochschätzung von Freiheit, sondern auch die Ansicht, dass Wertsetzung in der Verantwortung des Subjekts liegt. Sie räumt sogar ein: „si l'existentialisme était [...] un solipsisme, il devrait regarder l'aventurier comme son héros le plus achevé" (Beauvoir 2017: PMA, S. 75–76).

Der zweite Hauptaspekt von Beauvoirs Konzeption des Abenteurers liegt in seiner Einstellung zu anderen Menschen. Der Abenteurer ist „celui qui croit pouvoir affirmer sa propre existence sans tenir compte de celle d'autrui" (Beauvoir 2017: PMA, S. 78). Gemäß seiner Gleichgültigkeit gegenüber den Zielen, die er mit ihnen verbindet, betrachtet er andere Menschen mit Verachtung: „l'aventurier partage le mépris du nihiliste envers les hommes"; „rien ne l'empêche donc de sacrifier ces êtres insignifiants à sa propre volonté de puissance ; il les traitera comme des instruments" (Beauvoir 2017: PMA, S. 78). In politischer Hinsicht haben seine Gleichgültigkeit gegenüber Zielen sowie seine Geringschätzung anderer Menschen zur Folge, dass der Abenteurer Klassenprivilegien guthieße und autoritären Staatsformen wie dem Faschismus nahe stünde (Beauvoir 2017: PMA, S. 79).

Demgegenüber zeichnet sich der „homme authentiquement libre" dadurch aus, dass er die Freiheit anderer Menschen nicht nur anerkennt, sondern auch fördert (Beauvoir 2017: PMA, S. 77–78). Beauvoir zufolge existiert der Einzelne nur „en se transcendant et sa liberté ne peut s'accomplir qu'à travers la liberté d'autrui" (Beauvoir 2017: PMA, S. 193). Die Uneigentlichkeit des Abenteurers besteht folglich darin, dass er nicht erkennt, wie seine eigene Freiheit konstitutiv von anderen Menschen abhängt. Seine vermeintliche Unabhängigkeit (Freiheit) ist demnach eine Selbsttäuschung, „une fausse indépendance": „il refuse avec mauvaise foi de reconnaître que cette subjectivité se transcende nécessairement vers autrui" (Beauvoir 2017: PMA, S. 81).

In dieser Kritik kommt eine entscheidende Differenz zwischen Beauvoir und Sartre zum Ausdruck. Während der Andere für Sartre Freiheitsbegrenzung und Konflikt bringt, sieht Beauvoir eine Harmonisierung mit Anderen als nicht nur möglich, sondern notwendig für die volle Entfaltung der eigenen Freiheit. Aber wie ist diese Differenz zu verstehen? Auf welche Weise soll individuelle Freiheit von Anderen abhängen? Beauvoir selbst erläutert diese Abhängigkeit weder ausführlich noch systematisch, aber ich möchte hier vier Thesen über ihre Funktion unterscheiden, die sie andeutet und deren Tragweite im Folgenden (Abschnitt 3) ausgewertet wird:

i) *Faktisch-kausale Abhängigkeit:* Beauvoir hebt hervor: „l'aventurier sur son chemin rencontre toujours autrui [...] toute entreprise se déroule dans un monde humain et intéresse les hommes." (Beauvoir 2017: PMA, S. 77) Diese scheinbar triviale Feststellung hat wichtige Folgen. Denn sofern der Abenteurer seinen Projekten in der intersubjektiv geteilten Welt nachgeht, wirken andere Menschen mit, seinen Lebensunterhalt und die Bedingungen der Möglichkeit dieser Projekte zu konstituieren. Der gesamte Zusammenhang menschlicher Praxen (die Lebenswelt) bildet den Bezugsrahmen und Hintergrund für die „individuellen" Projekte des Abenteurers.

ii) *Abgrenzung:* Beauvoir erläutert weiter: „Ce qui distingue l'aventure d'un simple jeu, c'est que l'aventurier ne se borne pas à affirmer solitairement son existence ; il lui affirme par rapport à d'autres existences" (Beauvoir 2017: PMA, S. 77). Das heißt, er versteht sich selbst durch die Abgrenzung von anderen Menschen, die er sich als homogene konformgehende Angehörige der „ernsthaften Welt" vorstellt.

iii) *Anerkennung:* In gewisser Hinsicht genügt der Abenteurer sich selbst nicht. Damit ein Abenteuer (ggf. über sein Leben hinaus) „lebendig bleibt", bedarf es „la reconnaissance ou [...] l'admiration" von Anderen: „beaucoup souhaitent léguer à la postérité, à défaut d'une œuvre, leur propre figure ; au moins ont-ils besoin, de leur vivant, de l'approbation de quelques fidèles" (Beauvoir 2017: PMA, S. 80).

iv) *Fortsetzung:* Für Beauvoir gehört zur Transzendenz, dass unsere Projekte auch von Anderen übernommen und fortgesetzt werden. Ein Projekt, das nur ein Einzelmensch verfolgt, verfehlt in dieser Sichtweise echte Transzendenz: „ma liberté exige pour s'accomplir de déboucher sur un avenir ouvert : ce sont les autres hommes qui m'ouvrent l'avenir [...] une vie ne se justifie que si son effort pour se perpétuer est intégré dans son dépassement" (Beauvoir 2017: PMA, S. 103–104).

Beauvoir bespricht den „Abenteurer" typologisch, ohne zu sagen, gegen welche Autoren ihre Kritik sich richtet. Warum also sollte man diese Figur mit Nietzsche verbinden? Vor allem, wie der nächste Abschnitt zeigt, weil ihre zwei Hauptaspekte sowie ihre Bestimmung gegenüber dem konventionellen *homme sérieux* und dem Nihilisten denselben Motiven im Denken Nietzsches entsprechen. Weiterhin, auch ohne ihn namentlich zu erwähnen, legen Beauvoirs Ansichten, dass der Abenteurer andere Menschen als Instrumente seines „Willens zur Macht" behandelt und dem Faschismus nahesteht (Beauvoir 2017: PMA, S. 78–79), diese Verbindung nahe. Schließlich wird die Verbindung am Ende ihrer Typologie uneigentlicher Einstellungen (an einer *einzigen* Stelle) explizit, wo Beauvoir erneut gegen die Gleichsetzung von Existenzialismus mit „Solipsismus" mahnt. Während sie zuvor den Abenteurer als dessen vollkommensten Helden beschrieben hatte, nennt sie hier (Beauvoir 2017: PMA, S. 91) Nietzsche ausdrücklich als Vertreter dieses Solipsismus.

2 Nietzsches freier Geist

Es gibt mehrere Motive in Nietzsches Schriften, die seine Nähe zu Beauvoirs ‚Abenteurer' erkennen lassen: den „freien Geist", den „Wanderer", den „Versucher", das „souveraine Individuum" (Nietzsche, GM II 2), „große Gesundheit" usw.[4] Ich beschränke mich hier auf den freien Geist, nicht nur weil er die von Beauvoir anvisierten Züge leicht nachvollziehen lässt, sondern auch weil seine Gleichsetzung mit dem Abenteurer gelegentlich von Nietzsche selbst ausdrücklich nahegelegt wird. So beschreibt er freie Geister als „Abenteurer und Weltumsegler jener inneren Welt, die ‚Mensch' heisst" (Nietzsche, MA I Vorrede 7). Ähnlich spricht er vom „Meisterschafts-Vorrecht des freien Geistes", „a u f d e n V e r s u c h hin leben und sich dem Abenteuer anbieten zu dürfen" (Nietzsche, MA I Vorrede 4).

Nietzsche bestimmt den freien Geist – wie später Beauvoirs Abenteurer – im Gegensatz zu einer breiten Masse von Konformisten, die etablierte Meinungen und Werte passiv übernehmen, und die er verschiedentlich als „gebundene Geister", „Dogmatiker", „Fanatiker" bezeichnet:

> Man nennt Den einen Freigeist, welcher anders denkt, als man von ihm auf Grund seiner Herkunft, Umgebung, seines Standes und Amtes oder auf Grund der herrschenden Zeitan-

4 Nietzsches Werke werden im Folgenden nach der *Kritischen Studienausgabe* zitiert.

sichten erwartet. Er ist die Ausnahme, die gebundenen Geister sind die Regel [...]. (Nietzsche, MA I 225)

In *Menschliches, Allzumenschliches* zeichnet sich der freie Geist gerade als Erkennender aus: „er fordert Gründe, die Anderen Glauben" ohne Gründe (Nietzsche, MA I 225), und ist jemand, der „nur deshalb weiter lebt, um immer besser zu erkennen" (Nietzsche, MA I 34). In späteren Schriften wird Nietzsches Einstellung zum Erkennen ambivalenter infolge seiner Infragestellung des „Werthes der Wahrheit" (Nietzsche, JGB 1; GM III 24). Dementsprechend mahnt er gegen die Gefahr, dass der freie Geist seine Freiheit durch ein dogmatisches Festhalten an „Redlichkeit" (Nietzsche, JGB 227; GM III 27) als seine charakteristische Tugend oder an die Suche nach Wahrheit (Nietzsche, GM III 24) einschränkt. Ein wirklich freier Geist könne sich auch dem Erkennen nicht versklaven.[5] Die Erkenntnis sei nicht um ihrer selbst willen wertvoll, sondern diene vielmehr dazu, uns zu verwandeln (Nietzsche, JGB 64, 230 – 231).[6] Folglich wird der Freigeist primär nicht mehr als Erkennender, sondern als die Verkörperung von Freiheit als höchstem Wert konzipiert. Den damit potenzierten Freiheitsanspruch drückt Nietzsche aus, indem er nun von „‚freien, sehr freien Geister[n]'" (Nietzsche, GM III 24; JGB 44) spricht.[7]

So wichtig diese Entwicklung für ein volles Verständnis von Nietzsches Denkfigur sein mag, ändert sich nichts an den beiden von Beauvoir als Hauptaspekte aufgenommenen Zügen.[8] So ergibt sich das Verhältnis des Freigeistes zu Aktivität und Zielen daraus, dass die Loslösung von herkömmlichen Denkweisen nicht ein für alle Mal vollzogen wird, sondern immer wieder geschehen muss. Um Passivität sowie eine Bestimmung von außen zu vermeiden, muss der freie Geist seine Aktivität immer wieder unter Beweis stellen. Gegen die „Trägheit des Geistes", die unsere Gedanken zu bloßen „Ueberzeugungen erstarren" lässt, empfiehlt Nietzsche dem „freien, rastlos lebendigen Geist [...] beständigen Wechsel" (Nietzsche, MA I 637). Auf diese Weise verschreibt sich der freie Geist

5 Statt als Erkennender zeichnet er sich nun durch psychologische Stärke aus (vgl. Nietzsche, FW 347), wie Reginster (2003, S. 80 – 81) betont.

6 Somit wird Nietzsche in *Jenseits von Gut und Böse* der paradoxen Struktur dieses Freiheitsstrebens bewusst. Denn er lehnt nicht nur die Erkenntnis als unbedingten Wert ab, sondern auch ‚seine eigene Loslösung' (Nietzsche, JGB 41). Sofern es aber nicht sinnlos sein soll, setzt ein solches Streben immer irgendeine Motivation voraus – und sei es, um uns zu „verwandeln" –, die es nicht infragestellt.

7 Dieser Ausdruck markiert auch Nietzsches Abgrenzung von anderen zeitgenössischen „Freidenkern" (vgl. Nietzsche, JGB 44; GM III 24).

8 Für hilfreiche Erläuterungen zu Nietzsches Auffassung des freien Geistes siehe Mullin (2000); Reginster (2003).

einem dynamischen Verlauf, bei dem jeder erreichte Zustand wiederum überwunden werden muss.

Dieser Verlauf hat folgerichtig für Nietzsche kein Endziel, sondern bleibt ein offener und daher ungewisser Werdegang, den Nietzsche oft mit einer Reisemetaphorik beschreibt. Der freie Geist „ist unruhig und ziellos unterwegs wie in einer Wüste" (Nietzsche, MA I Vorrede 3). Wer „zur Freiheit der Vernunft gekommen ist", wird sich „nicht als Reisender n a c h einem letzten Ziele [fühlen]: denn dieses giebt es nicht [...] es muss in ihm selber etwas Wanderndes sein, das seine Freude an dem Wechsel und der Vergänglichkeit habe." (Nietzsche, MA I 638) Sein Freiheitsdrang zwingt ihn, immer weiter zu reisen:

> In vielen Ländern des Geistes zu Hause, mindestens zu Gaste gewesen; den dumpfen angenehmen Winkeln immer wieder entschlüpft [...]; voller Bosheit gegen die Lockmittel der Abhängigkeit, welche in Ehren, oder Geld, oder Ämtern, oder Begeisterungen der Sinne versteckt liegen [...]. (Nietzsche, JGB 44)

Der freie Geist wird somit durch eine gewisse Gleichgültigkeit gegenüber den eigenen Gedanken und Projekten charakterisiert. Er engagiert sich stets vorübergehend, im Bewusstsein ihrer Nichtendgültigkeit und Revidierbarkeit.[9] Und schließlich, obwohl er nicht mehr primär als Erkennender definiert wird, verkörpert der Freigeist Nietzsches erkenntnistheoretischen Perspektivismus. Seine „Reisen" machen ihn mit diversen Standpunkten vertraut, wodurch er als Erkennender überlegen wird, bewahren ihn aber zugleich davor, irgendeinen Standpunkt zu verabsolutieren.[10]

Wie steht es mit der Einstellung des freien Geistes zu anderen Menschen? Wie wir gesehen haben, wird er als das Streben nach Unabhängigkeit von den Denkweisen anderer Menschen und damit durch ein negatives Verhältnis zu Anderen definiert. Als Andersdenkender konstituiert er sich sowie sein eigenes Selbstverständnis gerade durch Absetzung von den Anderen. Es überrascht daher nicht, dass Nietzsche freie Geister als „die geborenen geschworenen eifersüchtigen Freunde der E i n s a m k e i t" beschreibt (Nietzsche, JGB 44). In seinen Schilderungen des freien Geistes werden die „Anderen" weiterhin als eine homogene Masse von Konformisten vorgestellt: *sie* werden auf Gleichheit reduziert, während der Freigeist *seine* Andersartigkeit behauptet. Gegenüber diesen Anderen bleibt der freie Geist allenfalls gleichgültig und häufig werden sie – als Herde, Fanatisten usw. – mit der „Verachtung" beschrieben, die Nietzsche für diejenigen

---

9 Darin wird er zum Vorbild für Rortys „Ironiker" (Rorty 1989, S. 73–108).

10 Vgl. Nietzsche, GM III 12. Der „r e i f e n Freiheit des Geistes" seien „die Wege zu vielen und entgegengesetzten Denkweisen erlaubt" (MA I Vorrede 4).

empfiehlt, die man nicht als seinesgleichen empfindet (z. B. Nietzsche, FW II, 379; JGB 260).

Man könnte vermuten, dass die Einstellung von freien Geistern zueinander positiver ausfällt und durch gegenseitige Anerkennung und Achtung bestimmt wird. Für Nietzsche aber wäre das ein Zeichen von Schwäche und in diesem Sinne tadelt er den „unvornehmen" „wissenschaftlichen Mensch[en]" dafür, dass er „seinesgleichen nöthig hat" und nach „Ehre und Anerkennung" strebt (Nietzsche, JGB 206). Für den freien Geist hingegen sind „Ehren, oder Geld, oder Ämtern" nur „Lockmittel der Abhängigkeit" (Nietzsche, JGB 44). Sofern dieser andere Menschen nötig hat, braucht er sie Nietzsche zufolge vor allem als Konkurrenten bzw. „Feinde" (Nietzsche, JGB 260), die er gerade als Feinde im Modus von „Ehrfurcht" würdigt (Nietzsche, GM I 10).

Dem freien Geist eignet also ein provisorisches Verhältnis zu Aktivitäten und Zielen sowie eine allenfalls gleichgültige Einstellung zu anderen Menschen. Diese zwei Züge, die Beauvoirs „Abenteurer" (als die oben identifizierten Hauptaspekte) zugrunde liegen, sind nicht unabhängig. Sofern es schwierig ist, Verachtung oder Gleichgültigkeit für Andere zu empfinden, mit denen man Ziele teilt, ergäbe sich eine solche Einstellung zu anderen Menschen viel leichter, wenn man jede Zielsetzung als eine nur vorübergehende ansieht.

Welche Art von Freiheit verkörpert der freie Geist? Er kann nicht, obwohl Nietzsche gelegentlich so redet,[11] Träger eines „freien Willens" sein, weil Nietzsche diesen konsequent und nachdrücklich ablehnt.[12] Weiterhin, obwohl er den Kausalitätsbegriff und die Vorstellung einer mechanistischen Welt ebenfalls ablehnt, hält Nietzsche an einer Art von Determinismus bzw. „Fatalismus" fest. „Unsere Bestimmung verfügt über uns, auch wenn wir sie noch nicht kennen" (Nietzsche, MA I Vorrede 7) – jeder Mensch ist ein „Stück Fatum" (Nietzsche, WS 61; GD Moral 6). Der freie Geist ist für Nietzsche also nicht in dem Sinne frei, dass er anders sein könnte als er ist. Aber was bedeutet dann „Selbstbestimmung" oder Freiheit von äußerer Bestimmung? Die Freiheit des freien Geistes, so möchte ich vorschlagen, muss anhand der zwei oben thematisierten Einstellungen – zu Aktivität und zu anderen Menschen – verstanden werden. Sie besteht in nichts anderem als der Fähigkeit, sich einerseits von bestehenden Denkgewohnheiten und -inhalten loszulösen, sowie sich andererseits als unabhängig von anderen Menschen zu erweisen. Damit ist diese Freiheit – nebenbei angemerkt – eine bloß

11 Z. B. Nietzsche, FW 347; MA I Vorrede 3; so wird auch das souveräne Individuum beschrieben in GM II 2.

12 Beispielsweise Nietzsche, MA I 39, 106; JGB 19; GM II 7; GD Irrthümer 7. – Für Mullin (2000, S. 388–391) gehört die Leugnung des freien Willens zum freien Geist als Vorbedingung.

negative: sie ist Freiheit, sich nicht oder nur mit Vorbehalt zu engagieren, Freiheit von ... statt Freiheit zu ...[13]

Mit Blick auf ihren jeweiligen Freiheitsbegriff sind sich Nietzsche und Beauvoir darin einig, dass Freiheit dem Einzelnen zukommt und dass individuelle Freiheit in vielen Fällen nicht verwirklicht wird. Aber sie erklären diese Tatsache auf unterschiedliche Weise: für Beauvoir handelt es sich um eine Unterlassung seitens des „uneigentlichen" Agens, der prinzipiell im vollen Besitz seiner Freiheit ist, auch wenn er sie aus „Faulheit" nicht einsetzt; für Nietzsche hingegen handelt es sich um eine von außen auferlegte Gebundenheit, von der sich der Einzelmensch lösen muss, um seine Freiheit erst zu erlangen. Beauvoir setzt sozusagen dort an, wo Nietzsche hinkommen will. Hinzu kommt, dass sich die Verwirklichung von Freiheit bei beiden in entgegengesetzte Richtungen vollzieht: Bei Beauvoir werden die zunächst vereinzelten Freiheitsinstanzen miteinander in einer Interessengemeinschaft harmonisiert; bei Nietzsche werden die zunächst gegebenen Schranken der gesellschaftlichen Konformität überwunden, um sich als Einzelner erst zu schaffen. Kurzum verwirklicht sich Freiheit für Beauvoir in einer Bewegung *zu den Anderen hin* und für Nietzsche *von den Anderen weg.*

3 Freiheit und die Anderen

Trotz dieser Unterschiede lässt sich Beauvoirs „Abenteurer" – wie wir gesehen haben – in seinen Hauptaspekten als Stellvertreter für Nietzsches „freien Geist" erkennen. Gerade weil diese Figur verlockend erscheinen kann, indem sie Unabhängigkeit sowie Überlegenheit als Erkennender verkörpert, stellt sich abschließend die Frage, inwiefern Beauvoir eine tragfähige Kritik derselben bietet. Unterliegt die Freiheit des freien Geistes tatsächlich den vier von Beauvoir angedeuteten Formen der Abhängigkeit von Anderen?

Am einfachsten verhält es sich mit der These, dass der freie Geist sich nur durch eine Abgrenzung gegenüber Anderen bestimmen lässt. Seiner Definition gemäß muss Nietzsches freier Geist *anders* sein und sich *als anders* verstehen. Beauvoirs Kritik weist zurecht nicht nur darauf hin, dass der Begriff des freien Geistes ohne den (verneinenden) Bezug zu anderen Menschen sinnlos wäre, sondern auch, dass diese Gegenüberstellung dem Selbstbild des freien Geistes zugrunde liegt. Entgegen Nietzsches Absicht erweist sich seine Freiheit dadurch als reaktiv und relativ statt als ursprüngliche Autonomie. Man könnte Beauvoir

13 Es stellt sich daher die Frage, auf die ich hier nicht eingehe, ob diese Freiheit die Ausbildung einer stabilen Identität erlaubt oder verhindert.

hinzufügen, dass dieses Selbstbild auf eine narzisstische Selbsttäuschung hinausläuft, weil es einseitig und unhaltbar annimmt, alle Anderen seien (als „Herdentiere“) auf Gleichheit zu reduzieren, während der freie Geist allein sich durch Unterschiede abhebt.[14]

Bei der Frage nach der faktisch-kausalen Abhängigkeit von Anderen empfiehlt es sich, vorab zu fragen, was der freie Geist überhaupt macht. In Bezug auf welche Projekte oder Tätigkeiten wird seine Freiheit konzipiert? *Prima facie* gibt es einen Unterscheid zwischen dem, was man mit bzw. ohne andere Menschen tun kann. So gibt es einerseits ‚individuelle‘ Praxen, an denen andere Menschen nicht *unmittelbar* teilnehmen müssen, wie beispielsweise bergwandern, ein Buch schreiben, malen, kochen usw. Bei seiner Beschreibung des freien Geistes scheint Nietzsche solche, von einer Einzelperson vollzogene Aktivitäten als Vorbild zu nehmen.[15] Andererseits gibt es ‚soziale‘ Praxen, deren Vollzug die unmittelbare Mitwirkung anderer Menschen einschließt, wie beispielsweise Fußball spielen, Handel oder Politik betreiben, reisen, einkaufen, eine Sprache sprechen, Bücher veröffentlichen, Gemälde ausstellen usw.

Dass es eine faktisch-kausale Abhängigkeit bei sozialen Tätigkeiten gibt, liegt auf der Hand, weil wir bei diesen definitionsgemäß mit Anderen im Rahmen einer geteilten Praxis interagieren. Aber auch wenn sie keinen *unmittelbaren* Umgang mit Anderen benötigen, weisen auch individuelle Praxen *hintergründig* eine faktische-kausale Abhängigkeit auf, sofern andere Menschen die Rahmenbedingungen, materiellen Voraussetzungen und oft den Sinn für solche Tätigkeiten schaffen. Gerade die verhältnismäßig einfache Leistung, (sich) anders zu denken, setzt eine Sprache oder Denkwelt – und mithin eine Sprachgemeinschaft – voraus.

In gewisser Hinsicht erkennt Nietzsche diese faktisch-kausale Abhängigkeit an. So betont er manchmal, dass Denken Muße statt Beschäftigung mit Lebensnotwendigkeiten voraussetzt (z. B. Nietzsche, MA I 283, 439, 462). Und, um diese Abhängigkeit zu vermeiden, empfiehlt er dem freien Geist an einer Stelle, seine „nothwendigen Bedürfnisse so viel wie möglich selber [zu] befriedigen“ (Nietzsche, WS 318). Vor allem aber kommt diese Anerkennung durch seine Thesen zum Ausdruck, dass die „Starken“ von den „Schwachen“ unterdrückt seien, oder dass Sklaverei „eine Bedingung jeder höheren Cultur“ sei (Nietzsche, JGB 239). Gleichwohl blendet sein Bild des Freigeistes diese Abhängigkeit aus und stellt

14 Heideggers (1993, S. 117–130) Begriff von uneigentlichem Dasein, das „zumeist und zunächst“ mit Anderen konform geht, vermeidet diesen Fehler.

15 Wie Daigle (2010–2011, S. 67) zu Recht feststellt: „[A] concept of intersubjective praxis seems to be absent from Nietzsche's philosophy“.

somit nicht in Rechnung, wie Gemeinsamkeiten bzw. Differenzen sowie unser Verhältnis zu Anderen durch Praxis konstituiert werden.

Beauvoirs weitere These, dass die Aufnahme bzw. Fortsetzung seiner Projekte durch Andere für die Freiheit des freien Geistes notwendig sei lässt sich aber nicht allgemein behaupten. Erstens übersieht Beauvoir bei ihrer Betonung von „Transzendenz", dass es durchaus Projekte gibt, die vollendet oder abgeschlossen werden und deren Fortsetzung daher entweder nicht notwendig oder nicht sinnvoll wäre. Beispielsweise wird auf individueller Ebene ein Buch oder Gemälde fertiggestellt, oder auf sozialer Ebene ein Fußballspiel beendet, ein Gebäude fertig gebaut usw. Solche Projekte würden vielleicht für Beauvoir an Wert verlieren, sofern sie nicht von Anderen fortgesetzt werden. Aber selbst unter dieser Annahme wäre zweitens damit nicht die *Freiheit* des freien Geistes betroffen, sondern sein Ruhm, Status oder Ruf, die alle von Nietzsche als „Lockmittel der Abhängigkeit" (Nietzsche, JGB 44) und damit Anzeichen von Schwäche getadelt wurden.[16]

Etwas komplexer verhält es sich bei der Frage, ob wir – wie Beauvoir meint – zur Verwirklichung unserer eigenen Freiheit eine Harmonisierung mit der Freiheit von Anderen und damit ihre Anerkennung benötigen. Diese Frage hat zwei Seiten. Muss man einerseits die Freiheit von Anderen respektieren, um die eigene Freiheit zu verwirklichen? Auf den ersten Blick könnte man meinen: nein, es steht uns frei, auf Andere herabzuschauen, den tyrannischen Chef oder zynischen Machtmenschen zu spielen, solange die Anderen machen, was wir brauchen. In diesem Sinne kritisiert Beauvoir die vom Abenteurer vollzogene Instrumentalisierung von Anderen, während Nietzsche die Notwendigkeit von Sklaven für ‚höhere Kultur' bejaht. Bedarf es andererseits zur Verwirklichung der eigenen Freiheit der Anerkennung von Anderen? Nietzsches Freiheitsbegriff, der im Verhältnis zu Aktivität sowie anderen Menschen liegt, legt wieder die Antwort „nein" nahe. Und für viele Aktivitäten, wo es auf individuelle Leistung ankommt, mag das auf den ersten Blick plausibel erscheinen: Ob man gut Tennis spielt oder Mathematik betreibt, hängt nicht unmittelbar von der Meinung Anderer ab.

Dieser erste Blick täuscht jedoch. Zum einen gibt es andere Aktivitäten, wo individuelle Exzellenz – z. B. als Schriftsteller, Musiker, Philosoph – durch die Anerkennung von Anderen mitkonstituiert wird. Zum anderen, auch in den Fällen, wo Leistungskriterien nicht unmittelbar von der Anerkennung durch Andere abhängen, verschränkt sich diese oft mit der faktisch-kausalen Abhängigkeit. Denn ohne die Anerkennung von Anderen bekommt man oft nicht die erforder-

16 Freilich ersehnte Nietzsche selbst Einfluss und Anerkennung. Insofern erfüllte er nicht konsequent die Anforderungen des Freigeistes.

lichen Gelegenheiten, seine „individuellen" Fähigkeiten zu entwickeln und sich von Anderen abzuheben. Ob es sich um Bildung, Übung, Zeit, finanzielle Unterstützung, Konkurrenz oder Betreuung handelt, haben auch solche Fähigkeiten Erfolgsbedingungen, zu denen man ohne die Anerkennung von Anderen keinen Zugang hätte. In solchen Fällen ergibt sich hervorragende ‚individuelle' Leistung mithin nur vor dem Hintergrund sozialer Ermöglichung.

In deskriptiver Absicht könnte man hinzufügen, dass *respektvolle Kooperation* oft dazu gehört. Vielleicht lässt sich auf theoretischer Ebene vorstellen, man könnte Andere für sich gewinnen, während man sie als Sklaven betrachtet oder instrumentalisiert. Aber in der wirklichen Welt verliert man schnell die Gunst oder Anerkennung von Anderen und damit ihre Unterstützung. Für viele Situationen und Projekte hat Beauvoir daher recht, dass eine Harmonisierung der eigenen Freiheit mit der Freiheit Anderer vonnöten ist. Nietzsche scheint diese Tatsache zu verkennen und orientiert sich in seiner Auffassung des freien Geistes ausschließlich an Situationen oder Praxen, in denen der Vollzug individueller Leistungen stärker als solche Kooperation in den Vordergrund hervortritt.

Man könnte dagegen einwenden, Nietzsche *verkennt* diese Tatsache nicht, weil sein Ziel normativ statt deskriptiv ist. Demnach will er nicht der bestehenden Wirklichkeit gerecht werden, sondern eine neue entwerfen. Freilich expliziert Nietzsche selbst nicht, und es lässt sich nicht ohne weiteres vorstellen, wie alltägliche Praxen sowie gesellschaftliches Leben ohne diese Harmonisierung einzurichten wären. Aber nimmt man an, eine solche Welt sei denkbar, so verdeutlicht Beauvoirs Einwand immerhin, was alles bei Nietzsches ‚anderer Welt' verloren ginge. Denn unsere bestehende Lebenswelt ist in dem Sinne liberal, dass unser Miteinandersein weitgehend von zwangloser Harmonisierung unserer Entscheidungen und Handlungen getragen wird. Auch wenn wir versuchen, Andere zu beeinflussen, erkennen wir sie – und sie uns – als Mithandelnde an, so dass gegenseitige Anerkennung durchgängig in die Struktur unserer lebensweltlichen Praxen eingewoben ist. Es ist schwer vorstellbar, was es heißen würde, diese durchgängige Liberalität aufzugeben, und was dabei gewonnen wäre. Wie Beauvoir aber erkennt, ist das die Alternative, die uns Nietzsche anscheinend vor Augen führen will.

Zusammenfassend also, auch wenn sie nicht in jeder Hinsicht überzeugt, macht Beauvoirs Kritik deutlich, dass Nietzsches Begriff des Freigeistes problematischer ist, als er auf den ersten Blick erscheinen mag. Zwar überzeugt ihre Behauptung nicht, dass alle Projekte von Anderen fortgesetzt werden müssen, um wertvoll zu sein, aber zurecht weist sie darauf hin, dass der freie Geist definitionsgemäß in seinem Sein wie in seinem Selbstverständnis von Anderen abhängt. Sie hat auch darin recht, dass Nietzsches Begriff des freien Geistes weder unserer faktisch-kausalen Abhängigkeit von Anderen noch der Rolle ihrer Anerkennung

Rechnung trägt. Man könnte zwar die Ansicht vertreten, dass Nietzsche diese zwei Faktoren nicht deskriptiv übersieht, sondern normativ umdeutet oder leugnet. Aber selbst wenn man Nietzsche im strengen Wortsinne auf diese Weise verteidigte, überzeugt Beauvoirs Argument, dass eine freie Harmonisierung mit Anderen vonnöten ist, sofern die Instrumentalisierung und Unterdrückung von Anderen nicht als Grundlage unseres Miteinanderseins dienen sollte.

Bibliographie

Beauvoir, Simone de (1976): *Le deuxième sexe.* Paris.

Beauvoir, Simone de (2017): *Pour une morale de l'ambiguïté.* Paris.

Daigle, Christine (2010–2011): „A Nietzschean Beauvoir? Becoming Who One Is As a Being With Others". In: *Simone de Beauvoir Studies* 27, S. 61–71.

Heidegger, Martin (1993): *Sein und Zeit.* Tübingen.

Mullin, Amy (2000): „Nietzsche's Free Spirit". In: *Journal of the History of Philosophy* 38, S. 383–405.

Parker, Emily Anne (2014): „Strange Freedom in Nietzsche and Beauvoir". In: *Trans-Humanities* 7, Nr. 1, S. 47–69.

Reginster, Bernard (2003): „What is a Free Spirit? Nietzsche on Fanaticism". In: *Archiv für die Geschichte der Philosophie* 85, Nr. 1, S. 51–85.

Rorty, Richard (1989): *Contingency, Irony, and Solidarity.* Cambridge.

Sartre, Jean-Paul (2006): *L'être et le néant.* Paris.

Thorgeirsdottir, Sigridur (2012): „Die Philosophie Nietzsches im Spiegel von Philosophinnen im 20. Jahrhundert". In: Renate Reschke und Marco Brusotti (Hrsg.): *„Einige werden postum geboren". Nietzsches Wirkungen.* Berlin, S. 97–115.

Tirrell, Lynne (1994): „Sexual Dualism and Women's Self-Creation". In: Peter Burgard (Hrsg.): *Nietzsche and the Feminine.* Charlottesville, S. 158–182.

Hans-Martin Schönherr-Mann

Das über sich Hinausgehen. Aber wohin? Nietzsches Wiederkünftige und der Übermensch als sich Transzendierende im Sinn von Sartre und de Beauvoir?

Abstract: Transcending Oneself. But To Where? Nietzsche's Returning and the *Übermensch* as Self-Transcending in the Sense of Sartre and Beauvoir? Sartre's and Nietzsche's philosophies are completely contrary. Nietzsche advocates a radical elitism that criticizes the contemporary elites for presenting themselves too indulgent to their subjects. Sartre develops a philosophy of resistance that anyone can participate in. Nietzsche's elites are supposed to create new elitist values. Sartre suggests that individuals shape their lives according to their own ideas. Accordingly, with Nietzsche, only the elites are responsible, in Sartre everyone is responsible for his or her own life and commitments.

„Niemals waren wir freier als unter der deutschen Besatzung", schreibt Sartre am 9. September 1944 in den *Lettres françaises* (Sartre 1980: RépSil, S. 37). Hatte der Mann den Verstand verloren? Am 26. Dezember 1888 schreibt Nietzsche an Franz Overbeck: „Ich habe nicht eher die Hände frei, bevor ich nicht den jungen Kaiser, sammt Zubehör in den Händen habe." (Nietzsche, Bf. an Franz Overbeck 26.12.1888, KGB III/5, Nr. 1212, S. 551) Das interpretiert Overbeck als Wirrnis, die eine Woche später nicht mehr zu leugnen sein wird. Lässt sich zwischen beiden Äußerungen mehr als eine scheinbare gemeinsame Wirrnis feststellen?

1 Wirrnis oder Größenwahn

Nun, Sartres Wirrnis lässt sich als Provokation entbergen, während sich die Sachlage bei Nietzsche komplexer darstellt. Um zu erklären, dass die Franzosen unter deutscher Besatzung frei waren, dazu braucht man nicht auf Sartres Philosophie zurückzugreifen. Denn mit der Besatzung gab es keine moralischen Autoritäten mehr. Vichy kollaborierte mit den Deutschen, der Papst auch, der mit diesen ein Konkordat geschlossen hatte. Niemand war gezwungen Widerstand zu leisten. In den ersten Jahren der Besatzung war der Widerstand nicht straff or-

https://doi.org/10.1515/9783110760163-012

ganisiert, so dass sich jeder selber ausdenken konnte, was er Widerständiges tun will.

Bei Nietzsche lässt sich die Wirrnis seiner Äußerung gegenüber Overbeck schwieriger entwirren. Wenn man die Briefe im November, Dezember durchblättert, findet man solchen Größenwahn an vielen Stellen. Entweder interpretiert man das dann als selbstironisch oder als sich ankündigende Verwirrnis. Solche Sätze ernst zu nehmen, erscheint absurd.

Ich möchte das aber versuchen, und fokussiere mich aber auf die Formulierung ‚den jungen Kaiser in den Händen zu haben'. Warum sollte man diesen Satz wie vergleichbare Äußerungen nicht ernst nehmen, wenn man die letzten Werke rezipiert, die von ihm teilweise noch unmittelbar vor seiner Abreise aus Turin bearbeitet wurden?

Es gibt wenig bezeugte Daten, wann Nietzsche zunehmend in die Verwirrung geriet. Am 3. Januar kann die Pferdegeschichte nicht stattgefunden haben, sondern nach dem Zeugnis der Familie Fino erheblich früher und führte womöglich dazu, dass die Finos einen berühmten Psychiater bestellten. Wann er das erste Mal kam, ist auch nicht bekannt. Aber Nietzsche hütete wahrscheinlich auf sein Geheiß hin bereits am 18. Dezember das Haus, als dort ein Leichenzug vorüberzog und er sich für den Toten hielt – das bezeugten die Finos – was sich nicht mehr mit übersteigertem Größenwahn erklären lässt, sondern mit Verwirrnis.

Größenwahn gehört durchaus zu häufigen Eigenschaften von Intellektuellen: Sartre und de Beauvoir beschließen am Ende ihres Studiums, die philosophische Welt neu zu erfinden. Rousseaus *Les Rêveries du Promeneur Solitaire* formulieren ähnliche Selbstbeweihräucherungen wie Nietzsches *Ecce Homo:* „So wurde durch einhelligen Beschluss, der geselligste und leutseligste Mensch von allen geächtet. [...] Ich hätte die Menschen geliebt, trotz allem. Meiner Zuneigung konnten sie sich nur entziehen, indem sie: keine mehr sind." (Rousseau 2012, S. 50) Auch in den Briefen der letzten Monate beschimpft Nietzsche die Deutschen, die ihn nicht zur Genüge würdigen.

Den Wunsch, ‚den jungen Kaiser in den Händen zu haben', formuliert er damit im Zustand fortschreitenden Größenwahns und geistigen Verfalls. Das lässt sich nicht leugnen. Das dreizehnte von neunzehn letzten Fragmenten unter dem Titel „Todkrieg dem Hause Hohenzollern" formuliert dasselbe noch schärfer: „Ich werde nicht eher die Hände frei bekommen, als bis ich den christlichen Husaren von Kaiser, diesen jungen Verbrecher s a m m t Zubehör in den Händen habe – mit Vernichtung der erbarmungswürdigsten Mißgeburt von Mensch, die bisher zur Macht gelangt ist." (Nietzsche, NL 1888, KSA 13, S. 643) Der Brief an Overbeck ist also abgemildert. Der Kontrollverlust kennt noch kommunikative Grenzen.

Am 31.12.1888 schreibt er an Strindberg sogar: „ich will den jungen Kaiser füsillieren lassen." (Nietzsche, Bf. an August Strindberg, 31.12.1888, NB 8,

Nr. 1229) Die letzten Briefe durchziehen weitere Fantasien der Größe. Am 26.11. schreibt er an Deußen: „Mein Leben kommt jetzt auf seine Höhe: noch ein paar Jahre, und die Erde zittert von einem ungeheuren Blitzschlage. – Ich schwöre Dir zu, daß ich die Kraft habe, die Zeitrechnung zu verändern. – Es giebt Nichts, das heute steht, was nicht umfällt, ich bin mehr Dynamit als Mensch." (Nietzsche, Bf. an Paul Deussen, 26.11.1888, NB 8, Nr. 1159)

Der Größenwahn macht auch nicht vor seinem Werk halt. Nietzsche, der den vierten Teil des *Zarathustras* selber drucken lassen musste, träumt von Millionenauflagen und Übersetzungen durch renommierte Intellektuelle. In einem Entwurf eines Briefes an Brandes heißt es Anfang Dezember: „Ich habe Übersetzungen in alle europäischen Hauptsprachen nöthig: wenn das Werk erst heraus soll, so rechne ich eine Million Exemplare in jeder Sprache als erste Auflage." (Nietzsche, Bf. an Georg Brandes, Anfang Dezember 1888, NB 8, Nr. 1170) Fast verwundert es nicht, dass er in seinem letzten Brief vom 3. Januar an Meta von Salis-Marschlins erklärt: „Gott ist auf der Erde. Sehen Sie nicht, wie alle Himmel sich freuen? Ich habe eben Besitz ergriffen von meinem Reich" (Nietzsche, Bf. an Meta von Salis, 3.1.1889, NB 8, Nr. 1239). Am Wahn lässt sich bei allen diesen Stellen nicht zweifeln und die Verwirrnis macht sich mindestens seit Anfang November bemerkbar.

Auch Sartre hat eine wahnsinnige Ader, die sich von den dreißiger Jahren sogar bis in die späten Vierziger zieht. In seinem ersten Roman *Der Ekel* aus dem Jahr 1938 widert den Helden Roquentin seine Umwelt, die bürgerliche Gesellschaft einfach an. In seiner Erzählung *Herostrat* steigert sich Sartres Ablehnung der Gesellschaft, indem er einen Amokläufer schildert, den er freilich nicht allzu viel Schaden anrichten lässt. Eine ähnliche private Rache nimmt Delarue im Roman *Der Pfahl im Fleische*. Denn Frankreich hat den Krieg bereits verloren, doch „er schoß auf den schönen Offizier, auf alle Schönheit dieser Erde, auf die Straße, auf die Blumen, auf die Gärten, auf alles, was er geliebt hatte. Die Schönheit machte einen obszönen Kopfsprung, und Mathieu schoß immer noch. Er schoß: er war rein, er war allmächtig, er war frei." (Sartre 1987: MorAme, S. 220) Als Sartre das publiziert, war er etwa so alt wie Nietzsche, als dieser den Kaiser in der Hand halten wollte.

2 Neue Werte schaffen oder sich transzendieren

Beide lehnen ihre Umwelt aus teilweise ähnlichen Gründen ab. Beide verachten den Untertan, Nietzsches letzten Menschen, wie den ‚Obertan', diesen aber aus gegensätzlichen Motiven. Nietzsche verachtet den Untertan, weil dieser sich mit seiner Rolle nicht zufriedengibt, Sartre, weil der Untertan sich nicht dagegen auflehnt. Den ‚Obertan' verachtet Nietzsche, weil er den Untertan nicht einfach unterwirft, sondern einen sozialen Ausgleich sucht, und Sartre, weil der ‚Obertan' ein Unterdrücker ist. Nietzsche entwickelt die Philosophie einer Elite, die die zeitgenössischen Eliten beseitigen und die Welt nach Nietzsches Vorstellungen gestalten soll. Sartre entwickelt die Philosophie des Widerstands jedes einzelnen gegen jegliche Elite, nicht nur die deutsche Terrorherrschaft.

Was haben sie dann miteinander noch gemeinsam außer dem Ekel und dem Dünkel? Nietzsche fordert dazu auf, die alten Werte des Christentums hinter sich zu lassen und neue Werte zu erfinden, die sich an der Erde orientieren und die das Leben feiern. Sartre attestiert dem Menschen ein Bewusstsein, das in der Lage ist, die eigene Situation zu reflektieren und sie zu verändern.

Was beide hier verbindet, ist nicht die Ausgangslage, sondern gewisse Perspektiven, deren Parallelität sich freilich ebenfalls verliert. Die Ausgangslage unterscheidet sich insoweit, als Nietzsche eher der Empiriker, Sartre der Theoretiker bleibt. Nietzsche hat im Anschluss an Schopenhauer vielerlei psychologische Einsichten, mit denen er manches traditionelle wie wissenschaftliche Verständnis brillant konterkariert. Aber Nietzsche beschreibt damit keine allgemeinen Fähigkeiten, sondern bietet spannende Einblicke in wissenschaftliche Probleme, wie in solche des täglichen Lebens.

Sartre bringt im Anschluss an Husserl das Bewusstsein auf einen theoretischen Begriff. Das Bewusstsein ist nie bei sich selbst, sondern immer bei etwas anderem, so dass es die Fähigkeit der Negation entwickelt. Daraus ergibt sich die Möglichkeit, die Faktizität der Existenz zu durchdenken, um sie dadurch transzendieren zu können. Diese Bewusstseinsstruktur eignet jedem Menschen.

Mit Hilfe seiner Kritik am Christentum wie an der Aufklärung, also mit einer Art Zerstörung, ebnet Nietzsche den Weg zu einer Neu-Erfindung der Welt, zu der indes nur gewisse Zeitgenossen in der Lage sein sollen, keinesfalls alle. Er führt vor, dass die herrschenden christlichen Werte keine mehr sind, sich jedenfalls heute verflüchtigen, weil die Zeitgenossen selbst ihnen nicht mehr huldigen. Wenn es keine gemeinsamen Werte mehr gibt, muss man neue erfinden.

Dagegen entdeckt oder erfindet Sartre mit der Transzendenz eine Fähigkeit des Menschen, d. h. jedes Menschen, sich selbst aus seiner Situation zu befreien, sich zu emanzipieren, was Simone de Beauvoir 1949 auf die Frauen überträgt und damit die spätere Emanzipation antizipiert. Frau muss nicht den traditionellen Lebenswegen folgen bzw. entsprechende Rollen ausfüllen. Also sowohl Nietzsche als auch Sartre und Beauvoir entwerfen andere Perspektiven des Lebens als die traditionellen. Beauvoir und Sartre entwickeln sie parallel zur theoretischen Einsicht aus der jeweiligen Situation heraus, in der sich Menschen befinden.

Bei Nietzsche stellt sich freilich die Frage, woher denn die Fähigkeit dazu kommen soll, neue Werte zu erfinden. Keinesfalls haben wie bei Sartre alle

Menschen diese Fähigkeit, eigentlich nur jene, die sich die Lehren Nietzsches aneignen, also die Verkünder der ewigen Wiederkunft, die er ja zu einem Bund vereinigen möchte. Er schreibt 1884: „Ich will einen neuen Stand schaffen: einen Ordensbund höherer Menschen [...].“ (Nietzsche, NL 1884, KSA 11, S. 195) Was aber zeichnet diese höheren Menschen aus? Was macht sie zu höheren Menschen?

Nietzsche gibt nur vage an, was sie tun und was sie für wahr halten müssen, z. B. die Lehre von der ewigen Wiederkehr des Gleichen und andererseits sollen sie dem Übermenschen den Weg bereiten, ihn also verkündigen. Ihre Triebfeder muss der Wille zur Macht sein.

Nur ist das die metaphysische Grundlage Nietzsches. Man kann sie positiv interpretieren als Mangel jeglicher Grundlage. Aber dann fragt sich, wie der Wille zur Macht Triebfeder sein kann. Die Lehren von Wiederkunft und Übermensch sind zwar phantastische Szenarien. Aber dass das so ist bzw. sein wird, dafür gibt es nicht allzu viele Argumente, handelt es sich eher um Mutmaßungen.

An die Stelle dieser Erzählungen in Nietzsches Lehren treten bei Sartre Strukturen, vor allem jene, dass die Existenz der Essenz vorausgeht. Wenn auch die Metaphysik das anders sieht, theoretisch wie empirisch lässt sich schwerlich behaupten, dass ein Mensch immer schon eine Essenz hätte, die ihn prägt. Dazu muss er überhaupt erst einmal existieren und erwachsen werden. Dazu sind keine bestimmten Lehren notwendig, wie es sich Nietzsche vorstellt. Hier ist Sartre moderner, während sich Nietzsche in die Tradition philosophisch essentieller Metaphysiken einreiht.

3 Genie oder Autor

Aber warum verkünden seine Jünger seine Lehren? Was befähigt sie dazu? Die Vernunft wäre sokratisch und aus dieser Tradition will Nietzsche aussteigen. Der Wille zur Macht, der Egoismus, den Nietzsche schätzt, wären indes z. B. vom *homo oeconomicus* kaum zu unterscheiden. Jedoch findet sich bei ihm noch eine andere Lehre, die den Willen zur Macht ergänzt, nämlich der Genius, der Nietzsches Anfänge nicht nur mit seinen letzten Bemerkungen verknüpft, der vielmehr wie ein roter Faden sein Werk durchzieht.

Bereits in *Die Geburt der Tragödie* entwickelt Nietzsche den Gedanken, dass der künstlerische Genius herausragende Werke schafft, die die antike Kultur begründen, indem sie Götter wie Menschen herausfordern. So schreibt er über Aischylos und Sophokles: „Der titanische Künstler fand in sich den trotzigen Glauben, Menschen schaffen und olympische Götter wenigstens vernichten zu können: und dies durch seine höhere Weisheit, die er freilich durch ewiges Leiden zu büssen gezwungen war. Das herrliche ‚Können‘ des grossen Genius [...], der herbe Stolz des Künstlers." (Nietzsche, GT, KSA 1, S. 68) Die Frage aber, woher das Genie kommt, was das Genie zum Genie macht, bleibt offen.

Bei Sartre spielt die Kunst, genauer die Literatur ebenfalls eine wichtige Rolle, freilich eine kommunikative zwischen Autor und Leser. Der Autor predigt nicht, sondern entwickelt zum Leser eine kommunikative Beziehung. Literatur sollte engagiert und kritisch sein und den Leser zum Denken anregen, was auch in umgekehrter Richtung passiert. Denn Sartre weiß, dass die zwischenmenschliche Beziehung eine rückkoppelnde Rolle spielt, ohne die das Engagement des Autors verhallt. Genialität allein würde nicht helfen, spielt im Existentialismus keine Rolle.

Nietzsche dagegen reagiert gereizt darauf, dass er nicht genügend rezipiert wird. Und er verlangt blinde Gefolgschaft, man denke an seine Aversion gegenüber Wagner, die er von Freunden fordert. Er bildet sich ein, durch Schriften im predigenden und aggressiven Tonfall den Erfolg erzwingen zu können. Ironischer Weise ist ihm das posthum über die Maßen gelungen.

Sartre konnte seinen Erfolg noch zu Lebzeiten genießen und den Nobelpreis ablehnen, was man sich von Nietzsche kaum vorstellen kann. Beauvoir avanciert zur Grande Dame der zweiten Frauenbewegung seit den siebziger Jahren, an der sie selbst teilnimmt. Vielleicht darf man sie als die wirkungsmächtigste Person in der Philosophie des 20. Jahrhunderts bezeichnen. Insofern muss man sie nicht nur neben Nietzsche stellen, mehr noch als Sartre, sondern auch neben Marx.

Zudem kritisiert Nietzsche bereits 1872 in seinen Basler Bildungsvorträgen die Bemühungen um eine allgemeine Schulbildung, die längst nicht realisiert war. Er beklagt:

> Hier wird jeder ohne Weiteres als ein litteraturfähiges Wesen betrachtet, das über die ernstesten Dinge und Personen eigne Meinungen haben dürfte, während eine rechte Erziehung gerade nur darauf hin mit allem Eifer streben wird, den lächerlichen Anspruch auf Selbständigkeit des Urtheils zu unterdrücken und dem jungen Menschen an einen strengen Gehorsam unter dem Scepter des Genius zu gewöhnen. (Nietzsche 1999, KSA 1, S. 680)

Gegensätzlicher kann es zwischen Nietzsche und dem französischen Existentialismus kaum aussehen. Überhaupt behindert für Nietzsche der zeitgenössische Staat mit seiner Schulstruktur, dass sich das Genie entfalten kann. Dazu trägt vor allem die Massenbildung bei, weil sie die dumme Masse und nicht das Genie fördert. Nietzsche fragt:

> Wozu diese auf die Breite gegründete Volksbildung und Volksaufklärung? Weil [...] man die aristokratische Natur der wahren Bildung fürchtet, weil man die großen Einzelnen dadurch zur Selbstverbannung treiben will, daß man bei den Vielen die Bildungsprätension pflanzt und nährt, weil man der strengen und harten Zucht der großen Führer damit zu entlaufen

> sucht, daß man der Masse einredet, sie werde schon selbst den Weg finden – unter dem Leitstern des Staates! (Nietzsche 1999, KSA 1, S. 710)

Letzteres postulieren Beauvoir und Sartre, freilich jenseits des Staates. An den Universitäten verhindert nach Nietzsche die wissenschaftliche Spezialisierung, die auch einem weniger erleuchteten Geist eine Hochschultätigkeit erlaubt, dass sich dort allein die wahren Genies tummeln, die doch als einzige an die Universität gehören. Denn die Universität hat für Nietzsche folgenden Sinn:

> Diese Einzelnen sollen ihr Werk vollenden, das ist der Sinn ihrer gemeinschaftlichen Institution – und zwar ein Werk, das gleichsam von den Spuren des Subjekts gereinigt und über das Wechselspiel der Zeiten hinausgetragen sein soll, als lautere Wiederspiegelung des ewigen und unveränderlichen Wesens der Dinge. Und alle, die an jenem Institute Theil haben, sollen auch mit bemüht sein, durch eine solche Reinigung vom Subjekt, die Geburt des Genius und die Erzeugung seines Werkes vorzubereiten. (Nietzsche 1999, KSA 1, S. 729)

Immerhin versucht er anzugeben, woraus sich das Genie speist, das sich vom wissenschaftlichen Geist abkehrt. Es handelt sich um das Ewige, was indes nicht für Klarheit, sondern Dunkelheit sorgt.

Unter dem Einfluss von Jacob Burckhardt, der die Idee eines politischen Genius entwirft, überträgt Nietzsche seine Genie-Vorstellung von der Kunst auf die Politik, was sich zunächst in den besagten Basler Bildungsvorträgen zeigt. Er kritisiert den wissenschaftlichen Geist des 19. Jahrhunderts, der sowohl an den Universitäten als auch an den Schulen vorherrscht. Dem stellt er einen ‚echten deutschen Geist' entgegen, der aus Reformation, Musik und Philosophie bestehen soll und der das Genie zum Genie zu machen scheint. Doch genauer erklärt ihn Nietzsche nicht und somit stellte sich weiterhin die Frage, wie das Genie zu ihm gelangt, um Genie zu werden. Auch der Geniebegriff in der Ästhetik des 18. Jahrhunderts füllt diese Leerstelle nur scheinbar, weil man eben nicht angeben kann, was das Genie befähigt, ein Kunstwerk zu schaffen, außer dass es ein Genie sein soll.

Außerdem verbreitet sich zu seinen Zeiten mit dem Journalismus ein aktualitätsheischender Trend, der „an die Stelle des großen Genius, des Führers für alle Zeiten, des Erlösers vom Augenblick getreten ist." (Nietzsche 1999, KSA 1, S. 671) Wenn sich doch das von der Ewigkeit umwehte Genie durchsetzt, dann passiert folgendes: „Es ist als ob dieses Genie in blitzartiger Seelenwanderung in alle diese halben Thierleiber gefahren sei und als ob jetzt aus ihnen allen wiederum nur das eine dämonische Auge herausschaue." (Nietzsche 1999, KSA 1, S. 751) Die Geführten stehen nur im Dienste des Führers, sieht letzterer indes das ewige Wesen der Dinge. Dann ergäbe sich „eine prästabilirte Harmonie zwischen Führer und Geführtem" (Nietzsche 1999, KSA 1, S. 752) somit eine festgefügte, weil von der

Ewigkeit her beseelte Ordnung des Lebens. Nur was diese Ewigkeit bedeutet, danach fragt Nietzsche nicht und damit verbleiben seine Genies im luftleeren Raum.

4 Unterwerfung oder Widerständigkeit

Damit lässt sich seine nur scheinbar paradoxe These erklären, warum er eine neue, eine andere Elite der Wiederkünftigen fordert, die herrschenden Eliten im kaiserlichen Deutschland aber ablehnt, seien es die politischen oder die wissenschaftlichen. Denn die herrschenden Eliten stützen sich auf die an den Universitäten sich verbreitende, zunehmend positivistischer werdende Bildung. Darauf darf sich Nietzsches Elite nicht berufen. Und mit profanen Tätigkeiten sollen sie sich auch nicht beschäftigen, wie Nietzsche knapp zehn Jahre später in der *Morgenröte* schreibt:

> Alle politischen und wirthschaftlichen Verhältnisse sind es nicht werth, dass gerade die begabtesten Geister sich mit ihnen befassen dürften und müssen: ein solcher Verbrauch des Geistes ist im Grunde schlimmer als ein Nothstand. Es sind und bleiben Gebiete der Arbeit für die geringeren Köpfe, und andere als die geringen Köpfe sollten dieser Werkstätte nicht zu Diensten stehen: möge lieber die Maschine wieder einmal in Stücke gehen! (Nietzsche, M, KSA 3, S. 157)

Damit deutet sich eine Antwort auf die naheliegende Frage an, warum er ‚den jungen Kaiser in der Hand haben' will, anstatt ihn zu beraten und derart zu lenken. Denn das war schon seit Jahrhunderten die politische Aufgabe von Gelehrten, unter anderem von Leibniz. Und Max Weber, der in vieler Hinsicht an Nietzsche anschließt, wird der Wissenschaft just die Funktion der Politikberatung zuschreiben. Leibniz wollte den Fürsten lenken, natürlich im rationalen Geist der Aufklärung. Er glaubte auch, dass sich mit der Macht des absoluten Fürsten seine Vorstellungen einfach durchsetzen lassen. Wie heißt es doch an einer berühmten Stelle in Schillers *Don Carlos:* „Gehen Sie Europens Königen voran./ Ein Federzug von dieser Hand, und neu/ Erschaffen wird die Erde. Geben Sie/ Gedankenfreiheit!" (Schiller 1976, S. 445) Ähnlich träumt Nietzsche vom Federzug, dass er mit einer absoluten Macht seine Ideen umsetzen kann, wenn er erst ‚den jungen Kaiser in der Hand' hätte.

Schwerlich hätte er sich von Leibniz' Enttäuschungen warnen lassen. Denn wie schreibt der Leibniz-Biograph Hirsch:

> Überglücklich ging Leibniz hinaus, gewiss auch voller Hoffnungen, weil ihm das gegebene Wort eines Herrschers immer auch das letzte Wort zu sein schien. Ein Befehl, ein Federstrich

würde genügen ...! So pflegte er zu sagen, es war eine feste Vorstellung bei ihm. Dass es meist ganz anders kam ..., konnte der Gelehrte nie begreifen. (Hirsch 2016, S. 411)

Letztlich musste die Bürokratie die Weisungen umsetzen. Wenn dazu kein Geld zur Verfügung stand und niemand da war, der das durchgesetzt hätte, nützt der königliche Federstrich nicht.

Also warum träumt Nietzsche nicht von Politikberatung? Seine politischen Thesen werden in der Schrift *Der griechische Staat* quasi untermauert. Hier formuliert Nietzsche seine politische Grundunterscheidung, dass Staat und Gesellschaft auf der Trennung von Oben und Unten beruhen – viel mehr hat er nicht zu bieten, um dem Genius auf seine genialen Sprünge zu verhelfen. Daher kann er nur mit dem ‚militärischen Genius' aufwarten, dem er die zentrale Rolle zuschreibt, um hierarchische Strukturen und absoluten Gehorsam durchzusetzen, was die Oberen von den Unteren scheiden soll. Er schreibt:

Wer den Krieg und seine uniformirte Möglichkeit, den Soldatenstand, in Bezug auf das bisher geschilderte Wesen des Staates betrachtet, muß zu der Einsicht kommen, daß durch den Krieg und im Soldatenstande uns ein Abbild, oder gar vielleicht das Urbild des Staates vor Augen gestellt wird. Hier sehen wir, als allgemeinste Wirkung der Kriegstendenz, eine sofortige Scheidung und Zertheilung der chaotischen Masse in militärische Kasten, aus denen sich pyramidenförmig auf einer allerbreitesten sklavenartigen untersten Schicht, der Bau der „kriegerischen Gesellschaft" erhebt. (Nietzsche, NL 1871, KSA 1, S. 775)

Und noch in seiner späten Schrift *Zur Genealogie der Moral* preist er dementsprechend den Krieg:

Eine Rechtsordnung souverän und allgemein gedacht, nicht als Mittel im Kampf von Macht-Komplexen, sondern als Mittel gegen allen Kampf überhaupt, [...] wäre ein lebensfeindliches Prinzip, eine Zerstörerin und Auflöserin des Menschen [...]. (Nietzsche, GM, KSA 5, S. 313)

Von dieser Grundlage ausgehend fasst er sein Verständnis vom Menschen noch präziser als in den Bildungsvorträgen. Es gibt weder naturgegebene Menschenrechte noch eine derartige Würde des einzelnen. Vielmehr ist der Mensch von Natur aus Untertan und nur als Untertan erhält er einen Wert, der nicht in ihm selbst liegt. Nietzsche schreibt:

Jeder Mensch, mit seiner gesammten Thätigkeit, hat nur soviel Würde, als er, bewußt oder unbewußt, Werkzeug des Genius ist; woraus sofort die ethische Consequenz zu erschließen ist, daß der „Mensch an sich", der absolute Mensch, weder Würde, noch Rechte, noch

> Pflichten besitzt: nur als völlig determinirtes, unbewußten Zwecken dienendes Wesen kann der Mensch seine Existenz entschuldigen. (Nietzsche, CV 3, KSA 1, S. 776)

In der Antike, bevor Sokrates den Rationalismus in die Welt setzte und damit das dionysisch geprägte Leben kanalisierte, sollen die Herrschenden die Masse einfach unterworfen haben, so die nietzschesche sicherlich falsche These, die wahrscheinlich nicht mal für das mythologische Zeitalter galt, wiewohl die Gefährten des Odysseus namenlos bleiben. Denn niemals konnten die Herrschenden auf die Kooperation der Masse verzichten. Und Nietzsche selbst hofft am Ende auf Millionen Leser.

Sartre hat dazu eine mögliche Begründung geliefert. Aus der Fähigkeit des Bewusstseins zur Transzendenz, aus dem Fürsich, das das Ansichsein überschreitet, indem es dieses reflektiert, ergibt sich für Sartre die Fähigkeit zum Widerstand, zu dem jeder auch unter dem grausamsten Tyrannen fähig ist – Sartres Philosophie der Résistance. Zuvor entwirft Camus in *Der Mythos des Sisyphos* sein berühmtes ‚Trotzdem', indem er die Auflehnung im Angesicht ihrer Aussichtslosigkeit propagiert.

Nach Nietzsche muss eine solche Widerständigkeit der Unteren, die für ihn keine natürliche Anlage sein kann, von den Eliten unbedingt gebrochen werden, lässt sich Nietzsche von der Arbeiterbewegung offenbar in Angst und Schrecken versetzen. So beobachtet er mit Argwohn, dass sich die preußisch deutschen Eliten darum bemühen, die Arbeiterklasse in die Gesellschaft zu integrieren, schreibt er im *Zarathustra:* „Und den Herrschenden wandt' ich den Rücken, als ich sah, was sie jetzt Herrschen nennen: schachern und markten um Macht – mit dem Gesindel!" (Nietzsche, Za II, KSA 4, S. 125) Je mehr er dieses festzustellen glaubt, umso mehr wird er diese Eliten hassen, wie den Staat, den sie prägen und zwar durch dessen Bürokratien, die sich der lebensfeindlichen Rationalität der Wissenschaften wie des Rechts bedienen. Denn Leben heißt Herrschen und Beherrschtwerden: „Die Menschen-Gesellschaft: die ist ein Versuch, so lehre ich's, – ein langes Suchen: sie sucht aber den Befehlenden!" (Nietzsche, Za III, KSA 4, S. 265) In diesen Zusammenhang gehört auch jener Satz, den man gerne individualistisch interpretiert: „Dort, wo der Staat aufhört, da beginnt erst der Mensch, der nicht überflüssig ist" (Nietzsche, Za I, KSA 4, S. 63). Das nicht überflüssige Individuum sind nicht die vielen, sondern die wenigen genialen Wegbereiter des Übermenschen. Also macht Beratung Wilhelms II. für Nietzsche keinen Sinn, bleibt somit nur seine Gefangennahme. Das ist natürlich Irrsinn, entspricht aber seinem Verständnis des Genius. Lange vor seinem Wahn will er eine Elite schaffen, in der sich die Genien zusammenfinden.

So schreibt er in der ersten *Unzeitgemäßen Betrachtung* den preußischen Sieg über Frankreich im gerade beendeten Krieg nicht einer deutschen Kultur zu, für die er nur Verachtung übrighat, sondern allein dem militärischen Geist:

> Strenge Kriegszucht, natürliche Tapferkeit und Ausdauer, Ueberlegenheit der Führer, Einheit und Gehorsam unter den Geführten, kurz Elemente, die nicht mit der Kultur zu thun haben, verhalfen uns zum Siege über Gegner, denen die wichtigsten dieser Elemente fehlten. (Nietzsche, DS, KSA 1, S. 160)

Aber Nietzsches Kritik an der zeitgenössischen Gesellschaft beschränkt sich keineswegs auf die Eliten. Wie fünfzehn Jahre später ertönt schon eine bittere Klage über die Deutschen im Allgemeinen. Überraschend nur, dass er plötzlich den Mangel gar nicht mehr bei den Führern, sondern bei den Untertanen diagnostiziert: „Denn an den einsichtigsten und kühnsten Führern und Feldherrn hat es den Deutschen nie gemangelt – nur dass diesen oftmals die Deutschen fehlten." (Nietzsche, DS, KSA 1, S. 161) Wahrscheinlich meint er jene Untertanen, die politische und soziale Ansprüche stellen. Es ist klar, dass die Untertanen das nicht dürfen, diese mit aller Gewalt unterdrückt werden müssen.

Für Sartre entsteht die soziale Frage erst nach dem zweiten Weltkrieg, je mehr er sich der Kommunistischen Partei Frankreichs annähert. Maurice Merleau-Ponty stellt die soziale Frage bereits 1945 in seinem frühen Hauptwerk *Phänomenologie der Wahrnehmung* und rückt damit den französischen Existentialismus nach links: die Eliten werden als konservativ und kapitalistisch kritisiert – Nietzsche in gewisser inhaltlicher Weise nicht mal so fern, aber mit gegensätzlicher Intention. Die Klassenunterschiede sollen nicht wie für Nietzsche unauflöslich festgezurrt, sondern für die französischen Existentialisten um Sartre abgebaut werden. Es geht dieser Gruppe um diverse Emanzipationsprozesse speziell der Arbeiter und der Frauen. Beides ist Nietzsches Sache gar nicht.

5 Wer trägt Verantwortung?

Nietzsche kritisiert aber nicht nur die politischen Vorstellungen der Aufklärung, sondern auch deren wissenschaftliche, was in eine fundamentale Ablehnung des Rationalismus der Aufklärung und damit dessen, was man im 20. Jahrhundert das naturwissenschaftlich technische Weltbild nennen wird. Nietzsches Kritik ist so treffend wie nachhaltig. Er weist den Naturwissenschaften nach, dass sie die Natur nicht erfassen, wie sie wirklich ist, sondern dass sie die Natur nur mit ihren Mitteln verstehen:

> Wir operiren mit lauter Dingen, die es nicht giebt, mit Linien, Flächen, Körpern, Atomen, theilbaren Zeiten, theilbaren Räumen –, wie soll Erklärung auch nur möglich sein, wenn wir Alles erst zum Bilde machen, zu unserem Bilde! (Nietzsche, FW, KSA 3, S. 472)

Für Nietzsche verlängert sich in den Naturwissenschaften nur der lebensfeindliche sokratische Rationalismus.

Eine solche wissenschaftskritische Einstellung findet sich im emanzipatorischen Existentialismus höchstens am Rande, wenn beispielsweise Camus das mittelmeerische Denken skizziert, das sich aber auf einen Relativismus beschränkt. Oder wenn Beauvoir vorführt, wie bestimmte wissenschaftliche Perspektiven das patriarchalische Denken verlängern. Aber grundsätzlich stehen Wissenschaften und Technologien nicht in ihrem Fokus. Das findet sich dagegen beim katholischen Existentialisten Gabriel Marcel und zwar teilweise vergleichbar mit Nietzsche.

Freilich spielt aber das Leben als solches eine durchaus orientierende Rolle, nicht im Sinne Nietzsches als eine dionysische dunkle, womöglich ewige Kraft, sondern konkret bei Camus als körperliches Erlebnis des Meeres am Strand von Algier oder als erotisch sexuelle Begegnung bei Beauvoir und Sartre. Verglichen mit Camus hat bei Nietzsche das Naturerlebnis weniger körperlichen als vielmehr romantischen Charakter, wenn die Natur abstrakt und fern bleibt und auf die Ewigkeit verweist.

Wenn Nietzsche den rationalistischen wissenschaftlichen Geist kritisiert und ihn aus der Bildung fernhalten will und wenn diese Bildung nicht allgemein verbreitet sein soll, dann betrifft das die Masse, die nicht verbildet werden soll, um dann womöglich soziale Ansprüche zu erheben anstatt sich zu unterwerfen. Die sozialistischen Bemühungen um Arbeiterbildung hatten in der Tat zum Ziel, die Arbeiter politikfähig zu machen. Dann aber würden sie anfangen, sich einen eigenen Sinn im Leben zu suchen, anstatt sich den Eliten zu unterwerfen und im Dienst für einen starken Genius den Sinn ihres Lebens zu erhalten.

Absolut gegensätzlich dazu stehen die emanzipatorischen französischen Existentialisten spätestens nach 1945. Je mehr sich Beauvoir und Sartre indes dem Marxismus annähern, umso mehr stellt sich ihnen auch die Frage der Gewalt, wie sie nicht nur in der proletarischen Revolution, sondern auch in den kolonialen Befreiungskriegen unvermeidlich scheint. Wenn Sartre in seinem Vorwort zu Frantz Fanons *Die Verdammten dieser Erde* die antikoloniale Gewalt als schöpferische Tätigkeit preist, dann scheint er Nietzsche nicht so fern. Freilich bleiben die Ziele gegensätzlich.

Mit Gewalt Unterworfene, so Nietzsches ziemlich abwegige Vorstellung, geben sich mit ihrer Unterwürfigkeit zufrieden, ja erkennen, dass sie selber zu gar keinem Sinn im Leben gelangen können, wenn nicht ein Genius, der sie unterworfen hat, sie in seine Ziele einspannt. Damit macht Nietzsche sicherlich eine richtige Beobachtung, dass sich viele Zeitgenossen nämlich gerne vor einen politischen Karren spannen lassen, wenn deren Führung als charismatisch erscheint und zugleich mit fetter Beute winkt.

Just dieses Modell übernimmt Max Weber in seinem Idealtypus des charismatischen Führers, zu dem auch Nietzsches Konzept des genialen politischen Führers passt. Der Rationalist Weber spricht denn von einer außeralltäglichen, übernatürlichen oder explizit sogar übermenschlichen Qualität dieser Person. Für den französischen Existentialismus ist das kein relevantes Problem.

Die Untertanen, die einem charismatischen Führer folgen, überlassen dabei die Verantwortung für ihr Leben wie für das Zutuende dieser Führung, eröffnet Nietzsche *en passant* die Debatte um die Verantwortungsethik, die im 20. Jahrhundert Karriere machen wird. Für Nietzsche ist Verantwortung eine Tugend des ‚souveränen', ‚autonom übersittlichen' folglich freien Individuums, das seine Wertvorstellungen aus sich selbst heraus entwickelt und dementsprechend denjenigen, die dazu nicht in der Lage sind, mit Herablassung und Verachtung begegnet. Er schreibt 1887 in *Zur Genealogie der Moral:*

> Das stolze Wissen um das ausserordentliche Privilegium der Verantwortlichkeit, das Bewusstsein dieser seltenen Freiheit, dieser Macht über sich und das Geschick hat sich bei ihm [dem souveränen Individuum] bis in seine unterste Tiefe hinabgesenkt und ist zum Instinkt geworden, zum dominierenden Instinkt [...]. (Nietzsche, GM, KSA 5, S. 294)

Sehr ähnlich spricht dann Max Weber explizit von Verantwortung der führenden Staats- und Wirtschaftsleute, für die absehbaren Folgen ihres Handelns, an der sich ihre Sittlichkeit misst, nicht daran, ob sie sich an obersten ethischen Normen orientieren. Dagegen haben die Untertanen nach Weber nur zu gehorchen und tragen dabei auch keinerlei Verantwortung: das Genie-Modell Nietzsches.

Sartre setzt als nächster die Debatte um die Verantwortungsethik fort, attestiert aber nicht nur jedem Menschen die Freiheit, sein Leben nach eigenen Vorstellungen zu entwerfen. Dafür muss er wie für seine ganze Existenz vielmehr selber die Verantwortung übernehmen – also längst nicht nur für die absehbaren Folgen seines Handelns. Aus der Freiheit folgt eine weitreichende individuelle Verantwortung, der niemand entgeht. Das sehen Nietzsche und Weber natürlich völlig anders.

Dagegen verbindet sich mit dem Privileg der Verantwortung bei Nietzsche am Ende eine Allmachtsvorstellung, die Leibniz' und Schillers Federstrich-Idee verlängert. Denn wenn der Kaiser und die Eliten nicht auf ihn hören – das war ihm trotz Wirrnis klar –, meint er vom Größenwahn getrieben, dass er selbst die Macht hätte, mit einem Federstreich durchzusetzen, was er für richtig hielt. So träumt er bereits lange vor dem Wahn in *Jenseits von Gut und Böse* davon, „durch das Mittel einer neuen über Europa herrschenden Kaste, einen langen furchtbaren eigenen Willen, der sich über Jahrtausende hin Ziele setzen könnte" (Nietzsche, JGB, KSA 5, S. 140) umzusetzen.

Natürlich muss er, um solche Ideen zu realisieren, ‚den jungen Kaiser in der Hand haben', der freilich Mitte der achtziger Jahre noch gar nicht regierte. Aber das junge Alter von Wilhelm II. könnte ihn dazu animiert haben, diesen zu seinem Spielball machen zu wollen. Der Wahnsinn, auf den der Zusammenbruch seines Denkens hinausläuft, der trotzdem jenem Satz über den jungen Kaiser einen Sinn verleiht, hat also mit seinem Genie-Verständnis doch eine lange Vorgeschichte in seinem Denken. An ihm scheiden sich von Anfang an seine Wege und die des emanzipatorischen Existentialismus.

Bibliographie

Hirsch, Eike Christian (2016): *Der berühmte Herr Leibniz. Eine Biographie.* München.

Nietzsche, Friedrich (1999): „Über die Zukunft unserer Bildungsanstalten. Vortrag II (1872)". In: *Kritische Studienausgabe* (KSA). Bd. 1. München, Berlin, New York, S. 641–752.

Rousseau, Jean-Jacques (2012): *Träumereien eines einsam Schweifenden* (1776–1778). Berlin.

Sartre, Jean-Paul (1980): „Die Republik des Schweigens" (1944). In: Jean-Paul Sartre: *Paris unter der Besatzung. Artikel und Reportagen 1944–1945.* Reinbek b. Hamburg.

Sartre, Jean-Paul (1987): *Der Pfahl im Fleische* (1949). In: *Gesammelte Werke. Romane und Erzählungen.* Bd. 4. Reinbek b. Hamburg.

Schiller, Friedrich (1976): *Don Carlos* (1787/88). In: *Werke.* Bd. 1. München.

Freiheit und Determinismus

Jaanus Sooväli

From Responsibility to Great Responsibility of Innocence: Nietzsche, Sartre, and Existentialism

Abstract: From Responsibility to Great Responsibility of Innocence: Nietzsche, Sartre, and Existentialism. The article sets out to determine Nietzsche's relation to Sartre's existentialism by analyzing their concepts of freedom and responsibility. The investigation shows that although Nietzsche and Sartre share the starting point, their conceptualizations differ considerably. Whereas Sartre defends the idea of absolute responsibility in his existentialist period, Nietzsche, throughout his work, puts forth the idea of complete unaccountability (*Unverantwortlichkeit*) and innocence. At the same time, though, Nietzsche keeps talking about responsibility, even great responsibility, and freedom. The article provides an interpretation of Nietzsche's reconceptualization of freedom and responsibility and relates it to Sartre's existentialist position.

I

Sartre's existentialism, as well as existentialism in general, takes shape in the context of what Nietzsche diagnosed as the death of God and European nihilism.[1] Nietzsche is the first name mentioned in Sartre's *Being and Nothingness*, in relation to abolishing the metaphysical dualism between being and appearance (Sartre 1993: EN, p. xlvi), and in a 1943 essay *A New Mystic* Sartre situates Nietzsche among existentialists like Jaspers and Heidegger, claiming that Nietzsche is an atheist who draws strongly and logically all the consequences from his atheism (Sartre 1947: NouvM, p. 166). Interestingly, in *Existentialism Is a Humanism* Sartre asserts that it is precisely his own existentialism that draws all the consequences from atheistic thought (Sartre 2007: EH, p. 53). Thus, in this respect, Sartre identifies himself with Nietzsche, or at least they are unmistakably close to one another, and Sartre's existentialism must be understood strictly in the context of Nietzsche's realm of thought. However, and

1 I believe it is not overly difficult to make the case that this is true also concerning religious existentialism (like Kierkegaard's or Shestov's). The "presence", or rather "absence" or distance of God (not complete death, though) motivates those discourses also.

https://doi.org/10.1515/9783110760163-013

on the other hand, Sartre certainly had an ambivalent and to some degree aversive relation to Nietzsche, at least retrospectively. When asked in a 1975 interview whether he was influenced by Nietzsche during his studies, he answered: "I remember giving a seminar paper on him in Brunschwicg's class, in my third year at the École Normale. He interested me, like many others; but he never stood for anything particular in my eyes." (Sartre/Rybalka/Pucciani/Gruenheck 1981: Schilpp, p. 9)[2]

II

The perhaps most significant categories of existentialism, particularly for Kierkegaard and Sartre, are freedom, responsibility, and choice. Hence, by examining these concepts or categories in Nietzsche as well as Sartre, we should be able to get a more accurate understanding of the former's relation to existentialism, and specifically of Sartre's version of it. At first sight, it seems as though there is an insurmountable gap between these two thinkers in this respect – Nietzsche rarely writes about making choices and decisions, and in the overwhelming majority of his work, he undermines the concepts of choice, responsibility, free will, and subject. In *Human, All Too Human* I 39, Nietzsche holds that "no one is accountable for his deeds, no one for his nature; to judge is the same thing as to be unjust. This also applies when the individual judges himself" (Nietzsche, HH-2 I 39/KSA 2, p. 64)[3]. A little further on, in HH-2 I 107, he talks of "complete unaccountability" (*die völlige Unverantwortlichkeit*) which he calls the most bitter drop that one has to swallow, and links it to the idea that "everything is innocence" (Nietzsche, HH-2 I 107/KSA 2, p. 105). Similarly, HH-2 I 133 (cf. Nietzsche, HH-2 I 588/KSA 2, p. 338) emphasizes "unconditional necessity of all actions and their complete unaccountability" (Nietzsche, HH-2 I 133/KSA 2, p. 128).

Human, All Too Human is, of course, a relatively early book by Nietzsche, so one might suspect that perhaps in his later writings, in his "mature thought", he gives up the idea of the absolute unaccountability and innocence of becoming, but that is not the case. In *Twilight of the Idols*, one of his very last works, he still believes that the individual is "a piece of fate" (*ein Stück Fatum*), and that demanding from him "change yourself" "means demanding that everything change, even retroactively" (Nietzsche, TI, Morality as Anti-Nature 6/KSA 6,

2 As for how much Sartre really read Nietzsche, see Daigle 2004, pp. 195–197.

3 All published works by Nietzsche are quoted from the editions of Cambridge University Press. For purpose of reference, the respective passage in the KSA is given additionally.

p. 87). He also asserts in the book that it is just an illusion to think that we (our I or will) is responsible for our actions (Nietzsche, TI, The Four Great Errors 3/KSA 6, p. 91). A little later, by summarizing his "teaching", he states:

> *Nobody* is responsible for people existing in the first place, or for the state or circumstances or environment they are in. The fatality of human existence cannot be extricated from the fatality of everything that was and will be. (Nietzsche, TI, Four Great Errors 8/KSA 6, p. 96)

At the end of this section, he also uses the famous expression "innocence of becoming". There are plenty of passages in the *Nachlass* from the same period where he makes a similar point. Hence, there is a continuity in Nietzsche's thought regarding the idea of the absolute unaccountability and innocence of becoming – we cannot be held responsible for our actions and punished for who we are for the simple reason that we cannot choose who we are, we cannot choose our essence or our character, we are "a piece of fate". Turning back to *Human, All Too Human* for a moment, you should, as Nietzsche puts it quite expressively there, "forgive yourself your own I" (*verzeihe dir dein eigenes Ich*) (Nietzsche, HH-2 I 292/KSA 2, p. 235).

III

When we turn to Sartre now, we notice immediately that in his existentialist period (approximately till 1952 when he writes *The Communists and Peace*)[4], he had a diametrically opposite position. For various reasons, I will mainly look at his famous essay *Existentialism Is a Humanism.* There are some arguments that would dissuade us focusing on that text, the first being that later in his career Sartre seems to be very critical of the ideas presented in that text. But in his later thought, he is also quite critical of *Being and Nothingness*, does that mean that we should have reservations concerning that work too (cf. Anderson, Fraser, Hoare, de Beauvoir 2006, p. 4)? Sartre's thought changes and develops

4 Dividing Sartre's thinking into clear distinct periods must be taken with a grain of salt. In the fifties, for example, Sartre tried to unite existentialism with Marxism, so he did not really give up some of the ideas of existentialism, though he does admit in a later interview that the latter is not easily compatible with the former (Gerassi 2009, p. 44). In any case, I think it can be shown that the space accorded to freedom and subjectivity in his existentialist period decreases significantly over the years. Perhaps *The Words* is the most obvious sign of that change – the tone of it is quite different from his existentialist works, it is a rather "fatalistic" description of his own development. Concerning the transformations and developments in Sartre's thought, cf. Busch 1990, pp. 90–100; Tirol 2017.

considerably over the years, this article is about his existentialist period. One could also try to make the claim that *Existentialism Is a Humanism* is just a popular simplification of the complex thought of *Being and Nothingness* and, therefore, rather insignificant. Let us keep in mind that the text is based on a public lecture, advertised in all the major journals of Paris, where Sartre was given an opportunity to clarify what he meant by existentialism, a very popular and notorious term and movement at the time, which had been under attack from various sides, including Marxists and Christians. In this sense, it may be understood as a kind of "manifesto" (in the sense of its Latin roots, that is, stating something clearly and conspicuously) of Sartre's existentialism. It is no doubt a simplification of the extremely complicated thought of *Being and Nothingness*, but it is not, I believe, contradictory to it, particularly concerning the notions of choice, freedom, and responsibility (cf. Sartre 1993: EN, pp. 553–556). The main reason why Sartre was later critical of *Being and Nothingness* and his existentialist period in general is that he thought that his understanding of freedom and responsibility at that time was not accurate[5].

As is well known, in *Existentialism Is a Humanism*, he puts forth the famous idea that "existence precedes essence" (*l'existence précède l'essence*), which itself, however, stems from the insight that God is dead (Sartre 2007: EH, pp. 20–21). The argument is beautiful, powerful, and attractive in its simplicity: since there is no God to conceive something like a human essence, humans first exist, that is, they are first thrown into the world, and create themselves and shape their own essence only *post factum*, by existing.[6] Hence, Sartre asserts that

5 Sartre's retrospective self-interpretations are somewhat inconsistent (at times he sees continuation, at times a break), but in many cases, he does clearly recognize the changes and transformations in his thinking. Cf. Sartre's words from a 1969 interview: "The other day, I re-read a prefatory note of mine to a collection of these plays – *Les Mouches, Huis Clos* and others – and was truly scandalized. I had written: 'Whatever the circumstances, and wherever the site, a man is always free to choose to be a traitor or not . . .' When I read this, I said to myself: it's incredible, I actually believed that!" – A little further on, he adds specifically about *Being and Nothingness:* "Thus, in *L'Etre et Le Neant*, what you could call 'subjectivity' is not what it would be for me now, the small margin in an operation whereby an interiorization re-exteriorizes itself in an act." (cf. Anderson, Fraser, Hoare, Beauvoir 2006: ItinT, pp. 4–8). See also Sartre 1976: ASD, pp. 98–99; Beauvoir 1984: EBS, pp. 361–363.

6 Here the respective quotation: "Atheistic existentialism, which I represent, is more consistent. It states that if God does not exist, there is at least one being in whom existence precedes essence – a being whose existence comes before its essence, a being who exists before he can be defined by any concept of it. That being is man, or, as Heidegger put it, the human reality. What do we mean here by 'existence precedes essence'? We mean that man first exists: he materializes in the world, encounters himself, and only afterward defines himself." (Sartre 2007: EH, p. 22)

> man is condemned to be free. Condemned, because he did not create himself, yet is nevertheless free, and from the moment that he is thrown into this world he is responsible for everything he does (Sartre 2007: EH, p. 29).

This is obviously quite a strong wording; he used a very similar wording, though, already in *Being and Nothingness*, including the expression "condemned to be free":

> The essential consequence of our earlier remarks is that man being condemned to be free carries the weight of the whole world on his shoulders; he is responsible for the world and for himself as a way of being. We are taking the word "responsibility" in its ordinary sense as "consciousness (of) being the incontestable author of an event or of an object." [...] It is therefore senseless to think of complaining since nothing foreign has decided what we feel, what we live, or what we are. Furthermore, this absolute responsibility is not resignation [...]. (Sartre 1993: EN, pp. 553–554)

Thus, Sartre's "absolute responsibility" *contra* Nietzsche's "absolute unaccountability" (*völlige Unverantwortlichkeit*). But let us continue with *Existentialism Is a Humanism.* A little later after having claimed that man is responsible for everything he does, Sartre utters what he calls "the first principle of existentialism":

> Man is not only that which he conceives himself to be, but that which he wills himself to be, and since he conceives of himself only after he exists, just as he wills himself to be after being thrown into existence, man is nothing other than what he makes of himself. This is the first principle of existentialism. (Sartre 2007: EH, p. 22)

Now, what is rather essential for Sartre is the idea that people have no right to make excuses for their misdeeds and unfortunate lives – there is nowhere to hide, they themselves are responsible for what they make of themselves. He calls those who want to flee their responsibility and invent deterministic doctrines self-deceivers, they are characterized by the famous *mauvaise foi*, bad faith (Sartre 2007: EH, p. 47; cf. Sartre 1993: EN, pp. 47–71). Bad faith is essentially a lie that we tell ourselves to flee our anguish. Thus, Nietzsche, according to Sartre, would be a self-deceiver, even the very worst kind – an absolute self-deceiver who does not want to take responsibility for anything at all, putting forth, as he does, a doctrine of pure necessity, pure innocence and absolute irresponsibility or unaccountability. It would follow from Sartre's position that Nietzsche simply did not have the courage to face the burden of his freedom and responsibility, he *chose* to flee his anguish.

As was pointed out above, Sartre claims about Nietzsche that he drew logically all the consequences from his atheism. Sartre asserted the same thing about his own existentialism. Thus, both draw rigorous logical consequences from the same idea, but the consequences end up being very different, even opposite to one another. What is happening here? Are there two different kinds of logic in play? Sartre, in his existentialist period, tends to jump from one extreme – that we are created by God with a predetermined essence which we are forced to realize in our existence – to another, similarly extreme position that we are free to create ourselves (not in the sense, of course, that we can give birth to ourselves, there is always some facticity surrounding us, but we can always choose how we relate to that) and hence absolutely responsible. But from the fact that God is dead does not follow – even less rigorously and logically – the position that we are free to choose and create our own "essence". There are obviously a number of things that can still determine us even if one adopts the idea of the death of God. Nietzsche, for instance, points out language, culture, climate, heritage, physiology, unconsciousness etc. – all these things, to a varying degree, determine who we are and limit considerably our choices and active stances (cf. BGE 21). A rather interesting case is, of course, the unconsciousness. Sartre rejects the theory of the unconscious mind precisely because it would limit one's freedom and open a way to avoid responsibility (cf. Sartre 1993: EN, pp. 557–574) Thus, what Sartre seems to do is to express an ethical outcry, a call for responsibility, a plea to feel the burden of your decisions and take charge of your life. One should not blame anybody else but oneself for one's mistakes and failures in life. One might commend this kind of ethics, but it is no way a "rigorous logical consequence". In his Marxist period, Sartre indeed recognizes and admits that human freedom is limited by various external factors (let me just say in passing that to some degree he seems to recognize it already in his existentialist period as well, but he does not take it seriously enough). For example, in a 1969 interview, Sartre expresses very clearly that after the Second World war life gradually taught him "*la force des choses* – the power of circumstances" (Anderson, Fraser, Hoare, Simone de Beauvoir 2006: p. 4). However, does that mean that he later approaches Nietzsche's praise for innocence? It is hardly the case. Note here, though, that Sartre and Nietzsche agree on the fact that one should not blame others for one's failures – doing so would be a kind of *ressentiment* for Nietzsche and *mauvaise foi* for Sartre; however, according to the latter, one should still blame oneself for one's failures, whereas for Nietzsche one should forgive oneself one's own ego. In short, Sartre retains the ego – or consciousness – as a sort of final center of gravity, the last piece of God in man, as one might call it, whereas Nietzsche clearly gives it up. His philosophy is not a philosophy of *consciousness*.

IV

Interestingly, Nietzsche who, as we saw, advocates absolute irresponsibility and innocence of becoming, keeps talking about freedom, "great decision" (*große Entscheidung*) (Nietzsche, NL 1884, 26[47], KSA 11, p. 160; Nietzsche, GM-1 II 24/KSA 5, p. 336), "the solitude of highest responsibility" (Nietzsche, NL 1883, 16[64], KSA 10, p. 522) and "great responsibility" (*große Verantwortung*) (Nietzsche, NL 1887, 11[363], KSA 13, p. 159), especially in his *Nachlass*, but also in the published writings. Similarly, he does not seem to have any problem using expressions like "will to responsibility" and "freedom of the will" (Nietzsche, GM-1 III 10/KSA 5, p. 361), "the most comprehensive responsibility" (Nietzsche, BGE-3 61/KSA 5, p. 79), the new philosophers' "willingness to great responsibilities" (Nietzsche, BGE-3 213/KSA 5, p. 149) and so on and so forth. The question that naturally arises is: How can he still do that and what would be the main differences from Sartre's notion of responsibility and freedom?

Firstly, in a sense, the death of God is a source of freedom for Nietzsche as well, namely in the sense of the new horizons (of knowledge, morality, art etc.) opening up. As Nietzsche puts it in *the Gay Science:* "The horizon has become open once more, every hazard is now permitted, and perhaps there has never been such an open sea before." (Nietzsche, GS-2 343/KSA 3, p. 574) But this freedom is not, of course, Sartre's freedom of choice. On the contrary, Nietzsche seems to believe that only by giving up the idea of the freedom of the will (in the traditional sense, that is) and choice, and fully accepting the necessity, we can reach the realm of freedom, *das Reich der Freiheit* (Nietzsche, HH-2 I 107/KSA 2, p. 105). Hence, and paradoxically, from the recognition of the depths of the darkest and thickest necessity, the light of freedom may be born. Negatively, this freedom is freedom *from* guilt, remorse, and punishment. Looking for the guilty party, for someone to blame, even if you put the guilt on yourself (and that is obviously different from Sartre), would still be a form of *ressentiment*. And positively, this is the freedom to engage in adventures of knowledge never deemed acceptable before, and more importantly, to accept and affirm yourself, your own self, as it is, without the evil eye of moral preconceptions (cf. Nietzsche, EH-1, Why I am so Wise 1/KSA 6, p. 278).

In one of the *Nachlass* notes from 1884, Nietzsche writes: "We work with all our powers to convince ourselves of nonfreedom: to feel that we are as free before ourselves as before the nature." (Nietzsche, NL 1884, 26[115], KSA 11, p. 180)[7]

---

7 In German: "Wir arbeiten mit allen Kräften, um uns von der Unfreiheit zu überzeugen: um uns so frei vor uns selber zu fühlen wie vor der Natur."

It is somewhat interesting that Nietzsche uses the word "convince" here but being free towards yourself as towards nature most likely means that we put ourselves at the mercy of the necessity that we are ("*ein Stück Fatum*"), that we remain in a certain sense passive towards the forces springing forth from the depths of our inside, without the feeling of guilt and bad conscience (cf. Filek 2001, pp. 128–129). The best formulation of this is, of course, "become who you are" (*werde, der du bist*) – open yourself up to the abyss of your own self. This freedom is in no way opposite to necessity, on the contrary, it is part of this necessity, it is this necessity itself, necessity let loose. One of the aphorisms that helps to explain this idea rather well I think stems from *Wanderer and his Shadow:*

> Over one man *necessity* stands in the shape of his passions, over another as the habit of hearing and obeying, over a third as a logical conscience, over a fourth as caprice and a mischievous pleasure in escapades. These four will, however, seek the *freedom* of their will precisely where each of them is most firmly fettered: it is as if the silkworm sought the freedom of its will in spinning. How does this happen? Evidently because each considers himself most free where his *feeling of living* is greatest; thus, as we have said, in passion, in duty, in knowledge, in mischievousness respectively. (Nietzsche, HH-2, WS 9/KSA 2, p. 545)

Thus, doing the things that one is, naturally and by necessity, good at doing, one feels oneself the freest because his feeling of living is the greatest there. From this, it is not difficult to understand what Nietzsche means by the idea that the self should open itself up to its own necessity. The strongest necessity and the highest freedom go hand in hand.[8] Similarly, in *Ecce homo*, in the course of explaining what is inspiration, Nietzsche states: "All of this is involuntary to the highest degree, but takes place as if in a storm of feelings of freedom, of unrestricted activity, of power, of divinity..." (Nietzsche, EH-1 Za III/KSA 6, p. 340) It is rather important that Nietzsche speaks here of the *feeling* of freedom. According to him it would be mythological to apply the notion of freedom (but also the concept of unfreedom, hence his criticism of traditional determinism) to being itself, it makes no sense to cherish ideas about some abstract absolute freedom that people supposedly have or do not have (Nietzsche, BGE-3 21/KSA 5, p. 36). It all comes down to the feeling and consciousness of freedom, which, in the end, is the feeling of power, of divinity, of inspiration, of creative forces etc.

---

8 Cf. Nietzsche, BGE 19: "'Freedom of the will' – that is the word for the multi-faceted state of pleasure of one who commands and, at the same time, identifies himself with the accomplished act of willing." (Nietzsche, BGE-3 19/KSA 5, p. 33)

Now, this freedom is also the source of, or at least directly intertwined with, (great) responsibility. In the aphorism titled "My Idea of Freedom" from the *Twilight of the Idols*, Nietzsche writes: "Because, what is freedom anyway? Having the will to be responsible for yourself". (Nietzsche, TI, Skirmishes of an Untimely Man 38/KSA 6, p. 139) This should be understood in the sense of unfolding yourself to yourself, listening to, and following, your creative instincts instead of some general morality and established values (cf. Nietzsche, BGE-3 212/KSA 5, p. 145). Steering away from the realm of the general also throws one to solitude and danger, and Nietzsche indeed links this new understanding of responsibility to solitude, as well as to overcoming hardships (Nietzsche, NL 1883, 16[64], KSA 10, p. 522). Listening to one's own self obviously involves some sort of passivity, and we know that some of the poststructuralist thinkers like Levinas and Derrida thematize responsibility precisely as being akin to a kind of "passivity" instead of activity of the subject, so it is not at all surprising that Derrida, for instance, refers to Nietzsche rather than to Sartre in this respect (cf. Derrida 2005, pp. 39–45; Pfeuffer 2008).

Thus, Nietzsche dismisses one notion of responsibility – the notion based on the feeling of guilt and remorse[9] and understood as an activity of the free autonomous subject, say, Sartrean notion to a considerable degree (cf. Filek 2001, p. 124) – and develops another which is closely intertwined with his new understanding of freedom. Freedom, as we saw, remains an essential part of responsibility, but not at all in the Sartrean sense. As it is to be expected, this new notion of responsibility is also directly linked to the innocence of becoming. An 1884 *Nachlass* note, titled "The Way to Wisdom", brings all these themes, although rather cryptically, together:

> *Great decision* whether you are fit for positive stance, for affirmation. No God, no human above me anymore. The instinct of the creator who knows where to lay his hands on. The *great* responsibility and innocence. (Nietzsche, NL 1884, 26[47], KSA 11, p. 160)[10]

We have here "affirmation", "freedom" ("no God"), the creative forces ("the instinct of the creator"), great responsibility and innocence – all the themes that form Nietzsche's new conceptualization of the phenomena in question.

9 Sartre does not say that one *should* feel oneself guilty, but the feeling of guilt and remorse would naturally follow from his conception, especially for those, of course, who are unsatisfied with their existence.

10 In German: "Große Entscheidung, ob tauglich zur positiven Stellung, zum Bejahen. Kein Gott, kein Mensch mehr über mir! Der Instinkt des Schaffenden, der weiß, wo er die Hand anlegt. Die große Verantwortung und die Unschuld."

As pointed out, Nietzsche often speaks of the feeling of freedom and responsibility, and he mainly seems to be interested in the mechanisms of how this kind of feeling could arise – under which circumstances and states, in whom (which "types") etc. His discourse is mainly genealogical, psychological, perhaps even "phenomenological" (even though it is, of course, Sartre who was heavily influenced by Husserl and Heidegger). In this sense, his interest is different from that of Sartre who seems to provide a more "moral" interpretation of the matter. So, how does this responsibility, particularly the great responsibility, understood as the extreme case of the former, arise according to Nietzsche? He applies it to "new philosophers", "free spirits" etc., that is, it is obviously not a matter of everybody (cf. Filek 2001, p. 139). But it still follows the same "logic" exhibited above: if one becomes free towards oneself, opens oneself up to the necessity that one is, thus liberating the "instincts of creation" from one's inside, he or she feels free, that is, powerful, self-assured, self-certain, godly. And out of this self-assuredness and godliness, one feels and *takes* the great responsibility, in some cases for the future of mankind. The more powerful the forces that spring forth from one's inside, the greater responsibilities one takes: "Is there enough pride, daring, courage, self-confidence, will of spirit [Wille des Geistes], will to take responsibility, freedom of will, for 'the philosopher' on earth to be really – possible? . . ." (Nietzsche, GM-1 III 10/KSA 5, p. 361) And in *Beyond Good and Evil*, he famously describes the philosopher, the way he understands him, "as the man with the most comprehensive responsibility, whose conscience bears the weight of the overall development of humanity" (Nietzsche, BGE-3 61/ KSA 5, p. 79).

A possible (moral) problem with Nietzsche's new conceptualization of responsibility concerns the fact that it is very difficult to evaluate and criticize somebody who, feeling powerful and godly enough, takes this great responsibility, say, for the future of humanity or mankind. What if that person is a serial killer or a murderous tyrant, or just a lunatic? Nietzsche mainly means, of course, philosophers – in a way, Plato shaped the humanity for thousands of years. But he also admires Caesar and Napoleon. The situation reminds us a little bit of Kierkegaard's "knight of faith" Abraham who, from the perspective of the general ethics, is nothing but a child killer (Kierkegaard 1983, p. 55). What "saves" him, though, is his direct, though unrepresentable and paradoxical, relation to the Absolute, to God. For Nietzsche, there is obviously no such Absolute. There is hardly any "objective" way[11] for him that would teach us that one

11 One could, of course, make the case that "affirmation of life" is some kind of "objective" criterium for Nietzsche, but it would probably be way too abstract as a criterium.

man is wrong taking this responsibility and the other one is right, but that does not mean that one could not come up with a number of "subjective" criteria based on, say, one's taste. Looking for the objective criteria already misses the point (but does not do away with the moral problem). This is already too far off from the main topic of the article, though.

Thus, Nietzsche and Sartre, although sharing the starting point (the death of God) do go their separate ways. In his Marxist period, Sartre comes closer to Nietzsche in the sense that he starts recognizing more "*la force des choses*", but with his new notions of responsibility and freedom, Nietzsche distances himself again and seems to approach (only that) some of the poststructuralist interpretations of responsibility. So, is Nietzsche an existentialist? It would be hard, if not impossible, to imagine existentialism without Nietzsche. Several existentialist themes run through his work, including the problem of the individual, the meaning of life, the problem of suffering, the solitude etc. However, he is certainly not an existentialist in the sense Sartre defines it in *Existentialism Is a Humanism*. In that sense, he would perhaps be the greatest enemy, the greatest critic of existentialism. The question whether Nietzsche is an existentialist is somewhat similar to the one whether Nietzsche is a postmodernist. There are a number of postmodernist themes in his work as well, but it would still be a great a number of stretch to call him one. It is in any case rather astonishing that two so different and distinct "movements" like existentialism and postmodernism can both justifiably regard Nietzsche as their forefather.

Bibliography

Anderson, Perry/Fraser, Ronald/Hoare, Quintin (2006): "Itinerary of a Thought". In: Perry Anderson/Ronald Fraser/Quintin Hoare/Simone de Beauvoir (ed.): *Conversations with Jean-Paul Sartre*. London, New York, Calcutta, pp. 1–66.

Beauvoir, Simone de (1984): *Adieux: A Farewell to Sartre*. London.

Busch, Thomas W. (1990): *The Power of Consciousness and the Force of Circumstances in Sartre's Philosophy*. Bloomington, Indianapolis.

Daigle, Christine (2004): "Sartre and Nietzsche". In: *Sartre Studies International*, Vol. 10, No. 2, pp. 195–210.

Derrida, Jacques (2005): *Politics of Friendship*. London, New York.

Filek, Jacek (2001): "Das Drama der Verantwortung bei Friedrich Nietzsche". In: *Archiv für Begriffsgeschichte*, 43, pp. 113–147.

Gerassi, John (2009): *Talking with Sartre: Conversations and Debates*. New Haven and London.

Kierkegaard, Søren (1983): *Fear and Trembling*. Princeton.

Pfeuffer, Silvio (2008): *Die Entgrenzung der Verantwortung: Nietzsche – Dostojewskij – Levinas. Monographien und Texte zur Nietzsche-Forschung*, Vol. 56. Berlin, New York.

Sartre, Jean-Paul (1947): "Un nouveau mystique". In: Jean-Paul Sartre (ed.): *Situations, I.* Paris, pp. 133–174.
Sartre, Jean-Paul (1976): "Autoportrait à soixante-dix ans". In: Jean-Paul Sartre (ed.): *Situations, X.* Paris, pp. 133–226.
Sartre, Jean-Paul (1993): *Being and Nothingness.* New York.
Sartre, Jean-Paul (2007): "Existentialism Is a Humanism". In: Jean-Paul Sartre: *Existentialism Is a Humanism.* New Haven, pp. 17–72.
Sartre, Jean-Paul/Rybalka, Michel/Pucciani, Oreste/Gruenheck, Susan (1981): "An Interview with Jean-Paul Sartre". In: Paul Arthur Schilpp (ed.): *The Philosophy of Jean-Paul Sartre.* LaSalle, pp. 5–51.
Tirol, Tarmo (2017): *Subjektiivsus Jean-Paul Sartre'i eksistentsialismis.* Dissertationes Philosophicae Universitatis Tartuensis 15. Tartu.

Anthony Kosar

„Alles ist *in actu*": Existentielles ‚Freiheitsbewusstsein' zwischen absoluter Verantwortung und völliger Unverantwortlichkeit

Abstract: „Everything is in action": Existential 'awareness of freedom' between absolute responsibility and complete unaccountability. Friedrich Nietzsche stresses early on in *Human, All Too Human I* that an historical sense can do a great deal to help undo the burden of responsibility for one's being, inasmuch as every 'deed' and 'doing' is shown thereby to be overdetermined and thus not really in one's hands after all (cf. Nietzsche, HH-2 I 17, 39, 70, 105–107, etc. etc.). Throughout *Being and Nothingness* – if most succinctly and forcefully in a relatively short subsection – Jean-Paul Sartre argues just the reverse, namely that an awareness of one's historical situatedness by no means 'frees' one of responsibility for one's being, but indeed impresses upon one an even fuller awareness of the absolute responsibility for oneself – and for the world at large – in light of the absolute freedom of choice with which one is damned from the start (cf. Sartre 1976: EN, Fourth Part, First Chapter, III. Freedom and responsibility). The following essay asks what Nietzsche and Sartre, when read together, might teach us about our everyday psychological notions of freedom and responsibility.

> Alles ist *in actu*. Hinter dem Akt gibt es weder Potenz noch „Hexis" noch Fähigkeit. Wir weigern uns zum Beispiel, unter „Genie" – in dem Sinn, wie man sagt, Proust hatte „Genie" oder „war" ein Genie – eine besondere Potenz zu verstehen, gewisse Werke hervorzubringen, die sich nicht genau in deren Hervorbringung erschöpfte. Prousts Genie ist weder das isoliert betrachtete Werk noch das subjektive Vermögen, es hervorzubringen: es ist das als die Gesamtheit der Manifestationen der Person betrachtete Werk. Deshalb können wir schließlich ebenso den Dualismus von Erscheinung und Wesen verwerfen. Die Erscheinung verbirgt nicht das Wesen, sie enthüllt es: sie *ist* das Wesen. (Sartre 1976/1991: EN, S. 12–13/11)[1]
>
> Ein Quantum Kraft ist ein eben solches Quantum Trieb, Wille, Wirken – vielmehr, es ist gar nichts anderes als eben dieses Treiben, Wollen, Wirken selbst, und nur unter der Verführung der Sprache (und der in ihr versteinerten Grundirrthümer der Vernunft), welche alles Wirken als bedingt durch ein Wirkendes, durch ein „Subjekt" versteht und missversteht,

1 Für alle Zitate aus *L'être et le néant* verwende ich die 1991 erschienene Übersetzung von Hans Schöneberg und Traugott König. Die Seitenzahlen vor dem Trennstrich entsprechen immer der korrigierten französischen Ausgabe von Arlette Elkaïm-Sartre (= Sartre 1976), die nach dem Trennstrich der deutschen Übersetzung von Schöneberg und König (= Sartre 1991).

https://doi.org/10.1515/9783110760163-014

> kann es anders erscheinen. Ebenso nämlich, wie das Volk den Blitz von seinem Leuchten trennt und letzteres als Thun, als Wirkung eines Subjekts nimmt, das Blitz heisst, so trennt die Volks-Moral auch die Stärke von den Äusserungen der Stärke ab, wie als ob es hinter dem Starken ein indifferentes Substrat gäbe, dem es freistünde, Stärke zu äussern oder auch nicht. Aber es giebt kein solches Substrat; es giebt kein „Sein" hinter dem Thun, Wirken, Werden; „der Thäter" ist zum Thun bloss hinzugedichtet, – das Thun ist Alles. Das Volk verdoppelt im Grunde das Thun, wenn es den Blitz leuchten lässt, das ist ein Thun-Thun: es setzt dasselbe Geschehen einmal als Ursache und dann noch einmal als deren Wirkung. (Nietzsche, GM I 13, KSA 5, S. 279)

Die beiden oben angeführten Zitate Jean-Paul Sartres und Friedrich Nietzsches bieten auf den ersten Blick eine fast gemeinsame Ontologie des menschlichen Handelns, die aber bei genauerer Betrachtung scheitert. Man mag fragen, inwiefern denn der Mensch wirklich handelt, wenn dem Subjekt seine Macht über die Umstände geraubt wird, ja ob es überhaupt noch sinnvoll ist, vom Tun und Handeln zu reden, wenn es einem offenbar nicht freisteht, zu handeln oder auch nicht. Aber solche Fragen sind weniger relevant, wenn man bedenkt, dass es sich bei den beiden Zitaten um unterschiedliche Motive handelt. Sartre fasst einige Grundzüge der Phänomenologie Edmund Husserls und Martin Heideggers zusammen, zu der der Rest seines Werkes *L'être et le néant* – mit dem Untertitel versehen *Essai d'ontologie phénoménologique* – bestenfalls einen an deren phänomenologische Arbeiten anknüpfenden Beitrag leisten soll, denn es handelt sich in Sartres ‚Versuch' nicht primär um das Sein des Phänomens als solchen, als vielmehr um die Möglichkeit der menschlichen Freiheit in konkreten Situationen.[2] Der Mensch – so stellt sich heraus – ist *tatsächlich* frei, zur wirklichen Tat frei, solange man seinen Blick nicht mehr nur auf die Vergangenheit, sondern vielmehr auf die Zukunft richtet, die noch nicht entschieden ist. Nietzsches Äußerungen über das Täter-Subjekt als einen sprachlichen ‚Grundirrt[u]m[] der Vernunft' sind weniger befreiend und schienen sogar die Wirklichkeit des Handelns selbst in Abrede zu stellen, hätte Nietzsche nicht schließlich am Primat eines täterlosen Tuns festgehalten. Aber wie unterscheidet sich dieses Tun von anderen Ereignissen, wenn es kein zugrundeliegendes Subjekt gibt? Der Primat der Akti-

2 Während Sartres phänomenologischer Versuch eher solipsistisch wirkt, gründet er dennoch auf der konkreten Situiertheit der einzelnen Perspektive. Zu Sartres Weg vom vor allem in *L'être et le néant* nahegelegten Solipsismus zu einer vom Marxismus geprägten Sozialphilosophie der konkreten Praxis vgl. Jacobi 1970. Alfred Betschart übt Kritik an der Widersprüchlichkeit des Terminus ‚existentialistischer Marxismus' in Betschart 2019.

vität, der ein Hauptanliegen der *Genealogie der Moral* sein soll, scheint dabei wie ins Wanken geraten.[3]

Um so erstaunlicher ist es daher, dass Sartre und Nietzsche das Phänomen des Tuns selbst fast gleich beschreiben, dessen Erscheinen den gewöhnlichen psychologischen Erklärungen widerspricht: Das Tun ist keine Wirkung eines hinter dem Phänomen steckenden Subjekts mehr, sondern mit dessen Antrieb als „integrierende[m] Bestandteil“ der Handlung selbst (Sartre 1976/1991: EN, S. 583/760) aufs engste verbunden; vielmehr, es unterscheidet sich kaum mehr vom tätigen Subjekt, sondern Tun und Täter sind nunmehr ein und dasselbe: „Die Erscheinung verbirgt nicht das Wesen, sie enthüllt es: sie *ist* das Wesen.“ (Sartre 1976/1991: EN, S. 12–13/11) Nicht ‚das Tun ist alles‘, sondern ‚alles ist im Tun‘ – der ‚Täter‘ mit eingerechnet.[4]

An dieser Stelle wird Nietzsches Anliegen in diesem kurzen Passus etwas klarer: Wenn es keinen ‚Täter‘ gibt, kann niemand mehr für schuldig befunden werden – niemand könnte für sein Handeln verantwortlich gemacht werden, auch fürs erfolgreiche Handeln nicht. Aber damit geht der Künstler-Stolz offenbar auch verloren. Sicherlich kann Nietzsche dies nicht gewollt haben. Die Frage stellt sich: Wie lässt sich das nietzschesche Subjekt verstehen? Als ein Ereignis etwa? Einen Prozess? Ist nicht Verantwortung ein notwendiger Bestandteil unserer Selbstheit? Und können sich Prozesse verantworten?

Sieht man näher hin, kommt das Problem der subjektiven Freiheit in den Fokus, die bisher zwei Arten der Verantwortung nahegelegt hat: die eine, die ein Künstler mit Stolz für sein Werk übernimmt – oben war schon von Prousts ‚Genie‘ die Rede – und die andere, mit deren Hilfe man einen Täter für schuldig erklärt. Aber die Freiheit des Subjekts setzt zuallererst ein Bewusstsein, genauer: das Selbstbewusstsein voraus, und dieses hat Nietzsche zufolge eine lange Geschichte. Der Hauptunterschied zwischen Sartre und Nietzsche mag folglich darin bestehen, dass für Sartre Freiheit und Bewusstsein – die dieser auch als eins versteht (vgl. Sartre 1976/1991: EN, S. 616/804) – ontologisch gegeben sind, während sie sich für Nietzsche sozio-historisch entwickeln: Erst innerhalb einer geschichtsspezifischen Gesellschaftsform – so führt Nietzsche in der *Genealogie* vor Augen, ohne aber freilich die genaue geschichtliche Zeit dieser Gesellschaft zu

3 Vgl. Nietzsche, GM II 12, KSA 5, S. 313–316. Zur Bedeutung der Aktivität für das ganze tierische Leben in der *Genealogie* vgl. insbes. Brusotti 2001.

4 So möchte jedenfalls Robert B. Pippin diesen Satz Nietzsches lesen. Vgl. Pippin 2010, insbes. S. 67–84.

bestimmen – tauchen Freiheit und Bewusstsein auf.[5] Das Bewusstsein wird, entwickelt sich und wird allmählich seiner Freiheit gewahr. Nun aber taucht ein weiteres Problem auf: Wie ist zu erklären, dass Nietzsche und Sartre beide scheinbar auf dem Primat der Aktivität bestehen, wenn für Sartre die Freiheit – mit der jeder Akt scheinbar verbunden ist – ontologisch gegeben ist, für Nietzsche aber erst sehr spät zutage tritt?

Eine erste These lautete, dass – für Nietzsche wie für Sartre – Aktivität in erster Linie ein psychologischer Begriff ist. Bei Sartre heißt es: „Ein erster Blick auf die menschliche-Realität lehrt, daß Sein sich für sie auf Handeln reduziert." (Sartre 1976/1991: EN, S. 632/824) Wir können uns gar nicht vorstellen, wie eine vollkommen *passive* Existenz aussehen würde, denn diese geht scheinbar über unsere tagtäglichen menschlichen Erfahrungen hinaus. Mehr noch, sie widerspricht nicht nur unseren menschlichen Erfahrungen im Besonderen, sondern auch unseren Grundvorstellungen des Lebens im Allgemeinen: „Wirklich undenkbar ist die passive Existenz, das heißt eine Existenz, die fortbesteht, ohne daß sie die Kraft hat, sich hervorzubringen oder zu erhalten." (Sartre 1976/1991: EN, S. 25/26) Damit ist aber noch nicht gesagt, dass diese Vorstellungen mit der Realität übereinstimmen, sondern nur dass unsere Psychologie an eine Grenze stößt, wenn wir versuchen, eine wahrhaft nichttätige Lebensform zu denken. Ein weniger bekannter Passus aus Nietzsches Spätschriften kann vielleicht an dieser Stelle helfen, Sartres „erste[n] Blick auf die menschliche-Realität" verständlicher zu machen, insofern *L'être et le néant* weder eine Streitschrift noch ein politisches Manifest, sondern vielmehr ein Werk existentieller Psychologie ist.[6] In Nietzsches *Götzen-Dämmerung* nämlich, die ursprünglich den Titel *Müssiggang eines Psychologen* tragen sollte,[7] versucht Nietzsche die älteste menschliche Psychologie als eine von Tun und Tätern zu entlarven. Da heißt es: „[A]lles Geschehen war ihr [der ältesten Psychologie, A.K.] ein Thun, alles Thun Folge eines Willens, die Welt wurde ihr eine Vielheit von Thätern, ein Thäter (ein ‚Subjekt') schob sich allem Geschehen unter." (Nietzsche, GD Irrthümer 3, KSA 6, S. 91) Die Sprache, sofern sie Subjekte nötig hat, schildert die Welt als wesentlich aktiv, sodass sogar nichtlebendige Gegenstände und Ereignisse tätig erscheinen, wenn sie zum Ausdruck kommen. (Früher war vom ‚Tun' des Blitzes die Rede, der eigentlich nichts als ein bestimmtes meteorologisches Geschehen ist.[8]) Um es noch einmal zu betonen,

5 Michel Foucault scheint es zu bedauern, dass Sartre, dessen erste Veröffentlichung ein „nietzscheanischer Text" zur Geschichte der Wahrheit war, letztlich zur Phänomenologie ging. Vgl. dazu Foucault 2005 [1983], S. 539.

6 Dazu Betschart 2010 und 2019, S. 91.

7 Zur Entstehungsgeschichte dieses Werkes vgl. insbes. Sommer 2012, S. 197–201 und 211–213.

8 Zu Nietzsches Kritik dieser Sprach-Mythologie vgl. Zavatta 2009.

muss das nicht besagen, dass diese Psychologie der Wahrheit entspricht, nur dass sie etwas Grundmenschliches zum Ausdruck bringt. Woher aber kommt diese menschliche Vorstellung?

Darin sind sich Sartre und Nietzsche wohl nicht einig. Wiederum gesagt ist dieses allgemeine Bewusstsein der Aktivität für Sartre ontologisch gegeben – wir können uns gar nicht vorstellen, wie es anders sein könnte –, während Nietzsche es auch historisch versteht. Es stellt sich somit die Frage, wie es laut Nietzsche zu einer derartigen Entwicklung des menschlichen Bewusstseins kommen konnte, dass es scheinbar nicht (mehr) imstande ist, sich eine vollkommen passive Existenz vorzustellen.

Die Welt als eine von Tätern Bewohnte wahrzunehmen, lässt sich zunächst thematisch mit dem Gedächtnis-Machen in der *Genealogie der Moral* verbinden. Dieser These zufolge wäre man nicht mit einer solchen Perspektive geboren, sondern hätte vielmehr gelernt, die Welt als eine Willenswelt zu begreifen, nachdem „ein eigentliches Gedächtniss des Willens" sich allmählich in einem herausgebildet hätte (Nietzsche, GM II 1, KSA 5, S. 292). Weil der „faselige[] Augenblicks-Verstand[]" der früheren Menschheit sich offenbar nicht in ein soziales Leben einfügen ließ – so Nietzsches *Genealogie* –, musste man sich mithilfe einer grausamen Mnemotechnik an das Nicht-zu-Vergessende erinnern lernen: „‚Man brennt Etwas ein, damit es im Gedächtniss bleibt: nur was nicht aufhört, weh zu thun, bleibt im Gedächtniss' – das ist ein Hauptsatz aus der allerältesten (leider auch allerlängsten) Psychologie auf Erden." (Nietzsche, GM II 3, KSA 5, S. 295) Am Ende dieser Geschichte würde man sich dann als einer, der frei handelt – oder der wenigstens von sich glaubt, dass er frei handelt , auf die ganze Außenwelt herausprojizieren, sodass diese Welt als eine von frei Handelnden wie ihm erscheint. Jedoch folgt ein Irrtum daraus, nämlich, dass der Wille etwas ist, das verursacht.

Sartre wiederrum versucht diesen Irrtum durch eine rein phänomenologische Analyse des Handelns zu beseitigen. Wenn dieses nicht mit anderen Ereignissen zu verwechseln sein soll, muss man die Frage beantworten können, wie es überhaupt in einer prädeterminierten, als ‚an sich schon gegeben' erscheinenden Situation entsteht. Sartres Erklärung folgt:

> Die Nichtung, durch die wir von der Situation Abstand gewinnen, ist eins mit der Ekstase, durch die wir uns auf eine Veränderung dieser Situation hin ent-werfen. Daraus ergibt sich, daß es tatsächlich unmöglich ist, eine Handlung ohne Antrieb zu finden, daß man daraus aber nicht schließen darf, der Antrieb sei Ursache der Handlung: er ist ihr integrierender Bestandteil. Denn da sich der auf eine Veränderung hin beschlossene Entwurf nicht von der Handlung unterschiedet, konstituieren sich Antrieb, Handlung und Zweck in einem einzigen Auftauchen. Jede dieser drei Strukturen erfordert die beiden anderen als ihre Bedeutung. Aber die organisierte Totalität der drei läßt sich nicht mehr durch eine Einzelstruktur er-

> klären, und ihr Auftauchen als reine verzeitlichende Nichtung des An-sich ist eins mit der Freiheit. Die Handlung entscheidet über ihre Zwecke und ihre Antriebe, und die Handlung ist der Ausdruck der Freiheit. (Sartre 1976/1991: EN, S. 583/760 – 761)

Die Entschlusskraft, eine Situation, in der man sich befindet, zu ändern, wird hier oben nicht als Ursache der Handlung selbst verstanden, sondern vielmehr als deren „integrierender Bestandteil". Aber damit eine Handlung erst sinnvoll als Handlung sein kann, muss sie zugleich Ausdruck der Freiheit sein, die es allein vermag, die Situation zu ändern, anstatt sie bloß als solche hinzunehmen. Sonst ist die Handlung für Sartre keineswegs von anderen Ereignissen zu unterscheiden. Wie er es an anderer Stelle formuliert: „Die Existenz der Handlung impliziert ihre Autonomie." (Sartre 1976/1991: EN, S. 632/825)

Nietzsche gibt selten zu verstehen, dass er ein großes Interesse an der Freiheit gehabt habe, jedoch redet er mitunter von einem „Freiheits-Gefühl"[9] und an einer wichtigen Stelle sogar von einem „Freiheits-Bewusstsein", das zugleich ein „Macht-[...]Bewusstsein" und „Vollendungs-Gefühl des Menschen überhaupt" sein soll (Nietzsche, GM II 2, KSA 5, S. 293). Es handelt sich hier nämlich um das Ende der grausamen Geschichte, in der die Mnemotechnik gegen jede Erwartung das ‚souveräne Individuum' hervorbringt:

> Stellen wir uns dagegen an's Ende des ungeheuren Prozesses, dorthin, wo der Baum endlich seine Früchte zeitigt, wo die Societät und ihre Sittlichkeit der Sitte endlich zu Tage bringt, wozu sie nur das Mittel war: so finden wir als reifste Frucht an ihrem Baum das souveraine Individuum, das nur sich selbst gleiche, das von der Sittlichkeit der Sitte wieder losgekommene, das autonome übersittliche Individuum (denn „autonom" und „sittlich" schliesst sich aus), kurz den Menschen des eignen unabhängigen langen Willens, der versprechen darf – und in ihm ein stolzes, in allen Muskeln zuckendes Bewusstsein davon, was da endlich errungen und in ihm leibhaft geworden ist, ein eigentliches Macht- und *Freiheits-Bewusstsein*, ein Vollendungs-Gefühl des Menschen überhaupt. (Nietzsche, GM II 2, KSA 5, S. 293, Kursivierung A.K.)

Interessant ist zunächst – vor allem angesichts des historischen Anliegens der *Genealogie* – die zeitliche Unbestimmtheit dieses Individuums.[10] Jedoch ist klar, dass es am Ende eines ‚ungeheuren Prozesses' steht. Sein Freiheitsbewusstsein wiederum ist kein Adelsbrief der Menschheit überhaupt, sondern das Vollendungsgefühl eines besonderen *Menschen*, der sich stellvertretend für seine Gattung glaubt; kein Gemeingut, sondern ein Privileg. Genauer: Es ist das Privileg der

9 Vgl. Nietzsche, JGB 260, KSA 5, S. 208 – 212; NL 1887, 9[152]; NL 1888, 14[98] und EH Z III, KSA 6, S. 339 – 340.

10 Vgl. dazu Brusotti 2019, S. 36 – 37 und Sommer 2019, S. 235 und 237.

Verantwortlichkeit, das Nietzsches souveränem Menschen eine gewisse ‚Festigkeit' verleiht, sodass er ohne das Versprechen den Anderen gegenüber noch nicht einmal zu denken wäre. Nietzsches Verantwortungsbegriff erweist sich somit als gesellschaftlich nützlich und tief persönlich in eins. Ein solcher Begriff der Verantwortung ist im Grunde soziologisch: Erst in einer Gesellschaft ist das Individuum überhaupt möglich.

„Aber wie kann das geschichtlich gewordene, souveräne Individuum wissen, ob es wirklich das Eigene tut und in dem, was es tut, nicht einfach durch seine Geschichte determiniert ist?" (Sommer 2019, S. 236) Mit dieser Frage bringt Andreas Urs Sommer eine Irritation der ‚Autonomie' des ‚souveränen Individuums' im Besonderen – und vielleicht auch eine Irritation des Handlungsbegriffs Nietzsches im Allgemeinen – zum Ausdruck. Ist Nietzsches großzügige Beschreibung dieses Individuums vielleicht nur rhetorisch gemeint? So scheint es manchmal, denn das Geschehen, auf das die Geschichte sich etymologisch zurückführt, scheint mitunter jede unserer Handlungen vorauszubestimmen. In *Menschliches, Allzumenschliches I* wird immer wieder betont, dass das Weltgeschehen einen strengen Verantwortungsbegriff unmöglich macht, insofern es jede Person überdeterminiert: In diesem Werk trägt der historische Sinn vor allem dazu bei, den Einzelmenschen von der Verantwortung für sein So- und So-sein zu befreien.[11] Aber schon in der *Genealogie der Moral* räumt Nietzsche mit diesem Determinismus auf, und zwar zugunsten der wirklichen Aktivität, die Freiheit und Spontaneität miteinbezieht. Nietzsches *Genealogie*, insofern sie zugleich eine Kontingenzlehre ist, räumt eine beinahe absolute Spontaneitätsmöglichkeit ein, wenn man die Gegenwart als die im Prinzip unvorhersehbare Zukunft einer fernen Vergangenheit betrachtet. Ein solches historisierendes Denken, das auf jede Teleologie überhaupt verzichtet, beschreibt Raymond Geuss folgendermaßen:

> Weder der immer wiederkehrende Wechsel der Jahreszeiten noch die logische Folge der Schritte in einem geometrischen Beweis, noch das Sichentwickeln des Kleinkindes zum Erwachsenen ist ein adäquates Vorbild für historisierendes Denken. (Voll) historisierendes Denken gibt es, so könnte man es paradoxerweise formulieren, nur wo eine Geschichtsphilosophie in emphatischen [sic] Sinne *fehlt*, nur wo es *kein* Schicksal, *keine* göttliche Fügung, *keine* Teleologie, *kein* vernunft- oder naturbedingtes übergreifendes Gesetz oder Entwicklungsschema gibt, auf das man sich beziehen kann; wo historisch vergesellschaftete Menschen, sozusagen, allein unter sich sind. (Geuss 2016, S. 198)

Geuss schließt ein solches ‚(voll) historisierendes Denken' an Nietzsches eigentliche Praxis der Genealogie an: Eine Genealogie nietzschescher Prägung – so

11 Siehe insbes. Nietzsche, MA I 17, KSA 2, S. 38; MA I 39, KSA 2, S. 62–64; MA I 70, KSA 2, S. 81; MA I 105–107, KSA 2, S. 102–106 usw. usf.

Geuss – ist mit dem Determinismus insofern unvereinbar, als „die kontingente Prozesshaftigkeit", die bei „der ‚historisierenden' Behandlung eines Sachverhaltes [...] im Mittelpunkt des Interesses [steht]", impliziert, dass dieser „Prozess [einer] sein [muss], bei dem Anfang, Verlauf und Ergebnis *nicht* als vorgeschrieben gelten, sondern auch hätte anders sein können." (Geuss 2016, S. 196) Aber legt diese absolute Kontingenz nicht eine absolute, wenn freilich auch ‚subjektlose' Freiheit nahe? Ist die Freiheit jeder und jedes Einzelnen vielleicht nicht nur auf die Menschheit einzuschränken? Könnte sie vielleicht sogar allen organischen Lebewesen überhaupt gelten, sofern es sich beim Leben immer um Aktivität handelt?

Eine solche Perspektive ist nur möglich, wenn man seinen Blick nicht mehr nur auf die abgeschlossene Vergangenheit, sondern vielmehr und vor allem auf die noch nicht entschiedene Zukunft richtet, und damit kehren wir zu Sartre zurück. Dass unsere jetzige Situation auch anders hätte sein können, als sie tatsächlich ist, ist ein Gedankenexperiment, das auf der Freiheit aller am Weltgeschehen Beteiligten beruht. Wenn diese prinzipiell immer anders hätten handeln können, als sie gehandelt haben, heißt das zugleich, dass sich unsere jetzige Welt jeder ihrer Handlungen verdankt, denn wenn jene tatsächlich anders gehandelt hätten, wäre die Welt auch anders. Aber wir können die Vergangenheit nicht mehr ändern, sondern nur auf die Zukunft hin arbeiten. Während manche Philosophen auf die Vergangenheit als Beweis für die Unfreiheit jeder und jedes Einzelnen zeigen mögen, argumentiert Sartre:

> Für uns dagegen ist die Reihenfolge der Interpretation streng *chronologisch*, sie sucht keineswegs die Zeit auf eine rein logische (*Grund*) oder logisch-chronologische (*Ursache*, Determinismus) Verkettung zurückzuführen. Sie wird also von der Zukunft her interpretiert. (Sartre 1976/1991: EN, S. 623/812–813)

Nimmt man die Eigenverantwortlichkeit ernst, wird man nunmehr nur mit Blick auf die Zukunft handeln. Nietzsche hat so viel zugestanden, als er in der neuen Vorrede zu *Menschliches, Allzumenschliches I* den folgenden Satz drucken ließ: „[E]s ist die Zukunft, die unserm Heute die Regel giebt." (Nietzsche, MA I Vorrede 7, KSA 2, S. 21)[12]

Eine solche zukunftsorientierte Perspektive führt für Sartre eindeutig zur absoluten Verantwortlichkeit jeder und jedes Einzelnen für die Welt im Ganzen:

---

12 Vanessa Lemm thematisiert diese Zukunftsorientierung der Verantwortung bei Nietzsche und Sartre mit dem Begriff des ‚Posthumen' in Lemm 2012.

> Die wesentliche Konsequenz unserer vorangehenden Ausführungen ist, daß der Mensch, dazu verurteilt, frei zu sein, das Gewicht der gesamten Welt auf seinen Schultern trägt: er ist für die Welt [*du monde*] und für sich selbst als Seinsweise verantwortlich. Wir nehmen das Wort „Verantwortlichkeit“ in seinem banalen Sinn von „Bewußtsein (davon), der unbestreitbare Urheber eines Ereignisses oder eines Gegenstandes zu sein“. In diesem Sinn ist die Verantwortlichkeit des Für-sich drückend, weil es das ist, wodurch geschieht, daß es eine Welt *gibt*; und weil das Für-sich auch das ist, das *sich sein macht*, muß es, was immer die Situation ist, in der es sich befindet, diese Situation gänzlich annehmen mit ihrem eigenen Widrigkeitskoeffizienten, und sei sie auch unerträglich. (Sartre 1976/1991: EN, S. 727/950)[13]

Wichtig ist, dass diese Verantwortlichkeit für Sartre im Grunde ein ‚Bewusstsein‘ ist, das den Einzelmenschen (das ‚Für-sich‘) mit dessen ‚Welt‘ verbindet, „ein Verbindungsnetz zwischen Mensch und Mensch“ (Nietzsche, FW 354, KSA 3, S. 591) gleichsam, solange man unter ‚Welt‘ ‚alle Welt‘ (*tout le monde*) versteht.[14] Aber dass man sich als Mensch a priori so fühlt, ist eine Annahme, die von der langen und langwierigen Vergesellschaftung der Menschheit absieht, mittels deren ein Freiheitsbewusstsein sich allmählich herausbilden konnte. Mit Recht bemerkt Sartre, dass seine Gedanken über die absolute Verantwortlichkeit einer und eines jeden wahrscheinlich nur den Moralisten angehen würden (vgl. Sartre 1976/1991: EN, S. 726–727/950), und nicht zuletzt deshalb, weil jedes Bewusstsein überhaupt schon – wenn wir Nietzsche trauen dürfen – samt der Sprache und jeder erst durch sie ermöglichten Interpretation soziomoralischen Charakter hat.[15] Wenn man sich eines Tuns bewusst wird, hat man es bereits in einen sozialen Nexus hineininterpretiert. Dass man sich auch dafür verantworten kann oder

---

13 Sartre gibt zu verstehen, dass sogar die alltäglichsten Gewohnheiten weltverändernd sind: „En ce sens on peut dire, si vous voulez, que chacun de nous fait l'absolu en respirant, en mangeant, en dormant ou en agissant d'une façon quelconque. Il n'y a aucune différence entre être librement, être comme projet, comme existence qui choisit son essence, et être absolu ; et il n'y a aucune différence entre être un absolu temporellement localisé, c'est-à-dire qui s'est localisé dans l'histoire, et être compréhensible universellement.“ (Sartre 1996: EH, S. 62)

14 Vgl. Sartre 1996: EH, S. 31–33. Es sei hier erwähnt, dass Sartre 1945 mit der Gründung der Monatszeitschrift *Les Temps modernes* die Verantwortung des Schriftstellers mit dessen politischem Engagement aufs engste verbunden sah und sich auch dazu äußerte. Vgl. dazu insbes. Münster 2007. Zum Verhältnis von Politik und Moral bei Sartre vgl. Betschart 2015.

15 Werner Stegmaier stellt fest, dass Nietzsches Begriff des Bewusstseins im Allgemeinen sich kaum vom Schuldbewusstsein im Besonderen unterscheidet. „Nach Nietzsches genealogischer Vermutung“ – so Stegmaier – „wäre [...] alles Bewusstsein schon soziales und, weil in Lebensnöten alles auf Verlässlichkeit und Verbindlichkeit ankommt, auch schon soziomoralisches Bewusstsein. Danach übersetzen wir alles, was wir zur Sprache und zu Bewusstsein bringen, schon in eine soziomoralische ‚Oberflächen- und Zeichenwelt‘, und, soweit wir voneinander einen verantwortlichen Sprachgebrauch erwarten, unweigerlich auch schon in eine Schuldbewusstseinswelt.“ (Stegmaier 2018, S. 405–406). Vgl. dazu auch Abel 2001, insbes. S. 22–27.

muss, macht nur in diesem Zusammenhang Sinn. Ohne die Sprache der Moral erübrigt sich ein Begriff des Tuns und der damit verbundenen Freiheit.

Auch Nietzsche schildert ein kosmisches Weltbild, nach dem jeder Einzelmensch einen wesentlichen Bestandteil des Ganzen ausmacht, jedoch kommt er in seiner Ab- und Einschätzung der Möglichkeit individueller Verantwortung in diesem Zusammenhang zum gegenteiligen Schluss. Er will uns nämlich belehren, dass

> [n]iemand [...] dafür verantwortlich [ist], dass er überhaupt da ist, dass er so und so beschaffen ist, dass er unter diesen Umständen, in dieser Umgebung ist. Die Fatalität seines Wesens ist nicht herauszulösen aus der Fatalität alles dessen, was war und was sein wird. (Nietzsche, GD Irrthümer 8, KSA 6, S. 96)

Nietzsches Ganzheitsrhetorik führt – wie bei Sartre auch – auf den Einzelmenschen zurück, der nun ein neues Selbstbild gewinnt als ein Unverantwortlicher jenseits der Moral von Schuld – und vielleicht auch jenseits der Gesellschaft, insofern für Nietzsche überhaupt keine Gesellschaft zu denken wäre ohne ein wie auch immer rudimentäres Schuldbewusstsein.[16]

Gewiss hat Nietzsche dieses neu gewonnene Bewusstsein der völligen „Unschuld des Werdens" (Nietzsche, GD Irrthümer 7, KSA 6, S. 96) als eine Überwindung des Schuldbewusstseins verstanden, jedoch ist und bleibt es bestenfalls nur ein Perspektivwechsel. Sartre mag freilich erwidern, Nietzsche unterschätze die weltverändernde Freiheit des Menschen und sei womöglich auch *de mauvaise foi*, und Nietzsche könnte Sartre beschuldigen, unsere menschliche Realität unkritisch auf das ganze Universum hinaus zu projizieren, sodass Sartre nie bewusst werden konnte, dass es bei unserer Teilnahme am Weltgeschehen gar nicht auf Verantwortung und Schuld ankommt. Jedoch scheint mir das, was die beiden Philosophen verbindet, aufschlussreicher zu sein. Sowohl Nietzsche als auch Sartre entwerfen schließlich ein kosmisches Weltbild, das auf der Interdependenz aller besteht, ohne das menschliche Dasein auf einen rein mechanistischen Determinismus zu reduzieren.[17] Während Nietzsche in *Menschliches, Allzumenschliches* einen gewissen Determinismus vertreten haben mag, gibt er diesen m. E. in den Werken ab 1886 dezidiert auf. Dem puren Determinismus wird wenigstens im

16 Siehe insbes. Nietzsche, GM II 4–15, KSA 5, S. 297–331.

17 Zu seinem Antimaterialismus in Verbindung mit seiner Determinismus-Kritik schreibt Sartre: „Tout matérialisme a pour effet de traiter tous les hommes, y compris soi-même, comme des objets, c'est-à-dire comme un ensemble de réactions déterminées, que rien ne distingue de l'ensemble des qualités et des phénomènes qui constituent une table ou une chaise ou une pierre." (Sartre 1996: EH, S. 58)

Nachlass der späten Achtzigerjahre insofern abgeschworen, als er von einem letzten Urheber ausgeht – von einem Gott etwa.[18] Aber wenn Gott tot ist, kann er nicht mehr wirken und sich für die Welt verantworten. Der selbst ernannte Atheist Sartre nimmt diese Verantwortung konsequent auf sich, aber die Gesellschaft bleibt noch als wirkungsmächtige Instanz: die anderen Menschen, die meine Hölle bedeuten.[19] Mit anderen Worten: die faktische Welt des An-sich, die mein Tun gewissermaßen mitbestimmt, beurteilt und verurteilt. Folglich kann es sein, dass auch die Aktivität, auf der Nietzsche wiederum noch in der *Genealogie* besteht, nicht minder ein soziomoralischer Begriff ist, den er wenigstens in dieser Streitschrift noch nicht bereit war aufzugeben.[20] Was den prozessualen Charakter des Subjekts angeht, erweist sich das Für-sich für Sartre auch als etwas Ereignishaftes, insofern es sich nur als Entwurf erfassen lässt, oder, um auf einen Ausdruck Nietzsches zu verweisen: „[E]s giebt kein ‚Sein' hinter dem Thun, Wirken, Werden" dieses Entwurfs, sondern er enthüllt sein Wesen vielmehr in jedem Augenblick, in dem er erscheint.[21]

So lassen sich die beiden Argumentationsweisen Sartres und Nietzsches wenigstens zeitweise vereinen: Wenn ein Individuum sich als ein Handelndes begreift, empfindet es zugleich ein gewisses Freiheitsbewusstsein und kann zudem nicht umhin, sich für sein Tun verantwortlich zu fühlen. Denkt es seine Freiheit konsequent zu Ende, versteht es auch, dass sein Handeln dazu beiträgt, die Welt so zu machen, wie sie ist. Nur ist dieses Bewusstsein für Sartre ontologisch gegeben, während es sich für Nietzsche historisch entwickelt (hat). Hier sei ausgeklammert, ob das existentielle Freiheitsbewusstsein, das wir alle mutmaßlich von Zeit zu Zeit empfinden, ein bloßes psychologisches Missverständnis ist, d. h. ob man nicht wirklich frei ist, sondern sich nur so glaubt. Nietzsche scheint mir hierauf keine klare Antwort zu geben, während Sartre – nicht überraschend – auf der absoluten Freiheit des Für-sich besteht. Aber insofern die *Genealogie* eine Grundlehre vom Primat der Aktivität enthält, legt sie zugleich diese absolute Freiheit – und eine ebenso schwere Verantwortung jeder und jedes Einzelnen für das Weltgeschehen – nahe. Nur stellt sich die Frage, ob diese Freiheit wiederum,

---

18 In Nietzsche, NL 1887, 9[91] hat es mit einem göttlichen Täter zu tun.

19 Zum Existentialismus als einem konsequenten Atheismus vgl. Sartre 1996: EH, S. 77.

20 Vgl. Nietzsche, GM II 12, KSA 5, S. 313 – 316. Allerdings verschwindet der Begriff der Aktivität in den Schriften von 1888, wie Marco Brusotti nachweist. Vgl. dazu Brusotti 2012.

21 Es gibt eine interessante Parallele zwischen Sartre und Nietzsche in diesem Punkt: Das Auffinden eines schuldigen Täters bei Nietzsche nämlich hat sein Korrelat im ‚Auf-sich-selbst-zurückgeworfen-Sein' des Schamgefühls bei Sartre, insofern das Tun in beiden Fällen kein Urheber- bzw. Täter-‚Subjekt' im Augenblick, da es sich ereignet, zu haben scheint, sondern erst im Nachhinein durch den Blick des Anderen ins Bewusstsein tritt. Vgl. dazu Strassberg 2004, S. 227.

die jede und jeder von uns mutmaßlich sind, sich auch zwangsläufig ihrer selbst bewusst sein muss.

Bibliographie

Abel, Günter (2001): „Bewusstsein – Sprache – Natur. Nietzsches Philosophie des Geistes". In: *Nietzsche-Studien* 30, S. 1–43.

Betschart, Alfred (2010): „Sartre and Adler: Existential Psychoanalysis and Individual Psychology". *17th Annual Conference*, UK Sartre Society. London.

Betschart, Alfred (2015): „Politik und Moral bei Jean-Paul Sartre". *Sartre und der Existentialismus.* Basisgruppe Philosophie der Universität Göttingen. Göttingen.

Betschart, Alfred (2019): „Sartre was not a Marxist". In: *Sartre Studies International* 25. Nr. 2, S. 77–91.

Brusotti, Marco (2001): „Wille zum Nichts, Ressentiment, Hypnose. ‚Aktiv' und ‚reaktiv' in Nietzsches *Genealogie der Moral*". In: *Nietzsche-Studien* 30, S. 107–132.

Brusotti, Marco (2012): „Reagieren, schwer reagieren, nicht reagieren. Zu Philosophie und Physiologie beim letzten Nietzsche". In: *Nietzsche-Studien* 41, S. 104–126.

Brusotti, Marco (2019): „Die Autonomie des ‚souveränen Individuums' in Nietzsches *Genealogie der Moral*". In: *Nietzsche-Studien* 48, S. 26–48.

Foucault, Michel (2005): *Schriften in vier Bänden. Dits et Ecrits*, Bd. IV: 1980–1988. Frankfurt a. M.

Geuss, Raymond (2016): „Historisierung, Aufklärung, Genealogie". In: *Filozofija i društvo* 27. Nr. 1, S. 189–202.

Jacobi, Klaus (1970): „J.-P. Sartres Weg zu einer Philosophie der konkreten Praxis". In: Paulus Engelhardt (Hrsg.): *Zur Theorie der Praxis: Interpretationen und Aspekte.* Mainz, S. 111–162.

Lemm, Vanessa (2012): „The Spectrality of Responsibility. The Posthumous in Nietzsche, Sartre and Derrida". In: Renate Reschke/Marco Brusotti (Hrsg.): *„Einige werden posthum geboren". Friedrich Nietzsches Wirkungen.* Berlin, Boston, S. 151–158.

Münster, Arno (2007): „Jean-Paul Sartre und die Verantwortung des Intellektuellen in der Gesellschaft". In: Horst Müller (Hrsg.): *Die Übergangsgesellschaft des 21. Jahrhunderts. Kritik, Analytik, Alternativen.* Norderstedt, S. 12–35.

Pippin, Robert B. (2010): *Nietzsche, Psychology, and First Philosophy.* Chicago.

Sartre, Jean-Paul (1976): *L'être et le néant. Essai d'ontologie phénoménologique.* Édition corrigée avec index par Arlette Elkaïm-Sartre. Paris.

Sartre, Jean-Paul (1991): *Das Sein und das Nichts. Versuch einer phänomenologischen Ontologie.* In: *Gesammelte Werke in Einzelausgaben, Philosophische Schriften.* Bd. 3. Reinbek b. Hamburg.

Sartre, Jean-Paul (1996): *L'existentialisme est un humanisme.* Paris.

Sommer, Andreas Urs (2012): *Kommentar zu Nietzsches* Der Fall Wagner. Götzen-Dämmerung. *Historischer und kritischer Kommentar zu Friedrich Nietzsches Werken.* Hrsg. v. d. Heidelberger Akademie der Wissenschaften. Bd. 6/1 (NK 6/1). Berlin, Boston.

Sommer, Andreas Urs (2019): *Kommentar zu Nietzsches* Zur Genealogie der Moral. *Historischer und kritischer Kommentar zu Friedrich Nietzsches Werken.* Hrsg. v. d. Heidelberger Akademie der Wissenschaften. Bd. 5/2 (NK 5/2). Berlin, Boston.

Stegmaier, Werner (2018): „Schuld und Rang. Nietzsches Vorschlag zur Überwindung des Schuldkomplexes". In: Andrea Christian Bertino (Hrsg.): *Europa im Geisterkrieg. Studien zu Nietzsche*. Cambridge, S. 401–419.

Strassberg, Daniel (2004): „Scham als Problem der psychoanalytischen Theorie und Praxis". In: *Schweizer Archiv für Neurologie und Psychiatrie* 155. Nr. 5, S. 225–228.

Zavatta, Benedetta (2009): „Die in der Sprache versteckte Mythologie und ihre Folgen fürs Denken. Einige Quellen von Nietzsche: Max Müller, Gustav Gerber und Ludwig Noiré". In: *Nietzsche-Studien* 38, S. 269–298.

Philosophie und Ästhetik

Josephina Bierl

Die ewige Wiederkunft und ihre Rezeption im *Mythe de Sisyphe* bei Albert Camus

Abstract: The eternal return and its reception in *The Myth of Sisyphus* by Albert Camus. The aim of this study is to shed new light on Camus's reception of Nietzsche and to examine the extent to which Nietzsche's philosophy of eternal return and Dionysian affirmation underlies Camus's early essay *The Myth of Sisyphus*. Not only does Nietzsche's doctrine constitute a theoretical basis for Camus's philosophy of the absurd but the affinity between the two philosophers also manifests itself on a philological level through their preference for aphorisms and essays and their recourse to poetic patterns and procedures. In both *Thus Spoke Zarathustra* and *The Myth of Sisyphus*, while rejecting any systematic and conceptualizing approach, criticism of language is combined with the search for a philosophical expression capable of accounting for the absurdity of human experience. In this sense, the eternal return is symptomatic of Nietzsche's philosophy, which does not rely on systematic argumentation, but rather, especially in his later works, proceeds on the basis of mytho-poetic images and patterns that cannot easily be translated into conceptual terms. Presumably, Camus's fascination with Nietzsche emanated precisely from these transitions between philosophy and literature.

Einleitung

Ausgangspunkt für die ewige Wiederkunft bei Friedrich Nietzsche und für Albert Camus' Interesse an diesem Gedanken ist der Bruch mit dem christlichen Glaubens- und Wertesystem. Sowohl in Nietzsches *Also sprach Zarathustra* als auch in Camus' *Mythe de Sisyphe* dient der Befund von dem Tod Gottes als Ansatzpunkt für eine Philosophie, die alternative Sinnfindungsmöglichkeiten in der Erhöhung des Diesseits und des gegenwärtigen Moments verortet. Deren Apotheose ist dabei im *Zarathustra* die Ausrufung der Lehre der ewigen Wiederkunft des Gleichen, die sowohl Absage an das eschatologische Weltbild als auch Symbol einer dionysischen Lebensbejahung ist.

Ziel des folgenden Beitrags ist es zu zeigen, in welchem Maß Nietzsches mythopoetische Denkfigur Camus' Philosophie des Absurden zugrunde liegt und bei ihm wie bei Nietzsche Teil des philosophischen Vorhabens ist, künstlerisch neue Werte zu schaffen. Zunächst sollen die Grundzüge vom *Mythe de Sisyphe* und

https://doi.org/10.1515/9783110760163-015

dessen Affinitäten zur Philosophie Nietzsches herausgestellt werden. Davon ausgehend gilt es, Camus' Auseinandersetzung mit dem Sisyphos-Mythos genauer zu beleuchten und seine prinzipiellen Eingriffe in den antiken Mythenstoff herauszustellen. In einem weiteren Schritt sei der Frage nach dem Verhältnis zwischen Literatur und Philosophie im Werk von Nietzsche und Camus nachzugehen. Abschließend soll Camus' spätere, weitaus kritischere Auseinandersetzung mit Nietzsche im *Homme revolté* mit einem Fokus auf die Diskussion der ewigen Wiederkunft angerissen werden.

Die Philosophie des Absurden und ihre Wegbereiter

Der Begriff des Absurden ist zentral in Camus' Frühwerk und soll deswegen im Folgenden kurz beleuchtet werden. Camus versteht das Absurde als unüberbrückbare Kluft und als Verlust der ursprünglichen Einheit zwischen dem Bewusstsein des Menschen und der ihn umgebenden Welt. Camus betont das Moment der Spaltung und der unüberbrückbaren Differenz: „Ce divorce entre l'homme et sa vie, l'acteur et son décor, c'est proprement le sentiment de l'absurdité" (Camus 2006: MS, S. 223) und „ce fossé ne sera jamais comblé" (Camus 2006: MS, S. 232). Das Absurde ist angelegt in dem menschlichen Verlangen nach Einheit, Harmonie und Klarheit der Welt. Diese erweisen sich jedoch durch die Abwesenheit einer Transzendenz, die dem menschlichen Leben Sinn verleihen könnte, auf den ersten Blick als unmöglich: „Ce qui est absurde, c'est la confrontation de cet irrationnel et de ce désir éperdu de clarté dont l'appel résonne au plus profond de l'homme" (Camus 2006: MS, S. 233).

Obwohl das Absurde erst im Zuge des französischen Existentialismus zu einem philosophischen Zentralbegriff geworden ist,[1] enthält bereits Nietzsches Philosophie deutliche Spuren eines absurden Weltbilds. Es ist jedoch zu betonen, dass dem Begriff des Absurden keine zentrale Rolle im Denken Nietzsches zukommt und er nur vereinzelt eingesetzt wird.

1 Begriffsgeschichtlich sei auf die unterschiedliche Verwendung des Begriffs des Absurden im Kontext der Logik und der Theologie hingewiesen. So bedient sich die christliche Theologie des Arguments, dass Absurdes, insofern es die menschliche Vernunft übersteigt, Gegenstand des Glaubens sei. Dieses Argument wird u. a. von Blaise Pascal und Søren Kierkegaard wieder aufgegriffen, deren Texte wiederum prägend für den französischen Existentialismus und die Umformung des Begriffs bei Albert Camus und Jean-Paul Sartre sind. Vgl. dazu Frackowiak 2007, S. 4–6.

Bereits in der Tragödienschrift behandelt Nietzsche die Gefahr der lähmenden Erkenntnis des Absurden in Zusammenhang mit der schopenhauerschen Willensresignation. Nietzsche selber leitete die Vorstellung eines absurden Daseins vermutlich von Arthur Schopenhauer ab, denn obgleich auch bei diesem der Begriff wenig Erwähnung findet, wird er genauso wie bei Nietzsche am Beispiel des Monologs von Hamlet diskutiert.[2] Im zweiten Band von *Die Welt als Wille und Vorstellung* erläutert Schopenhauer zudem die Absurdität in Bezug auf das „Fatum" und die „Nothwendigkeit" des menschlichen Daseins, denen nur in der Überwindung des Willens entronnen werden kann (Schopenhauer 2018b, S. 375).

Ähnlich wie für Camus enthält für Nietzsche der durch die Tragödie hervorgerufene dionysische Zustand ein „l e t h a r g i s c h e s Element", das eine „Kluft" zwischen „der alltäglichen und der dionysischen Wirklichkeit" öffnet (Nietzsche, GT 7, KSA 1, S. 56).

Das Eintauchen in die dionysische Welt erlaubt einen Blick in den Abgrund der Dinge, der menschliches Handeln lächerlich erscheinen lässt. Wie Hamlet verspürt der Mensch Ekel gegenüber der Welt und verfällt in eine „asketische, willenverneinende Stimmung", vor der ihn nur die ‚Zauberkraft' der Kunst bzw. die Kraft der apollinischen Illusion zu retten vermag:

> Hier, in dieser höchsten Gefahr des Willens, naht sich, als rettende, heilkundige Zauberin, die K u n s t; sie allein vermag jene Ekelgedanken über das Entsetzliche oder *Absurde des Daseins* in Vorstellungen umzubiegen, mit denen sich leben lässt. (Nietzsche, GT 7, KSA 1, S. 57, Kursivierung J.B.)

Durch die Synthese von Apollinischem und Dionysischem Prinzip führt die Tragödie den Zuschauer zu einer lustvollen dionysischen Bejahung des Lebens zurück.

In seiner späten Schaffensperiode greift Nietzsche die Thematik des Absurden im Kontext des Nihilismus wieder auf, den er mithilfe der Wiederkunftslehre zu entkräften versucht. So heißt es in seinem nachgelassenen Lenzer Heide-Fragment:

---

2 Siehe auch Schopenhauer 2018a, S. 404–405: „Jeder, welcher aus den ersten Jugendträumen erwacht ist, [...] wird [...] wohl das Resultat erkennen, daß diese Menschenwelt das Reich des Zufalls und des Irrthums ist, die unbarmherzig darin schalten, im Großen, wie im Kleinen, neben welchen aber noch Thorheit und Bosheit die Geißel schwingen: daher es kommt, daß jedes Bessere nur mühsam sich durchdrängt, [...] aber das Absurde und Verkehrte im Reiche des Denkens [...] eigentlich die Herrschaft behaupten; [...] Was aber das Leben des Einzelnen betrifft, so ist jede Lebensgeschichte eine Leidensgeschichte: [...] Der wesentliche Inhalt des weltberühmten Monologs im *Hamlet* ist, wenn zusammengefaßt, dieser: Unser Zustand ist ein so elender, daß gänzliches Nichtseyn ihm entschieden vorzuziehn wäre."

> Die Dauer, mit einem „Umsonst“, ohne Ziel und Zweck, ist der lähmendste Gedanke […].
>
> Denken wir diesen Gedanken in seiner furchtbarsten Form: das Dasein, so wie es ist, ohne Sinn und Ziel, aber unvermeidlich wiederkehrend, ohne ein Finale ins Nichts: „die ewige Wiederkehr“.
>
> Das ist die extremste Form des Nihilismus: das Nichts (das „Sinnlose“) ewig! (Nietzsche, NL 1886, 5[71], KSA 12, S. 213)

Obgleich im Fragment der Begriff des Absurden nicht fällt, erinnert die Darstellung der ewigen Wiederkunft als „der lähmendste Gedanke“ an die Beschreibung des Absurden in der *Geburt der Tragödie* und erlaubt somit die Assoziation von Absurdem und Nihilismus im Werk Nietzsches. Tatsächlich lässt sich das Absurde, auch im Sinne Camus', als spezifische Ausdrucksform des Nihilismus verstehen, die der Befindlichkeit der Zeit eigen ist.Der Begriff des Nihilismus hingegen bezieht sich auf eine Konstante in der westlichen Denktradition und hat somit eine globalere Ausrichtung (vgl. Weyembergh 2009, S. 9).

In dem vorliegenden Passus zeigt Nietzsche perspektivisch, dass die Vorstellung einer ewigen Wiederkehr in ihrer „furchtbarsten Form“ die lähmende Erkenntnis der Absurdität und Sinnlosigkeit des menschlichen Daseins auslösen und somit eine Quelle des Nihilismus darstellen kann. Ähnlich verfährt Camus, wenn er das Absurde in all seinen möglichen Konsequenzen, wie z. B. dem Suizid, durchdenkt. Dem Nihilismus und der Resignation ist gemäß beiden Philosophen nur durch eine lebensbejahende Einstellung entgegenzuwirken. Das Dionysische ist somit als Voraussetzung für das Leben im Diesseits zu verstehen, buchstäblich als Treiber des Apollinischen.

An dieser Stelle sei auf Nietzsches Unterscheidung zwischen einem aktiven und einem passiven Nihilismus hingewiesen: Der zitierte Passus bezieht sich auf den passiven Nihilismus, der sich als Äquivalent von Willensresignation, Leiden und Lebensmüdigkeit verstehen lässt. Der aktive Nihilismus hingegen ist Zeichen von Stärke und beinhaltet die Möglichkeit einer Neupositionierung.[3] Durch die Erzeugung neuer Werte erfolgt eine Überwindung des nihilistischen Befunds des Sinnmangels.

Roquentin in Sartres *Nausée* entspricht in diesem Hinblick dem passiven Nihilismus, Meursault und Caligula hingegen, die Protagonisten von Camus' Roman *L'Étranger* und seinem Drama *Caligula*, aber auch der absurde Mensch im *Mythe de Sisyphe* verkörpern den aktiven Nihilismus (vgl. Weyembergh 2009, S. 9).

Vergleichbar mit dem „größte[n] Schwergewicht“ (Nietzsche, FW 341, KSA 3, S. 570), als das Nietzsche die ewige Wiederkunft beschreibt, handelt es sich bei dem Bewusstsein um die Absurdität laut Camus um „[d]es vérités écrasantes“

3 Vgl. dazu Nietzsche, NL 1886, 5[71], KSA 12, S. 350 – 351.

(Camus 2006: MS, S. 303), eine erdrückende, oder wie auch Nietzsche in *Die Fröhliche Wissenschaft* sagt, eine „zermalmen[de]" Erkenntnis (Nietzsche, FW 341, KSA 3, S. 570). An dieser Stelle knüpft Camus mittels der Gewichtsmetapher sowohl an den Aphorismus aus *Die fröhliche Wissenschaft* als auch an den antiken Mythos des Sisyphos an. Dessen Felsen assoziiert er metaphorisch mit der Last des absurden Lebens. Anhand einer kreativen ‚Mythenkorrektur' vollzieht Camus jedoch eine entscheidende Umwertung und schafft neue Sinnpotenziale.[4] Der üblichen Darstellung des Sisyphos als die eines bestraften Büßers und Leidenden stellt Camus die These entgegen, man müsse sich Sisyphos glücklich vorstellen (vgl. Camus 2006: MS, S. 304). Das Wissen um die Vergeblichkeit seines Tuns ermögliche ihm, sich über sein Schicksal zu erheben und seine Arbeit als sein eigenes Werk zu proklamieren. Anhand von diesem Gedankengang plädiert Camus für eine Affirmation des Lebens im vollen Bewusstsein seiner Sinnlosigkeit. Indem er seinem Schicksal bejahend entgegentritt, entzieht er seiner Strafe und der göttlichen Macht außerdem jegliches Fundament.

Obgleich Camus' Sisyphos-Text keine explizit genannten Bezüge zu Nietzsches Lehre aufweist, stellt die ewige Wiederkunft meines Erachtens den Ausgangspunkt seiner Philosophie des Absurden dar. So liegt die zyklische, antiteleologische und ästhetisch ausgerichtete Denkfigur Nietzsches dem Essay als strukturelle Folie zugrunde. Der Essay gipfelt in der Korrektur des antiken Mythos, welche die theoretische Auseinandersetzung mit dem Absurden synthetisiert, erzählerisch gestaltet und bildlich vor Augen führt. Noch offensichtlicher als Meursault in *L'Étranger*, der jeden Tag dieselbe Alltagsroutine von „Lever, tramway, quatre heures de bureau ou d'usine [...]" (Camus 2006: MS, S. 227–228) wiederholt,[5] verkörpert Sisyphos geradezu das in der *Fröhlichen Wissenschaft* formulierte Gedankenexperiment Nietzsches, nämlich die unendliche und exakte Wiederholung aller Dinge.

Mit der Abkehr von der christlichen Jenseitsvorstellung vollziehen beide Denker eine Aufwertung des gegenwärtigen Moments, des Lebens und der Erde,

---

4 Für die Begriffsverwendung sei auf den von Martin Vöhler und Bernd Seidensticker herausgegebenen Sammelband verwiesen, in welchem der Begriff der ‚Mythenkorrektur' am Beispiel konkreter literarischer Texte definiert und so für die Literaturwissenschaft fruchtbar gemacht wird. Die Mythenkorrektur zeichnet sich durch eine stoffliche oder semantische Modifikation des narrativen Kerns sowie durch einen kritischen Rückbezug auf die mythische Vorlage aus. Vgl. insb. Hühn 2005, S. 355–368.

5 „Lever, tramway, quatre heures de bureau ou d'usine, repas, tramway, quatre heures de travail, repas, sommeil et lundi mardi mercredi jeudi vendredi et samedi sur le même rythme [...]" (Camus 2006: MS, S. 227–228).

die einerseits in *Also sprach Zarathustra* und anderseits gegen Ende des *Mythe de Sisyphe* in lyrischer Form zelebriert wird.

Nach dem Vorbild Zarathustras besingt der Dichterphilosoph die Schönheit der Erde:

> Dans l'univers soudain rendu à son silence, les mille petites voix émerveillées de la terre s'élèvent. Appels inconscients et secrets, invitations de tous les visages, ils sont l'envers nécessaire et le prix de la victoire. Il n'y a pas de soleil sans ombre, et il faut connaître la nuit. [...] Cet univers désormais sans maître ne lui paraît ni stérile ni futile. Chacun des grains de cette pierre, chaque éclat minéral de cette montagne pleine de nuit, à lui seul, forme un monde. (Camus 2006: MS, S. 303–304)

Der Passus erinnert an das Ende vom vierten Buch des *Zarathustra:* „Also sprach Zarathustra und verliess seine Höhle, glühend und stark, wie eine Morgensonne, die aus dunklen Bergen kommt" (Nietzsche, Za IV, KSA 4, S. 408). An dieser Stelle erlangt Zarathustra das höchste Bewusstsein seines Daseins und beschwört die Stunde des großen Mittags herauf, eine Metapher, die Camus übrigens im Schlusskapitel seines *Homme révolté* aufgreift.

Die Hell-dunkel-Metapher, die durch den Kontrast zwischen der „Morgensonne" und den „dunklen Bergen" erzeugt wird, übernimmt Camus ebenfalls in seinem Sisyphos-Kapitel. Sein Sisyphos zelebriert „[c]hacun des grains de cette pierre, chaque éclat minéral de cette montagne pleine de nuit" (Camus 2006: MS, S. 304).

Eine Ästhetik der Wiederholung

Bei Camus stellt das absurde Kunstwerk eine Möglichkeit dar, den Mangel an Kohärenz und Einheit im menschlichen Dasein auszugleichen und sich gleichzeitig, wie sein Sisyphos, über das Schicksal zu erheben: „Créer, c'est ainsi donner une forme à son destin" (Camus 2006: MS, S. 299). Der nihilistischen Resignation stellt Camus kreatives Schaffen entgegen: „Le sort de [l]a pensée n'est plus de se renoncer mais de rebondir en images" (Camus 2006: MS, S. 300).

Ähnlich wie bei Nietzsche zeichnet sich Camus' frühe Ästhetik durch eine bildhaft-sinnliche Sprache aus. Denn für ihn gilt es nicht mehr „d'expliquer et de résoudre, mais d'éprouver et de décrire" (Camus 2006: MS, S. 284). Anstatt zu versuchen, die absurde Situation zu erklären oder in ihr eine höhere Finalität zu finden, soll das absurde Kunstwerk sie lediglich reproduzieren. In dieser Hinsicht stellt Kunst eine produktive Wiederholung und Intensivierung des Lebens dar: künstlerisches Schaffen bedeutet zweimal zu leben: „Créer, c'est vivre deux fois" (Camus 2006: MS, S. 283–284), heißt es bei ihm.

Damit etabliert Camus im *Mythe de Sisyphe* eine Ästhetik der Wiederholung. Kunst sei eine Wiederholung und ein Spiegel der Welt. Von dieser quasi realistischen Ästhetik löst er sich jedoch in seinen späteren Schriften, insbesondere im *Homme révolté*. Dem Künstler kommt im *Mythe de Sisyphe* als ‚homme absurde' par excellence, eine zentrale Rolle zu. Wie Nietzsches Zarathustra erschafft er seine eigenen Werte in dem „[u]nivers magnifique et puéril du créateur" (Camus 2006: MS, S. 284).

Auch für Nietzsche ist der schöpferische Akt von großer Bedeutung. Die Erzeugung neuer Werte, die Nietzsche als Antwort auf den Nihilismus fordert, ist stark ästhetisch konnotiert. Die Emphase, mit der er, besonders im *Zarathustra*, die Bedeutung des künstlerischen Schaffens betont, zielt auf die Möglichkeit einer kreativen sinnstiftenden Umwertung, durch die die Welt und der Mensch in ihr einen neuen Sinn erlangen. Im *Zarathustra* ist der Schaffende „[d]er, welcher des Menschen Ziel schafft und der Erde ihren Sinn giebt und ihre Zukunft" (Nietzsche, Za III, KSA 4, S. 247). Künstlertum steht im Kontext einer Vision der Erhöhung, deren Symbol und Apotheose der Übermensch ist (vgl. Schacht 2020, S. 125). In diesem Sinne gilt für Nietzsche wie für Camus Kunst als ein Prinzip der Lebensbejahung, weil sie die Möglichkeit bietet, der Absurdität entgegenzutreten und den Nihilismus zu überwinden. Die Kunst verleihe Sinn in einem sinnlosen und leidgeprägten Leben, meint Camus mit Verweis auf Nietzsche: „L'art et rien que l'art dit Nietzsche, nous avons l'art pour ne point mourir de la vérité" (Camus 2006: MS, S. 283).[6]

Es ist unumstritten, dass Camus Nietzsche vor allem in der frühen Schaffensperiode zu seinem philosophischen Vorbild erhebt. Weniger beachtet wurde jedoch, dass er ihm auch als künstlerisches Vorbild dient. So bezeichnet er Nietzsche als den einzigen Künstler, der eine absurde Ästhetik entwickelt habe:

> Nietzsche paraît être le seul *artiste* à avoir tiré les conséquences extrêmes d'une esthétique de l'Absurde, puisque son ultime message réside dans une lucidité stérile et conquérante et une négation obstinée de toute consolation surnaturelle. (Camus 2006: MS, S. 314, Kursivierung J.B.)

Das Verhältnis von Dichtung und Philosophie

Von diesem Befund ausgehend stellt sich die Frage nach dem Verhältnis zwischen Literatur und Philosophie im Werk von Camus, eine Frage, die auch in Bezug auf Nietzsche immer wieder diskutiert wird. Camus' essayistisches Werk steht in ei-

6 Siehe auch Nietzsche, NL 1888, 16[40], KSA 13, S. 500.

nem produktiven Verhältnis zu seinem literarischen Schaffen. Jeder der drei von Camus geplanten Zyklen[7] enthält essayistische, narrative und dramatische Texte, die sich auf unterschiedliche Weise mit dem übergeordneten Thema auseinandersetzen. Es ist zu vermuten, dass gerade die Übergänge zwischen Philosophie und Literatur Camus' Begeisterung für Nietzsche ausmachten.

Sowohl bei Nietzsche als auch bei Camus geht die Verweigerung der systematischen Argumentation im Sinne der traditionellen Philosophie einher mit der Wahl einer philosophisch-poetischen Sprache, die zwischen philosophischem Essay und literarischem Schreiben oszilliert und von lyrischen, dramatischen und mythischen Elementen durchzogen ist.[8] Wie der Aphorismus erlaubt das offene Genre des Essays eine stilistische und inhaltliche Freiheit bei der Gedankenführung.

Mit seinen Essays wirkt Camus der traditionellen Gegenüberstellung von Philosophie und Literatur entgegen und überführt die Philosophie in die Dichtung. Er postuliert eine Kunst, in der sich Denken und künstlerische Schöpfung verbinden. Für ihn gilt, „[l]e philosophe, même s'il est Kant, est créateur", denn „[p]enser, c'est avant tout vouloir créer un monde" (Camus 2006: MS, S. 287). Wie ein Schriftsteller habe der Philosoph seine Symbole, Figuren und Handlung. Gleich wie für Camus die großen Philosophen diejenigen sind, bei denen die Philosophie in die Dichtung übergeht, sind die großen Schriftsteller die „romanciers philosophes" (Camus 2006: MS, S. 288). Diese gehen in ihren Texten anhand von Bildern und Szenarien philosophischen Fragen nach: „Un roman n'est jamais qu'une philosophie mise en images. Et dans un bon roman, toute la philosophie est passée dans les images" (Camus 2006: MS, S. 794). Der Anspruch an die Philosophie, in Bildern zu schreiben, zeugt von einem größeren Vertrauen in die sinnliche Erfahrung als in die philosophische Logik und Konzepte.

In dieser Hinsicht besteht eine starke Affinität zu Nietzsche, der ebenfalls sein Misstrauen gegenüber der Sprache und den in die Sprache eingegangenen Konzepten bekundet. In seinem frühen Aufsatz *Ueber Wahrheit und Lüge im aussermoralischen Sinne* argumentiert er, dass Konzepte das Resultat eines dem Men-

7 Camus konzipierte sein Werk in drei aufeinanderfolgenden Zyklen, die jeweils narrative, dramatische und essayistische Texte vereinen. Der Zyklus des Absurden steht im Zeichen des antiken Sisyphos-Mythos und enthält neben dem hier besprochenen Essay u. a. auch den berühmten Roman *L'Étranger* und den Dramentext *Caligula*. Der darauf aufbauende Zyklus der Revolte erhebt Prometheus zu seinem Vorbild und umfasst u. a. *L'Homme Revolté*, *La Peste* und *Les Justes*. Der dritte Zyklus hätte, wie Camus in seinen Carnets und Briefen mehrfach betonte, im Zeichen der Liebe stehen sollen, blieb aufgrund des frühen Todes des Schriftstellers allerdings unvollendet.

8 Weiterführende Lektüren zur lyrischen und dramatischen Gestaltung des *Mythe de Sisyphe:* Mailhot 1980 und Ferrage 1997.

schen eigenen Prozesses der Metaphernbildung seien. Begriffe sind für Nietzsche folglich pseudo-objektive Wahrheiten, deren vorkonzeptueller Ursprung vergessen worden sei. Wie die Denkfigur der ewigen Wiederkunft verdeutlicht, stützt Nietzsche sein Denken auf nichtbegriffliche Substrate und potenziert insbesondere im *Zarathustra* den poetischen Aspekt der Sprache. Auf dieser Grundlage untersucht er die Konzepte in Bezug auf ihre bildlich-figurativen Ursprünge.

In diesem Zusammenhang sei auf Nietzsches frühe Schrift *Richard Wagner in Bayreuth* verwiesen, in der er nichtbegriffliches und bildhaftes Sprechen mit mythischem Denken gleichsetzt:

> Das Dichterische in Wagner zeigt sich darin, dass er in sichtbaren und fühlbaren Vorgängen, nicht in Begriffen denkt, das heisst, dass er mythisch denkt, so wie immer das Volk gedacht hat. Dem Mythus liegt nicht ein Gedanke zu Grunde, wie die Kinder einer verkünstelten Cultur vermeinen, sondern er selber ist ein Denken; er theilt eine Vorstellung von der Welt mit, aber in der Abfolge von Vorgängen, Handlungen und Leiden. (Nietzsche, WB 9, KSA 1, S. 485)

Diese Beschreibung lässt sich eins zu eins auf den *Zarathustra* übertragen. Zentral ist dabei die Vorstellung eines Denkens in „Vorgängen", bei denen es sich, wie auch Claus Zittel argumentiert, um die Verknüpfung einzelner Bilder und Eindrücke unterhalb der Bewusstseinsschwelle handelt. Diese unbewusst verknüpften Bildketten stehen am Ursprung von den später gebildeten Begriffen (Zittel 2000, S. 95–96). Auch in *Ueber Wahrheit und Lüge im aussermoralischen Sinne* bekräftigt Nietzsche bereits, er bekämpfe die „Herrschaft der Abstractionen" (Nietzsche, WL 1, KSA 1, S. 881) zugunsten einer intuitiven schöpferischen Bildersprache. Diese lasse sich, so Theo Meyer, in der „lyrischen Kunstprosa" des *Zarathustra* erkennen, in der Nietzsche durch den kreativen Einsatz metaphorischer, rhetorischer und syntaktischer Mittel, die Aufhebung von Gattungsgrenzen und die Erzeugung einer eigenen Zeichensprache die Möglichkeiten von Sprache ausschöpft (Meyer 1993, S. 131).

Das spielerisch-kreative Schaffen von bildlicher Sprache stellt für Nietzsche eine poetische Strategie dar, dem Bann der christlichen Werte und des Nihilismus zu entkommen. Spielerisch gilt es, neue Werte und Ausdrucksformen zu schaffen. Somit lässt sich die Poesie als Grundlage eines neuen tragisch-dionysischen Zeitalters jenseits von Gott und Götzen verstehen. Nietzsche erhofft sich von seiner Lehre eine Erneuerung der europäischen Gesellschaft sowie eine Überwindung des dekadenten Gegenwartsmenschen. Der Kunst, die Nietzsche bereits in *Geburt der Tragödie* als ‚Stimulans' und Rechtfertigung des Lebens qualifiziert, kommt dabei eine erhebliche Rolle zu, wie u. a. an der poetischen Form des *Zarathustra* zu erkennen ist. In dem Versuch einer Selbstkritik an der *Geburt der Tragödie* kritisiert er noch: „Sie hätte s i n g e n sollen, diese ‚neue Seele' – und nicht reden!

Wie schade, dass ich, was ich damals zu sagen hatte, es nicht als Dichter zu sagen wagte: ich hätte es vielleicht gekonnt!" (Nietzsche, GT Selbstkritik 3, KSA 1, S. 15)[9] Genau das scheint Nietzsche im *Zarathustra* realisiert zu haben: Er stellt seine Lehre „singend", tanzend und „als Dichter" dar, wie an dem poetisch-lyrischen Stil und an den Tanzliedern deutlich wird. Auf diese Weise verleiht er dem Text einen performativen Charakter, der im weitesten Sinne an den kultischen Ursprung des griechischen Mythos erinnert.

Anders als bei Camus äußert sich bei Nietzsche die ‚Arbeit am Mythos' nicht im Rückgriff auf oder in der Variation tradierter Mythen. Mit Hans Blumenberg kann man Nietzsches Wiederkunftslehre jedoch als ‚formale Mythenrezeption' erkennen, insofern diese die zyklische Mythenstruktur aufgreift und somit die Form selbst zu ihrem Thema erhebt (Blumenberg 1979, S. 355). *Also sprach Zarathustra* weist zwar grundlegende Merkmale mythischen Erzählens auf. Angesichts der Definition des Mythos als kollektives Produkt einer ganzen Gemeinschaft wird jedoch deutlich, dass es sich um keinen Mythos im eigentlichen Sinne handelt, sondern das Mythische lediglich fingiert wird.

Indem Camus in *Mythe de Sisyphe* ebenfalls philosophische Konzeptualisierungen ablehnt und nicht-konzeptuelle Gesichtspunkte in den Vordergrund rückt (vgl. Champigny 1980, S. 54), knüpft er an Nietzsches Verfahren an. Wie schon bei Nietzsche lässt sich auch bei Camus eine Bewegung weg vom Begrifflichen hin zu Szenarien und Bildern beobachten. Somit setzt Camus seine Forderung nach einer Philosophie, die in sinnlich greifbaren Bildern verfährt, selbst um. Damit rückt er die poetische und dramatische Gestaltung seiner *Essais* in den Vordergrund.

Der Einsatz von mythopoetischen Elementen, die bei Camus zwar nicht wie bei Nietzsche überhandnehmen, sondern nur in geringem Maße auftreten, zeugt von der Suche nach einer neuen philosophisch-literarischen Sprache jenseits systematischer Argumentation. Die intertextuelle und mythologische Verweisstruktur sowie die Heraufbeschwörung mythisierender Bilder ermöglichen Camus ein bildhaftes und nicht-konzeptuelles Vorgehen. Damit geht er einen Schritt weiter als seine Zeitgenossen Jean-Paul Sartre, André Gide und Jean Anouilh, um nur einige zu nennen, die den antiken Mythos in erster Linie zur Untermauerung ihrer Argumentation verwenden. Camus bezieht sich auf den Mythos als Denkform und ist in diesem Hinblick Ernst Cassirer sehr verbunden. Cassirer beschreibt den Mythos in Rahmen seiner umfassenden Phänomenologie und Strukturanalyse als autonome Denkform. Hinsichtlich seiner Fähigkeit, menschliche Eindrücke und Wahrnehmungen zu strukturieren, stehe der Mythos auf einer Ebene

---

9 Damit hätte Nietzsche gewissermaßen eine Übereinstimmung vom Inhalt der Tragödie und ihrer mythisch-musikalischen Form erzielt.

mit Sprache und Wissenschaft sowie anderen symbolischen Formen wie Kunst und Religion. Ihm zufolge zeichnet sich das mythische Denken vor allem durch eine symbolische Sinnkonstruktion aus (vgl. Cassirer 1964, insb. S. 51–52).

Wie Nietzsche beruft sich Camus auf den Mythos als Form einer sinnlich-ästhetischen Weltwahrnehmung. Er macht deutlich, dass das Gefühl des Absurden nicht empirisch-wissenschaftlich, sondern allein sinnlich erfasst werden kann. Das Absurde ist bei ihm ein Gefühl oder gar eine Leidenschaft.[10] Das sinnlich-ästhetische Erleben der Welt, das im *Mythe de Sisyphe* suggeriert wird, vermag den Menschen angesichts des Absurden vom Nihilismus abzuwenden. Für den absurden Menschen gilt es, in der immanenten Erfahrung und der Lebensbejahung zu der „vérité d'être et de sentir" (Camus 2006: MS, S. 215) zurückzufinden.

Darüber hinaus ermöglicht bei Camus der Mythos dank der ihm eigenen Rückbindung an das Ursprüngliche eine Strukturierung der absurden Welt (vgl. Cassirer 1964, S. 130). Der Verweis auf den antiken Mythos vermittelt einerseits die Erfahrung des Absurden am Beispiel des Sisyphos, anderseits verleiht er ihr den Status der Allgemeingültigkeit und des Immer-Schon-Dagewesenen und macht sie dadurch erträglich.

Die ewige Wiederkunft als Ansatzpunkt von Camus' Kritik im *Homme révolté*

Abschließend sei auf Camus' späte kritischere Auseinandersetzung mit Nietzsche und dessen Wiederkunftslehre im *Homme révolté* verwiesen. Wird im *Mythe de Sisyphe* noch jegliche explizite Erwähnung der ewigen Wiederkunft vermieden, greift Camus im *Homme révolté* die Denkfigur dezidiert auf. Er vollzieht Nietzsches Affinität zu vorsokratischen Kreisfiguren und dem heraklitischen Prinzip des spielenden Weltkindes kritisch nach:

> Nietzsche retourne alors aux origines de la pensée, aux présocratiques. Ces derniers supprimaient les causes finales pour laisser intacte l'éternité du principe qu'ils imaginaient. Seule est éternelle la force qui n'a pas de but, le ‚Jeu' d'Héraclite. Tout l'effort de Nietzsche est de démontrer la présence de la loi dans le devenir et du jeu dans la nécessité. (Camus 2008: HR, S. 123)

---

10 Camus spricht vom „sentiment de l'absurdité" (Camus 2006: MS, S. 223). Wer dem Absurden anhand einer lebensbejahenden Einstellung entgegentritt, lebe es als eine „passion sans lendemain" (Camus 2006: MS, S. 300).

Indem er betont, die Vorsokratiker hätten jegliches Ziel und jegliche Finalität geleugnet, relativiert er die Legitimität ihres philosophischen Verfahrens. Es ist augenfällig, dass er sich von dem Glauben an eine ewige Wiederkunft hier distanziert. Er dementiert sie als Fiktion, die auf der ‚Imagination' der Präsokratiker und dem „effort", der Argumentationskraft, Nietzsches beruhe. Die Vorstellung der Unschuld und der Notwendigkeit des Werdens, die Nietzsche von Heraklit ableitet, durchzieht Nietzsches Spätwerk und bildet in Form der leitmotivischen Spielmetapher die Grundlage der antiteleologischen Wiederkunftslehre.

Im *Homme révolté* beruht Camus' kritische Einstellung gegenüber der Denkfigur darauf, dass er die Bejahung derselben als „la soumission absolue de l'individu au devenir" (Camus 2008: HR, S. 123) betrachtet. Sie führe zu einer „divinisation de la fatalité" (Camus 2008: HR, S. 122) und der Welt. Während Camus im *Mythe de Sisyphe* eine absolute Bejahung des Lebens und des Schicksals im Sinne Nietzsches noch als einzige mögliche Antwort auf den Nihilismus postuliert, relativiert er im *Homme révolté* seine frühere Position und kritisiert Nietzsches *amor fati:* „Dire oui à tout suppose qu'on dise oui au meurtre" (Camus 2008: HR, S. 126). Da eine absolute Bejahung auch die Zustimmung zum Bösen bedeuten würde, wären dadurch auch Mord und Gewalt legitimiert.

Geprägt von den Schrecken des Zweiten Weltkriegs wirft Camus nun Nietzsche vor, durch seine Schriften die Schreckensherrschaft der Nationalsozialisten vorbereitet zu haben. Er habe ihnen philosophisches Material geliefert, in dem sie eine Legitimation für ihr Vorgehen finden konnten. In Nietzsches Wiederkunftslehre erkennt Camus eine Rechtfertigung für den Holocaust, da sie das Leben einzelner Individuen oder ‚Rassen' gegenüber der Ewigkeit der menschlichen Spezies abwerte:

> Quand il demandait que l'individu s'inclinât devant l'éternité de l'espèce et s'abîmât dans le grand cycle du temps, on a fait de la race un cas particulier de l'espèce et on a plié l'individu devant ce dieu sordide. (Camus 2008: HR, S. 124)

Es stellt sich die Frage, inwiefern Camus' teils sehr vehemente Kritik gerechtfertigt ist. Die Vorstellung einer ewigen Wiederkunft identischer Inhalte und Vorgänge grenzt zwar an eine gewisse Fatalität, doch steht bei Nietzsche genauso wie bei Camus die ästhetische Bewältigung der unabänderlichen Situation im Vordergrund. Es mutet außerdem problematisch an, den Philosophen für seine Rezeption *post mortem* verantwortlich zu machen.

Dessen scheint sich Camus bewusst zu sein, wenn er seine Kritik bereits selbst entschärft, indem er betont, es sei trotz allem unmöglich, Nietzsche mit dem führenden NS-Theoretiker und -Ideologen Alfred Rosenberg gleichzusetzen. Aus diesem Grund plädiert er: „Nous devons être les avocats de Nietzsche“ (Camus 2008: HR, S. 125).

Fazit

Trotz der Präsenz eines zyklischen Denkens, das der Philosophie Camus' zugrunde liegt und sich vermutlich an Nietzsches Figur der ewigen Wiederkunft orientiert, konstatiert man ebenso eine frappierende Diskrepanz zu Nietzsches Lehre. So tut man gut daran, sich zu vergegenwärtigen, dass Camus die ewige Wiederkunft keineswegs wörtlich versteht. Der ewigen Wiederholung stellt er die Vergänglichkeit des menschlichen Lebens entgegen: „À bout de tout cela, malgré tout, est la mort. Nous le savons. Nous savons aussi qu'elle termine tout“ (Camus 2006: MS, S. 281). Sogar Sisyphos werde durch den Tod erlöst, was der Logik des Mythos respektive jener der unendlichen Strafe widerspricht. Camus' Interesse an Nietzsches Wiederkunftslehre lässt sich somit weniger in einer Diskussion des objektiven Wahrheitsgehalts erkennen, sondern beruht vielmehr auf den Implikationen, die sie mit sich führt. Die Wiederholung enthält für Camus einen Moment künstlerischer Produktivität. Der Notwendigkeit der dionysischen Bejahung, die Camus von Nietzsche herleitet, wohnt bei ihm ebenfalls eine kreative Kraft inne. In der lebensbejahenden Einstellung und im künstlerischen Schaffen kann das Absurde unterminiert und der Nihilismus überwunden werden (vgl. Rosenthal 1977, S. 99). Insgesamt zeigt sich, dass bei Camus keine wörtliche, sondern eine primär ästhetische Deutung der ewigen Wiederkunft vorliegt.

Bibliographie

Blumenberg, Hans (1979): *Die Arbeit am Mythos.* Frankfurt a. M.

Camus, Albert (2006): „Le mythe de Sisyphe“. In: Jacqueline Lévi-Valensi/Raymond Gay-Croisier (Hrsg.): *Œuvres* Complètes. Bibliothèque de la Pléiade. Bd. 1. Paris, S. 217–315.

Camus, Albert (2008): „L'Homme révolté“. In: Raymond Gay-Croisier (Hrsg.): *Œuvres* Complètes. Bibliothèque de la Pléiade. Bd. 3. Paris, S. 60–324.

Cassirer, Ernst (1964): *Philosophie der symbolischen Formen.* Bd. 2. Darmstadt.

Champigny, Robert (1980): „Compositions philosophiques et concepts“. In: Raymond Gay-Crosier (Hrsg.): *Albert Camus 1980. Second International Conference Gainesville.* Gainesville, S. 49–54.

Ferrage, Hervé (1997): „Lyrisme et histoire“. In: Jacqueline Lévi-Valensi/Agnès Spiquel (Hrsg.): *Camus et le lyrisme.* Mayenne, S. 9–18.

Frackowiak, Ute (2007): „Absurd“. In: Harald Fricke (Hrsg.): *Reallexikon der deutschen Literatur.* Bd. 1. Berlin, New York.

Hühn, Helmut (2005): „Revolte gegen das Absurde: Sisyphos nach Camus“. In: Martin Vöhler/Bernd Seidensticker (Hrsg.): *Mythenkorrekturen. Zu einer paradoxalen Form der Mythenrezeption*. Berlin, S. 345–368.

Mailhot, Laurent (1980): „Aspects théâtraux des récits et essais de Camus“. In: Raymond Gay-Crosier (Hrsg.): *Albert Camus 1980. Second International Conference Gainesville*. Gainesville, S. 153–160.

Meyer, Theo (1993): *Nietzsche und die Kunst*. Tübingen, Basel.

Rosenthal, Bianca (1977): *Die Idee des Absurden. Friedrich Nietzsche und Albert Camus*. Bonn.

Schacht, Richard (2020): „Nietzsche. After the death of God“. In: Steven Crowell (Hrsg.): *The Cambridge Companion to Existentialism*. Cambridge, S. 111–136.

Schopenhauer, Arthur (2018a): „Die Welt als Wille und Vorstellung I“. In: *Werke in zehn Bänden*. Zürcher Ausgabe. Der Text folgt der historisch-kritischen Ausgabe von Arthur Hübscher (3. Aufl. Brockhaus, Wiesbaden 1972). Bd. 1–4. Angelika Hübscher (Hrsg.). Zürich.

Schopenhauer, Arthur (2018b): „Die Welt als Wille und Vorstellung II“. In: *Werke in zehn Bänden*. Zürcher Ausgabe. Der Text folgt der historisch-kritischen Ausgabe von Arthur Hübscher (3. Aufl. Brockhaus, Wiesbaden 1972). Bd. 1–4. Angelika Hübscher (Hrsg.). Zürich.

Weyembergh, Maurice (2009): „L'Absurde“. In: Jeanyves Guérin (Hrsg.): *Dictionnaire Camus*. Paris.

Zittel, Claus (2000): *Das ästhetische Kalkül von Friedrich Nietzsches „Also sprach Zarathustra“*. Würzburg.

Jens Bonnemann

Die Geburt der Kunst aus dem Ekel am Dasein. Die philosophische Relevanz ästhetischer Erfahrungen bei Nietzsche und Sartre

Abstract: The Birth of Art from the Nausea Against Existence. The Philosophical Relevance of Aesthetic Experiences in Nietzsche and Sartre. Comparisons between Nietzsche and Sartre are a rather unusual subject in philosophical study, all the more in regard to the field of aesthetics. In this paper, however, I lay the focus on the early thinking of both philosophers, namely on Nietzsche's *The Birth of Tragedy* and Sartre's *Nausea*. The central question I envisage in this essay is to what extent both Nietzsche's and Sartre's notions of art are grounded on a nihilistic-aesthetic worldview. As to that, I will point out the similarities and differences between these two positions within philosophical aesthetics, in both of which art takes over the function of religion.

Was die Beiträge Nietzsches und Sartres zur philosophischen Ästhetik betrifft, so sind ihre Verwandtschaften und Differenzen bisher fast gar nicht, geschweige denn im gebührenden Maße, zum Gegenstand der Forschung gemacht worden.[1] Das ist umso mehr verwunderlich, weil ein gemeinsamer Grundzug ihrer Ansätze ganz entschieden dazu einlädt: Denn für beide Philosophen wurzelt das künstlerische Schaffen in einer ausgesprochen düsteren Erfahrung, welche gleichzeitig nicht weniger als die Keimzelle des philosophischen Denkens Nietzsches wie auch Sartres bildet. So sind beide Philosophen nicht von ungefähr konsequente Vertreter eines atheistischen Nihilismus; und beide stimmen darin überein, dass dasjenige, was ihr eigenes Denken in Atem hält, auch das Grundmotiv einer jeden Kunst sei.

Der vorliegende Aufsatz versteht sich nun als Beitrag dazu, eine alles andere als selbstverständliche Leerstelle der Forschungsliteratur zu füllen. Um die Rolle der ästhetischen Erfahrung für das gesamte Denken im einen wie im anderen Fall

---

1 Studien über das Verhältnis zwischen Nietzsche und Sartre nehmen in der philosophischen Forschungsliteratur ohnehin eher den Status einer Nischenexistenz ein. Das Thema der philosophischen Ästhetik wird dabei allenfalls am Rande gestreift: Siehe hierzu Beam 1998; Daigle 2004; Louette 1996; Mitchell 2020; Noudelmann 2014.

https://doi.org/10.1515/9783110760163-016

mit der nötigen Schärfe zu konturieren, konzentrieren sich die Ausführungen im Folgenden auf das jeweilige Frühwerk, d. h. zum einen auf Nietzsches *Geburt der Tragödie* (1872) und zum anderen auf Sartres ersten Roman *Der Ekel* (1938). Hier wie dort geht es um verstörende intuitive Einsichten, denen einerseits der Status von metaphysischen Offenbarungen zugeschrieben wird und die andererseits das Fundament für ein Kunstverständnis bereitstellen, mit dem sich eine Art von Heilsversprechen verknüpft.

Vorab sind allerdings ein paar Worte über Sartres Verhältnis zu Nietzsche angebracht: Während andere Existenzphilosophen wie Martin Heidegger oder Karl Jaspers mit umfangreichen Monografien Nietzsche gewürdigt haben, bringt Sartre diesem deutschen Denker offensichtlich – zumindest in seinen philosophischen Werken – kaum ein nennenswertes Interesse entgegen. Es finden sich lediglich hin und wieder verstreute Erwähnungen und gelegentliche Zitate, wobei Sartre noch am ausführlichsten in seinem Buch *Saint Genet, Komödiant und Märtyrer* (1952) auf Nietzsche zu sprechen kommt. Aber auch dort geschieht das nur, um dessen Weltsicht mit der des zeitgenössischen Schriftstellers Jean Genet zu vergleichen. Wenn Sartre also keine Verwendung für Nietzsche als Stichwortgeber für die eigene Theorie hat, zieht er ihn jedenfalls nur als Thema einer existentiellen Psychoanalyse in Betracht, *über* das man spricht, keineswegs aber als einen Philosophen, *mit* dem man spricht bzw. diskutiert (Sartre 1982: SG, S. 539–546).

1 Nietzsche und die ästhetische Rechtfertigung des menschlichen Daseins

Im ersten Teil des Aufsatzes soll nun aufgewiesen werden, auf welche Weise für Nietzsche die Kunst eine metaphysische Erkenntnis über das Wesen der Welt bietet und gleichzeitig dazu verhilft, eine solche Erkenntnis überhaupt ertragen zu können. In seinem ersten größeren Werk, *Die Geburt der Tragödie aus dem Geiste der Musik*, bezieht sich Nietzsche auf die griechische Mythologie, wenn er an eine Sage anknüpft, die sich unter anderem bei Ovid findet: Der phrygische König Midas fängt Silen, Begleiter des Dionysos und Anführer des Chors der Satyrn, weil er von ihm eine Auskunft darüber erhalten will, was das Beste für den Menschen sei. Zur Antwort bekommt er, das Beste sei, überhaupt gar nicht erst geboren worden zu sein, und dementsprechend das Zweitbeste, möglichst bald zu sterben (Nietzsche, GT, KSA 1, S. 35). Wie Nietzsche hervorhebt, ist allein der dionysische Mensch dazu berufen, das wahre Wesen der Dinge – die dionysische Wahrheit schlechthin – zu erfassen. Aber hierbei handelt es sich schließlich um eine

„grauenhafte Wahrheit" (Nietzsche, GT, KSA 1, S. 57), die das „Entsetzliche oder Absurde des Menschseins" (Nietzsche, DW, KSA 1, S. 566) in aller Schärfe offenbar werden lässt.

So ist bei Nietzsche nun – wie später auch bei Sartre – von einem Ekel die Rede (Nietzsche , GT, KSA 1, S. 18, 56), der sich nicht auf einzelne Objekte, sondern auf das gesamte menschliche Leben richtet. Angesichts eines solchen Ekels vor dem Entsetzlichen und Absurden des Menschseins drängt sich, wie Nietzsche fortfährt, das Theodizee-Problem auf, also die Frage, warum Gott eine solche schreckliche, ja einfach nur ekelhafte Welt erschaffen hat und inwiefern sich die Existenz einer solchen Welt überhaupt rechtfertigen lässt. Aus der griechischen Antike entnimmt Nietzsche hierauf eine Antwort, die er sich offensichtlich auch selbst zu eigen macht: So taucht mehrfach in *Die Geburt der Tragödie* die schillernde These auf, dass das menschliche Dasein ausschließlich als ästhetisches Phänomen gerechtfertigt sei (Nietzsche, GT, KSA 1, S. 17, 47, 152). Das Einzige, was das Leben also für die Menschen erträglich machen kann, ist der Schein, die Täuschung – und eben die Kunst. So grauenhaft dieser Ekel auch sein mag, er ist also für Nietzsche nichtsdestotrotz zugleich die Inspirationsquelle der abendländischen Kunst – wenn nicht der Kunst überhaupt.

Während sich die klassische deutsche Ästhetik von Winckelmann bis Hegel ausschließlich auf die von Nietzsche so bezeichnete *apollinische* Kunst beschränkt, weist er mit Nachdruck auf eine zweite, bisher ignorierte Form hin: nämlich die *dionysische* Kunst. Beim Apollinischen und dem Dionysischen soll es sich, wie es nun weiter heißt, ursprünglich um zwei außermenschliche Naturtriebe handeln, die erst im Nachhinein durch Nachahmung zu menschlichen Kunsttrieben werden. Wenn also von der apollinischen und der dionysischen Kunst die Rede ist, dann geht es hierbei um die künstlerische Nachahmung von Zuständen, die sich bereits in der Natur entdecken lassen.

Diese Naturzustände und ihre ästhetischen Transformationen korrespondieren nun auf Seiten der Rezipierenden mit bestimmten vorherrschenden Erlebnisweisen, d. h. sie rufen beim Menschen entweder traumhafte oder rauschhafte Erfahrungen hervor. Da Nietzsche im Grunde eine wirkungsästhetische Differenzierung vornimmt, wenn er den Rausch der dionysischen und den Traum der apollinischen Kunst zuordnet, ist es angebracht, dementsprechend von dionysischer und apollinischer Rezeption zu sprechen. Eine Kunst des Rausches soll die dionysische Kunst deswegen sein, weil sie das *principium individuationis* auflöst und dadurch den Menschen mit anderen Menschen und der gesamten Natur in eine umfassende Einheit bringt. Was so schrecklich sein soll, klingt zunächst einmal eher erstrebenswert:

> Unter dem Zauber des Dionysischen schliesst sich nicht nur der Bund zwischen Mensch und Mensch wieder zusammen: auch die entfremdete, feindliche oder unterjochte Natur feiert wieder ihr Versöhnungsfest mit ihrem verlorenen Sohne, dem Menschen. (Nietzsche, GT, KSA 1, S. 29)[2]

Als Naturzustand entspricht diesem Rauschcharakter die euphorisierende Wirkung des Frühlings oder der Genuss des Alkohols. Der dionysischen Rauschkunst, die sich vor allem in der Musik verwirklicht, wird von Nietzsche nun die apollinische Kunst an die Seite gestellt, welche genauer als eine Kunst des Traums und des Scheins ausgewiesen wird. Hierdurch wird im Gegenzug die dionysische Kunst offensichtlich *ex negativo* durch ihren fehlenden Scheincharakter, also im Grunde als eine nicht-fiktionale Kunst bestimmt. Unterschieden wird damit zwischen einer „Kunst des Bildners" und einer „unbildlichen Kunst" (Nietzsche, GT, KSA 1, S. 25), in der „die Zerreissung des principii individuationis ein künstlerisches Phänomen" (Nietzsche, GT, KSA 1, S. 33) wird.

Während die dionysische Kunst durch den Rausch die Individuen versöhnt, indem sie ihre Formen und damit ihre Grenzen zur Auflösung bringt, besteht die apollinische Kunst ganz im Gegenteil darin, schöne Formen zu gestalten, ja sie lebt geradezu von der Freude an der sinnlichen Gestalt der Individuen. Weit entfernt von Verzückung und Rausch der dionysischen Rezeption, zeichnet sich die apollinische Rezeption daher durch Distanz und Begrenzung, Maß und Ruhe aus. In einer solchen Kunst, für die die Malerei oder die Skulptur Paradebeispiele sind, lösen sich die Konturen der Individualität nicht auf, sondern kommen vielmehr umso schärfer und vollendeter zum Vorschein.[3]

Die apollinische Kunst besteht im Wesentlichen aus einer Nachahmung und Vervollkommnung von einzelnen Formen und Gestalten, die sich in der Natur finden lassen. Ein Ausdruck der ganzen Welt ist für Nietzsche hingegen die Musik als Krönung der dionysischen Kunst. Wie er fortfährt, legen wir beim Hören von Musik das Ohr an die Herzkammer des Weltwillens (Nietzsche, GT, KSA 1, S. 135). Während die apollinische Kunst als ein Spiegel einzelner Erscheinungen in der Welt zu verstehen ist, blickt die dionysische Kunst *hinter* den Schleier all dieser Erscheinungen, d.h. sie ist ein Spiegel des Lebens, das all diese Formen aus sich hervortreibt (Nietzsche, GT, KSA 1, S. 51).

---

2 Wie Reuber 1988, S. 102 betont, fallen in der dionysischen Kunst dann entsprechend auch die Grenzen zwischen Künstler, Rezipient und Werk.

3 An dieser Stelle ist allerdings die zutreffende Bemerkung von Steinmann 2000, S. 18 zu beachten: „Diese jeweilige Beschränkung auf eine Kunstart ist jedoch nur paradigmatisch zu verstehen, denn an sich bedeuten beide zwei bestimmte Formprinzipien, die sich in Plastik und Musik nur besonders deutlich zeigen."

Es wäre jedoch ein Irrtum, wenn man Nietzsche so interpretieren würde, dass diese beiden Kunsttriebe in einem antagonistischen Verhältnis stehen, denn ganz im Gegenteil finden sie – so lautet eine Hauptthese der *Geburt der Tragödie* – in der griechischen Tragödie, aber auch im Gesamtkunstwerk Richard Wagners in einer, wenn auch spannungsvollen Einheit zueinander. Genauer stellt sich Nietzsche diese Kooperation so vor, als würde sich der Chor, das dionysische Element der Tragödie, in der apollinischen Bilderwelt der dramatischen Handlung konkretisieren (Nietzsche, GT, KSA 1, S. 62). In diesem Sinne sollen nach Nietzsche die apollinischen Darstellungen von leidenden Menschen allesamt Exemplifikationen der dionysischen Weisheit des Silens sein (Nietzsche, GT, KSA 1, S. 151). Durch die Vermittlung mit dem apollinischen Schein wird auf diese Weise eine Distanz zur dionysischen Wahrheit möglich, welche nun nicht mehr ungefiltert in ihrer ganzen Brutalität über die Zuschauenden hereinbricht:

> Wo uns das „Naive" in der Kunst begegnet, haben wir die höchste Wirkung der apollinischen Cultur zu erkennen: welche immer erst ein Titanenreich zu stürzen und Ungetüme zu tödten hat und durch kräftige Wahnvorstellungen und lustvolle Illusionen über eine schreckliche Tiefe der Weltbetrachtung und reizbarste Leidensfähigkeit Sieger geworden sein muss. (Nietzsche, GT, KSA 1, S. 37)

Vor dem Hintergrund dieser Ausführungen stellt sich nun die Frage, wie man sich eigentlich die ästhetische Rechtfertigung des menschlichen Daseins genauer vorstellen soll. Ist sowohl Apollo als auch Dionysos jeweils auf seine Weise hierzu imstande? Tröstet nur einer von ihnen über die Wahrheiten des menschlichen Daseins hinweg oder ist hierzu ihre Kooperation erforderlich? Einerseits ist es die Kunst, die nach Nietzsche überhaupt erst die Scheußlichkeit des Lebens enthüllt, andererseits bietet sie allein die Möglichkeit, eine Zuflucht vor diesem Ekel zu finden. Wenn das, was die Wunde reißt, auch das Einzige sein soll, was sie heilen kann, ist damit natürlich die Frage aufgeworfen, ob man sich nicht von vornherein den Umweg ersparen und lieber gleich auf die Kunstrezeption verzichten sollte.

Was die näheren Fragen zur ästhetischen Rechtfertigung der Welt betrifft, so ist Nietzsche leider an keiner Stelle so eindeutig, wie man es sich wünschen würde. Hier und da finden sich Textpassagen, denen zufolge die dionysische Kunstform für sich allein bereits durch die versöhnende Wirkung des Rausches einen Trost bietet (z. B. Nietzsche, GT, KSA 1, S. 56). Wenn Nietzsche jedenfalls in Übereinstimmung mit Schopenhauer behauptet, dass die Kunst vom individuellen Willen befreit, dann kann damit an dieser Stelle eigentlich nur die dionysische Kunst gemeint sein; sie ist es, „die den Willen in seiner Allmacht gleichsam hinter dem principio individuationis, das ewige Leben jenseits aller Erscheinung und trotz aller Vernichtung zum Ausdruck bringt" (Nietzsche, GT, KSA 1, S. 108). Die Musik versöhnt den Einzelnen mit der Welt, weil sie ihn dazu bringt, in einer sympathetischen Liebesempfindung im ewigen Leben und Leiden das eigene Leben und Leiden wiederzuerkennen.

Nietzsche macht im Grunde auf die Unvollständigkeit der dionysischen Kunst aufmerksam, indem er betont, dass das menschliche Leiden, in das uns der dionysische Zustand den tiefsten Einblick gewährt, in letzter Konsequenz überhaupt erst durch die Schönheit der Erscheinungen zu überwinden ist: Die Urgewalt der Musik muss durch die Kunst geformt werden (Nietzsche, GT, KSA 1, S. 108). Wenn es sich so verhält, dann liegt das Schwergewicht der ästhetischen Rechtfertigung jedoch auf der apollinischen Kunstform: Es sind die Traumgestalten der Götter und Helden im Epos und in der Tragödie, die den Ekel über das menschliche Dasein verklären, weil sie ihn in Vorstellungen kleiden, mit denen sich leben lässt: Dieser Zwischenbereich, in dem eine „Umschleierung der Wahrheit" durch die „Vereinigung von Dionysos und Apollo" möglich wird, „offenbart sich in einem Spiel mit dem Rausche, nicht in einem völligen Verschlungensein durch denselben." (Nietzsche, GT, KSA 1, S. 567)

Also ist es weniger der dionysische Rausch als vielmehr der apollinische Schein, von dem zwar keine logische, aber immerhin doch eine emotionale bzw. ästhetische Auflösung des Theodizee-Problems zu erwarten ist. Nach Nietzsche steht sowohl die griechische Tragödie als auch die Oper von Richard Wagner, in der Nietzsche ein Wiedererwachen des dionysischen Geistes zu erkennen glaubt (Nietzsche, GT, KSA 1, S. 127), vor der Aufgabe, dionysische Wahrheit und apollinische Schönheit in ein Gleichgewicht zu bringen, weil von dem dionysischen Untergrund der Welt nur so viel im Kunstwerk bewusst werden darf, wie die „apollinische Verklärungskraft" (Nietzsche, GT, KSA 1, S. 155) zu bewältigen imstande ist.

Damit legt Nietzsche zwei Deutungen nahe, von denen zumindest die erste ausgesprochen kontraintuitiv ist: Denn zum einen sieht es bei ihm so aus, als wäre die Schönheit der Musik selbst eigentlich wesensfremd. Zum anderen folgt aus seinen Erläuterungen, dass erst das Dionysische der Kunst einen ergreifenden Gehalt gibt, wobei das Apollinische im Gegenzug die Aufgabe übernimmt, das, was dem Kunstwerk Tiefe und Gewicht verleiht, durch den Schein eine ästhetische Form zu geben: Das „ewig Leidende und Widerspruchsvolle" braucht „die entzückende Vision, den lustvollen Schein, zu seiner steten Erlösung" (Nietzsche, GT, KSA 1, S. 38). Plakativ gesprochen, gerät das Dionysische in seiner Nacktheit und Rohheit für sich allein in die Gefahr, hässlich und unerträglich zu sein, während das Apollinische für sich allein zur Seichtigkeit eines unverbindlichen schönen Formenspiels tendiert.

Wie sich gezeigt hat, ist es letztlich die apollinische Kunst, auf der die ästhetische Rechtfertigung des Daseins beruht: Das Erlöstwerden durch die Kunst ist ein „Erlöstwerden durch den Schein" (Nietzsche, GT, KSA 1, S. 38). Aber ohne das

Dionysische würde es der Kunst an einem existentiell bedeutsamen Gehalt fehlen, weil ihm allein jene verstörenden Einsichten zu verdanken sind, die nach einer Rechtfertigung verlangen und an denen sich das Formgestaltungsvermögen des Apollinischen abarbeitet.

2 Sartre und die Kunst als Heilsweg aus der Kontingenz

Genau wie bei Nietzsche steht nun auch bei Sartre am Ursprung des Kunstverständnisses eine existentielle Erfahrung, die in seinem ersten Roman *Der Ekel* anhand der Erlebnisse des Ich-Erzählers Antoine Roquentin ausführlich geschildert wird. Der Einwand liegt natürlich auf der Hand, dass man die Position eines Romanautors nicht mit derjenigen seiner Romanfigur identifizieren sollte, selbst wenn es sich dabei um einen Ich-Erzähler handelt. In diesem besonderen Fall ist jedoch eine solche Gleichsetzung erlaubt, denn Sartre selbst gesteht Jahrzehnte später in seiner Autobiografie *Die Wörter* (1964): „Ich war Roquentin, ich zeigte an ihm ohne Gefälligkeit das Muster meines Lebens." (Sartre 1983: Mots, S. 193)

Die Kontingenz ist zwar eine philosophische Idee, wie Sartre in einem späteren Interview erklärt (Sartre 1978: SFilm, S. 39), aber im Roman stößt man auf sie eben nicht dadurch, dass Logik und Argumentation, also die Werkzeuge des diskursiven Denkens, ins Spiel gebracht werden. Vielmehr handelt es sich um eine rein *intuitive* Erkenntnis, d. h. um den Gehalt einer konkreten Erfahrung, die als Ekel bezeichnet wird. Dieser Ekel ist für Sartre demzufolge keine bloß subjektive Befindlichkeit, weil ihm darüber hinaus nämlich eine welterschließende Qualität zukommt, insofern sich im Ekel – und nur in ihm – eine fundamentale Verfasstheit des Seins zeigt, eben jene Kontingenz, die für Sartre gewissermaßen der Teig ist, aus dem jedes einzelne Ding und die ganze Welt geknetet ist.

Allerdings fährt der frühe Sartre zweigleisig: Während die Kontingenz im Roman affektiv erlebt ist, soll sie in seinem philosophischen Hauptwerk *Das Sein und das Nichts* (1943) durch eine Beweisführung rational erkannt werden. Die Kontingenz wird damit einmal als ein phänomenaler Gehalt, ein anderes Mal als ein abstrakter Gedanke verstanden. Man könnte es auch so sagen: Sartre traut sich in *Das Sein und das Nichts* zu, dass er einen Beweis für den Atheismus erbringen kann, während er sich in *Der Ekel* vornimmt, jene existentielle Erfahrung zu beschreiben, in der der Atheismus verwurzelt ist.

Mit dem so genannten ontologischen Beweis wendet Sartre sich also einerseits gegen den subjektiven Idealismus, andererseits gegen die Idee einer göttlichen Schöpfung und versucht zu begründen, warum das Sein nicht aus einer Subjektivität – nicht einmal aus einer göttlichen – abgeleitet werden kann. Lässt sich das Sein nicht auf eine Subjektivität zurückführen, dann ist es ungeschaffen und unabhängig. Hieraus folgt weiterhin, dass es nicht darin aufgeht, einen von einer Subjektivität gesetzten Zweck zu erfüllen und sich als völlig überflüssig und sinnlos, also kontingent herausstellt (vgl. Sartre 1994a: EN, S. 33–45).[4] Würde sich die Welt hingegen einer göttlichen Schöpfung verdanken, dann wären die Menschen, Tiere, Pflanzen und Dinge zwar immer noch insofern kontingent, als die Notwendigkeit der Existenz allein dem göttlichen Schöpfer zukäme. Aber in dem Fall hätte zumindest Gott gewollt, dass es sie gibt, und darum wären sie in ihrer Existenz gerechtfertigt. Sie wären da, weil sie da sein sollen.

Was in *Das Sein und das Nichts* bewiesen werden soll, enthüllt sich in *Der Ekel* in einer Art von Offenbarungserlebnis, das an die christliche Tradition, aber vor allem auch an die ebenso plötzliche wie überwältigende Enthüllung des Wesens der Welt erinnert, von der im Zen-Buddhismus die Rede ist.[5] So heißt es bei Sartre in der berühmten Kastanienwurzelszene: „Und mit einem Schlag, mit einem einzigen Schlag zerreißt der Schleier, ich habe verstanden, ich habe gesehen." (Sartre 1989: N, S. 144) Diese individuelle konkrete Kastanienwurzel in ihrer einzigartigen Erscheinungsweise lässt sich weder aus dem Begriff der Kastanienwurzel noch aus ihrer allgemeinen biologischen Funktion oder einem göttlichen Willen ableiten. Insofern die Dinge jenseits ihrer vertrauten Klassifizierung und jenseits ihrer instrumentellen Verfügbarkeit zur Erscheinung kommen, gerät die menschliche Ordnung mit ihren Sinnzusammenhängen gefährlich ins Wanken:

> Ich sehe, was darunter ist! Die Lackschichten schmelzen, die glänzenden Samthäutchen, die Pfirsichhäutchen des lieben Gottes platzen unter meinem Blick, sie reißen auf und klaffen auseinander. (Sartre 1989: N, S. 142)

Angesichts einer von Grund auf kontingenten Natur stoßen wissenschaftliche Erkenntnis und technische Kontrolle an ihre Grenzen, weil selbst die Naturgesetze in letzter Konsequenz unzuverlässig bleiben, so dass jede Prognose immer nur auf wackeligen Beinen stehen wird. Nietzsche meint im Grunde dasselbe, wenn er von einem „Grausen" spricht, das aufkommt, sobald der Satz vom Grunde eine Ausnahme erleidet (Nietzsche, GT, KSA 1, S. 28). In Sartres Roman finden sich Horrorszenarien, die äußerst drastisch eine aus den Fugen geratene Natur veranschaulichen, die zweifellos das von Nietzsche erwähnte Grausen hervorrufen würde:

---

4 Sehr aufschlussreich ist die Diskussion von Sartres ontologischem Beweis bei Damast 1994.
5 Siehe hierzu z. B. die immer noch lesenswerte Monografie Schüttler 1974, insbesondere S. 35.

> Zum Beispiel ein Familienvater wird beim Spazierengehen einen roten Fetzen durch die Straße auf sich zukommen sehen, wie vom Wind getrieben. Und wenn der Fetzen ganz nahe bei ihm ist, wird er sehen, daß es ein Brocken verfaultes Fleisch ist, staubbeschmutzt, das sich kriechend und hüpfend dahinschleppt, ein Stück geschundenes Fleisch, das sich durch die Gosse wälzt und in Krämpfen Blutspritzer ausstößt [...]. Und ein anderer wird merken, daß ihn etwas im Mund kratzt. Er wird vor einen Spiegel treten, den Mund öffnen: und seine Zunge wird ein riesiger, äußerst reger Tausendfüßler geworden sein, der mit den Füßen strampelt und ihn am Gaumen kratzt. Er wird ihn ausspucken wollen, aber der Tausendfüßler wird ein Teil seiner selbst sein, und er wird ihn mit den Händen herausreißen müssen. (Sartre 1989: N, S. 178–179)

Genauso wie bei Nietzsche kommt nun ebenfalls bei Sartre die Kunst als Zuflucht vor einer gänzlich absurden Welt ins Spiel. Es gibt sogar noch mehr Parallelen: Denn es ist auch hier ausgerechnet die Musik, die dem kontingenzgebeutelten Roquentin ein Refugium bietet (vgl. Sartre 1989: N, S. 31–34, 194–199). Was Roquentin beim Hören der Musik erfährt, ist eine, wie es heißt, „unbeugsame Ordnung" (Sartre 1989: N, S. 31). Für wenige Momente lässt ihn die „Notwendigkeit der Musik" (Sartre 1989: N, S. 32) die Kontingenz vergessen. Was aber ist notwendig an der Musik? Notwendig ist sie nach Sartre in dem Sinne, dass es in ihr keinerlei Kontingenz und Zufälligkeit gibt, denn jeder einzelne Ton ist da, weil eben genau dieser Ton und zwar genau an dieser Stelle da sein muss, d. h. er ist durch den Zusammenhang mit allen anderen Tönen der Melodie gerechtfertigt. Angesichts eines Ekels, der sich auf den gesamten Bereich der Realität bezieht, soll jedenfalls Roquentin der Verzweiflung entkommen können, weil ihm mit der Kunst ein Heilsweg aufgewiesen wird, der sich auf die folgende Weise charakterisieren lässt: Alles ist kontingent, absurd, ekelhaft – bis auf die Kunst, denn in einem Kunstwerk ist jedes Wort, jeder Ton, jeder Farbstrich gewollt und damit notwendig und gerechtfertigt.[6]

Ein kurzer Seitenblick in Sartres phänomenologische Studie *Das Imaginäre* (1940) macht noch einmal deutlicher, was es mit diesem schroffen Antagonismus zwischen Kunst und Kontingenz auf sich hat.[7] Am Ende jener Schrift, die zwei Jahre nach dem Roman erschienen ist, bringt Sartre seine Forschungsresultate für den Bereich der Ästhetik zur Anwendung: Die Kunst ist ein Imaginäres, die all-

---

6 Sartre zufolge würde der Tachismus oder das Action Painting seinem Kunstverständnis nicht widersprechen, denn hier wäre die Kontingenz der Farbaufträge selbst noch gerechtfertigt, die Zufälligkeit als solche gewollt. Selbst für jene zufälligen Farbflecken auf der Leinwand gilt, dass es sie ohne die schöpferische Subjektivität des Künstlers nicht geben würde. Es handelt sich also um eine Kontingenz innerhalb des Bereichs der Notwendigkeit.

7 Siehe ausführlich zum Verhältnis zwischen Imagination und Ästhetik bei Sartre: Bonnemann 2007, insbesondere Kap. 2, 6, 10.

gemeinen Wesensgesetze des Imaginären gelten darum auch für die Kunst (vgl. Sartre 1994b: I-aire, S. 296). Aus der Perspektive von Nietzsche wird damit allerdings die Eigengesetzlichkeit der dionysischen Kunst ignoriert und der Bereich der Kunst auf den der apollinischen Kunst reduziert. Und während für Nietzsche gerade in der Musik die Besonderheit der dionysischen Kunst ins Auge – bzw. ins Ohr – springt, sieht Sartre überhaupt keinerlei Veranlassung, ihr einen Sonderstatus zuzusprechen: So wie die realen Farben dazu da sind, damit sich ein imaginäres Objekt im Gemälde manifestiert, so sind ganz analog die einzelnen Töne real, aber das Musikstück selbst gehört zur Imagination. Das soll sich nach Sartre darin zeigen, dass man zwar die *Aufführung* der 9. Symphonie, aber niemals die 9. Symphonie selbst unterbrechen kann (Sartre 1994b: I-aire, S. 302).

Die strikte Trennung zwischen dem Bereich der Wahrnehmung und dem der Imagination fällt hier mit der Trennung zwischen dem Bereich der Kontingenz und dem der Notwendigkeit zusammen. Während die Realität ungeschaffen und daher dem Subjekt fremd, unverdaulich und ekelhaft ist, gilt all dies eben gerade nicht für das Imaginäre: Das Imaginäre wird als solches immer von einem Subjekt hervorgebracht und deswegen entgeht es aufgrund dieser Subjektabhängigkeit der Kontingenz des Realen. Wenn ich mir beispielsweise einen großen weißen Hasen vorstelle, dann ist dieser Hase nur da, weil und solange ich ihn mir vorstelle, d. h. ihn in meinem Bewusstsein erschaffe. Während ein realer Hase subjektunabhängig ist, weil er auch dann noch da ist, wenn ich ihn nicht wahrnehme, wird der imaginäre Hase von meinem imaginierenden Bewusstsein im Sein gehalten. Mit anderen Worten, gerade weil er keinerlei Subjektunabhängigkeit und darum auch keinerlei Kontingenz besitzt, löst er sich ins Nichts auf, sobald ich von ihm kein Bewusstsein mehr habe (vgl. Sartre 1994b: I-aire, S. 34). So ist der reale Stein auf dem Weg zufällig da und kontingent, aber der imaginäre Stein in einem Gemälde ist von einem Maler, also dem Schöpfer des Gemäldes, erschaffen worden, weil er will, dass der Stein an dieser Stelle des Gemäldes zu sehen ist. Daraus folgt, dass in einem Kunstwerk alles seinen Ursprung in einer schöpferischen Subjektivität hat, wobei es sich hierbei eben nicht um Gott, sondern um einen Künstler handelt.

Sartre räumt schließlich in *Die Wörter* ein, dass er selbst wie Roquentin das Sakrale aus der Religion genommen und in die Literatur verschoben hat (vgl. Sartre 1983: Mots, S.191–196). Diese Kunstreligion beruht auf dem Glauben, dass derjenige, der schreibt, bleibende Werke schafft und seine eigene Existenz durch diese bleibenden Werke rechtfertigt. Zu rechtfertigen ist das menschliche Leben also allein dadurch, dass es sich einer rein selbstzweckhaften Kunst zum Opfer bringt: Das Reale ist nur dafür da, damit das Imaginäre erscheinen kann.

Fazit – gemeinsame Grundintuitionen und getrennte Wege

Nietzsche und Sartre sind sich wohl darin einig, dass nur die Kunst gegen die Erkenntnis der Absurdität des menschlichen Lebens hilft; und beide würden darum wohl die folgende Äußerung von Michel Houellebecq unterschreiben:

> Wer das Leben liebt, liest nicht. Und geht erst recht nicht ins Kino. Was immer auch darüber gesagt wird, der Zugang zum künstlerischen Universum ist mehr oder weniger für jene reserviert, die ein wenig *die Schnauze voll* haben! (Houellebecq 2002, S. 13)

Zwar gehen die Übereinstimmungen zwischen Nietzsche und Sartre hinsichtlich der Erlösungsfunktion der Kunst überraschend weit, dennoch ist nicht zu übersehen, in welchem Ausmaß beide Denker wiederum bei diesem Thema auch völlig getrennte Wege gehen. Erstens ist für Nietzsche die Kunst nicht nur das Therapeutikum, sondern zugleich auch die Quelle der dionysischen Weisheit, wohingegen Sartre in seinem Roman zwar eine literarische Beschreibung der Kontingenzerfahrung liefert, aber keineswegs behaupten würde, dass die Einsicht in die Sinnlosigkeit des Seins sich nicht auch außerhalb von Literatur und Kunst vollziehen könnte. Zweitens bestehen entschieden unterschiedliche Auffassungen von der Absurdität des menschlichen Lebens: Mit Schopenhauer teilt auch der frühe Nietzsche die Überzeugung, dass jedes Leben im Wesentlichen ein fortwährendes Leiden ist. Sartres Ekel besteht zwar in der Einsicht in die Kontingenz der Welt, aber damit ist keineswegs einer tragischen Weltsicht das Wort geredet. Während Nietzsche das menschliche Dasein als eine Abfolge sinnlosen Leidens versteht, ist es für Sartre zwar sinnlos, aber damit wird keineswegs ausgeschlossen, dass es nichtsdestotrotz weitgehend glücklich verlaufen kann – zumindest insoweit, als Glück und Sinnlosigkeit sich nicht ausschließen.

Drittens sind sowohl Nietzsches dionysische Weisheit als auch Sartres Ekel dahingehend als mystische Intuitionen zu charakterisieren, dass jeweils hinter der Vielheit der Dinge ein allgemeiner Zusammenhang – sei es das Leben, sei es das Sein – entdeckt wird. Aber keine Übereinstimmung besteht schließlich darin, wie diese fundamentale Einheit zu verstehen ist: Was sich der dionysischen Weisheit offenbart, ist ähnlich wie der *élan vital* in Henri Bergsons Lebensphilosophie ein dynamisches Prinzip, sozusagen der Urwille der Welt, das ewig dahinströmende Leben, das die einzelnen Gestalten mit sich reißt. Während Nietzsches Gedanke eines wildbewegten Weltgrundes an Heraklit und Anaxagoras erinnert, schlägt Sartre, der in jungen Jahren sehr von Spinoza begeistert war, mit der Rede vom Sein völlig andere Töne an: Sartres Sein ist jedenfalls alles andere als dynamisch, sondern vielmehr unveränderlich und zeitlos und jegliches Werden ist ihm völlig fremd. Nicht von ungefähr hat z. B. Hans-Georg Gadamer auf die Nähe zwischen Sartre und Parmenides hingewiesen und die Frage aufgeworfen, ob man Sartre als einen Eleaten bezeichnen könnte (Gadamer 1999, S. 117–118).

Sowohl Nietzsche wie auch Sartre haben in ihrer späteren Philosophie mit jenem explizit herausgestellten Pathos der Selbstüberwindung, das für den einen so gut wie für den anderen Denker charakteristisch ist, jene pessimistische bzw. fatalistische Weltsicht hinter sich lassen wollen, in der ihr anfängliches Kunstverständnis verwurzelt ist. Gegen den Nihilismus und die Lebensabwertung, die auf Schopenhauers Einfluss zurückgehen, will Nietzsche das menschliche Dasein nun gerade in seiner ganzen Tragik leidenschaftlich bejahen und dafür macht er das Leben selbst – „jene dunkle, treibende, unersättlich sich selbst begehrende Macht“ (Nietzsche, GT, KSA 1, S. 269) – zum alleinigen Maßstab. Was nach Nietzsches Wertezertrümmerung übrig bleibt, ist dann nur noch die Frage, ob ein bestimmtes kulturelles Phänomen das Leben stärkt oder schwächt.

Eine vollkommen entgegengesetzte Richtung schlägt Sartre ein, wenn er die *L'art pour l'art*-Kunstkonzeption aus *Der Ekel* und *Das Imaginäre* überschreitet, um – nur wenige Jahre später – in seiner Schrift *Was ist Literatur?* (1947) für eine Literatur des moralischen und politischen Engagements zu argumentieren. Sartre hängt den Ästhetizismus zugunsten der Moral an den Nagel. Kurz darauf werden jedoch auch die moralphilosophischen Ambitionen wieder aufgegeben, weil er zu der Überzeugung kommt, dass die schlechten sozialen Verhältnisse individuelle Handlungen unvermeidlich zum Scheitern verurteilen, selbst wenn ihnen die besten Absichten zugrunde liegen sollten. Aus Solidarität mit den Unterdrückten und Ausgebeuteten, denen mit Moral alleine nicht zu helfen ist, sympathisiert Sartre nun viele Jahre mit dem Marxismus, den er allerdings existentialistisch uminterpretiert. So oder so stellt er sich in Nietzsches Augen damit natürlich in den Dienst einer Sklavenmoral.

Für die Fragestellung dieser Untersuchung entscheidend ist jedenfalls, dass für den späten Sartre der Ekel und die Sinnlosigkeit des menschlichen Lebens nur noch ein Luxusproblem ist, das bis auf Weiteres verschoben wird, bis die tatsächlichen Probleme gelöst sind: „[I]ch habe plötzlich entdeckt, dass die Entfremdung, die Ausbeutung des Menschen durch den Menschen, die Unterernährung usw. das metaphysische Elend, das Luxus ist, in den Hintergrund drängen. Hunger, das ist wirkliches Elend.“ (Sartre 1985: Mon64, S. 64) Und er fügt hinzu: „Ich habe Kinder verhungern sehen. Gegenüber einem sterbenden Kind hat *Der Ekel* kein Gewicht.“ (Sartre 1985: Mon64, S. 65) Dieses Mitgefühl, das nach Sartres eigenem Bekunden seinen Sinneswandel motiviert hat, ist aus Nietzsches Perspektive nun wenig mehr als ein heimtückischer Trick, mit dem das schwache das starke Leben bekämpft und das Leben selbst in letzter Konsequenz dem Verfall preisgegeben wird. Zumindest in seinen philosophischen Schriften hat Nietzsche hierfür nur Hohn und Spott übrig:

> [D]er Glaube an sich selbst, der Stolz auf sich selbst, eine Grundfeindschaft und Ironie gegen „Selbstlosigkeit" gehört eben so bestimmt zur vornehmen Moral wie eine leichte Geringschätzung und Vorsicht vor den Mitgefühlen und dem „warmen Herzen". (Nietzsche, JGB, KSA 5, S. 210)

So vertritt Nietzsche schließlich einen Aristokratismus, verteidigt eine Herrenmoral und träumt vom Übermenschen, wohingegen Sartre sich in einer vorbehaltlos egalitären Haltung vehement gegen Hierarchie und Privilegierung ausspricht:

> Was bleibt, wenn ich das unmögliche Heil in die Requisitenkammer verbanne? Ein ganzer Mensch, gemacht aus dem Zeug aller Menschen, und der soviel wert ist wie sie alle und soviel wert wie jedermann. (Sartre 1983: Mots, S. 196)

Hier wird unübersehbar, wie weit sich schließlich die Schere zwischen zwei Denkern auftut, die – wie der vorliegende Aufsatz zeigen wollte – zumindest in ihren Anfängen und trotz ihrer völlig unterschiedlichen philosophischen Herkunft auf überraschend ähnliche Weise über das Verhältnis zwischen Kunst und Leben nachgedacht haben.

Bibliographie

Beam, Craig (1998): „Sartre vs. Nietzsche: Will to Power, Platonism, and Pessimism". In: *The Carleton University Student Journal of Philosophy.* http://docs.sartre.ch/Sartre%20vs%20Nietzsche.pdf, zuletzt abgerufen am 15. 06. 2021.

Bonnemann, Jens (2007): *Der Spielraum des Imaginären. Sartres Theorie der Imagination und ihre Bedeutung für seine phänomenologische Ontologie, Ästhetik und Intersubjektivitätskonzeption.* Hamburg.

Daigle, Christine (2004): „Sartre and Nietzsche". In: *Sartre Studies International* 10, S. 195–210.

Damast, Thomas (1994): *Jean-Paul Sartre und das Problem des Idealismus. Eine Untersuchung zur Einleitung in L'être et le néant.* Berlin.

Gadamer, Hans-Georg (1999): „‚Sein und das Nichts' (J. P. Sartre)". In: Gadamer, Hans-Georg: *Gesammelte Werke 10. Hermeneutik im Rückblick.* Tübingen.

Houellebecq, Michel (2002): *Gegen die Welt, gegen das Leben.* Köln.

Louette, Jean-François (1996): *Sartre contra Nietzsche.* Grenoble.

Mitchell, David (2020): *Sartre, Nietzsche and Non-Humanist Existentialism.* Cham.

Noudelmann, François (2014): *Le toucher des philosophes. Sartre, Nietzsche et Barthes au piano.* Paris.

Reuber, Rudolf (1988): *Ästhetische Lebensformen bei Nietzsche*. München.
Sartre, Jean-Paul (1978): *Sartre. Ein Film von Alexandre Astruc und Michel Contat*. Reinbek b. Hamburg.
Sartre, Jean-Paul (1982): *Saint Genet, Komödiant und Märtyrer*. Reinbek b. Hamburg.
Sartre, Jean-Paul (1983): *Die Wörter*. Reinbek b. Hamburg.
Sartre, Jean-Paul (1985): „Was bedeutet Literatur, in einer Welt die hungert? Interview mit Jacqueline Piatier, Le Monde, 18. April 1964". In: ders.: *Was kann Literatur? Interviews. Reden. Texte 1960–1976*. Reinbek b. Hamburg, S. 63–67.
Sartre, Jean-Paul (1989): *Der Ekel*. Reinbek b. Hamburg.
Sartre, Jean-Paul (1994a): *Das Sein und das Nichts. Versuch einer phänomenologischen Ontologie*. Reinbek b. Hamburg.
Sartre, Jean-Paul (1994b): *Das Imaginäre. Phänomenologische Psychologie der Einbildungskraft*. Reinbek b. Hamburg.
Schüttler, Günter (1974): *Die Erleuchtung im Zen-Buddhismus. Gespräche mit Zen-Meistern und psychopathologische Analyse*. Freiburg, München.
Steinmann, Michael (2000): *Die Ethik Friedrich Nietzsches*. Berlin.

Selbstwerdung, Ökologie, Leiblichkeit

Marco Brusotti

"Choice" *or* "Gift"? "Choice" *and* "Gift"? Nietzsche, Jaspers and Ricœur in a Text by Georges Canguilhem on Becoming Oneself

Abstract: "Choice" *or* "Gift"? "Choice" *and* "Gift"? Nietzsche, Jaspers and Ricœur in a Text by Georges Canguilhem on Becoming Oneself. In the early post-war period, Georges Canguilhem deals with Nietzsche's maxim "Become who you are!". In his analysis of this "apparently contradictory formula of a philosopher full of contradictions", the concept of "gift" plays a pivotal role, as it can be understood either in the naturalist sense of then contemporary psychology (giftedness, talent) or in the existential sense it takes on in Jaspers's philosophy ("*Sichgeschenktwerden*"). Canguilhem's philosophical inquiry into "Become who you are!" not only sheds new light on the ambivalence in his privileged relationship to Nietzsche and on the latter's specific importance for him, but also reveals Canguilhem's engagement with Jaspers's philosophy of *Existenz* as well as with its French interpreters, Mikel Dufrenne and Paul Ricœur. Canguilhem's hitherto unpublished text, which involves a sustained discussion with Jaspers's *Nietzsche*, reveals a perhaps somewhat unexpected consonance between Canguilhem's vitalism-friendly standpoint and the philosophy of *Existenz*.

Becoming oneself is what is at stake in Karl Jaspers's critical confrontation with Nietzsche: "The *task* is to become oneself by appropriating Nietzsche."[1] "Faced with this enigmatic and, to tell the truth, incomprehensible figure, it must be said: The task is to become oneself by appropriating Nietzsche. An infinite and almost impracticable task."[2] As Mikel Dufrenne and Paul Ricœur point out, Jaspers, who also takes the concept of "task" from Nietzsche, is not as confident as the author of *Ecce homo* about his ability to fulfill this challenging task. Nevertheless, becoming oneself is the very core of his own philosophy: "*Existenz*-philosophy is the way of thought by which man seeks to become himself; [...]

1 "Die *Aufgabe* ist, selbst zu werden in der Aneignung Nietzsches." (Jaspers 1981, p. 31; Engl.: Jaspers 1965, p. 23; translation modified.)

2 Dufrenne/Ricœur 1947, pp. 259–260; my translation. Dufrenne and Ricœur, both in their mid-thirties at the time of publication, had met each other as prisoners of war in a prison camp for officers and it was there that they studied Jaspers. On *Karl Jaspers et la philosophie de l'existence*, cf. Dosse 2008, pp. 121–130; on their imprisonment, pp. 75–105.

https://doi.org/10.1515/9783110760163-017

This way of thought does not cognise objects, but elucidates and makes actual the being of the thinker."[3]

1. Canguilhem on becoming oneself. – Shortly after the publication of *Karl Jaspers and the Philosophy of Existence*, another French philosopher, Georges Canguilhem, addressed the issue of becoming oneself in a manuscript text on Nietzsche, where he also discusses Jaspers and the interpretation of his thought by Dufrenne and Ricœur. The title of Canguilhem's hitherto unpublished text consists of the name "Nietzsche" to which the maxim "Become who you are<!>" is premised, first in French and then in the original German: *Deviens celui que tu es / Werde<,> der du bist / Nietzsche.*[4] *Deviens celui que tu es*, as I will call it succinctly, is the third of eight sections ("components") of a manuscript from 1947/48 and belongs to the second and last phase of Canguilhem's philosophical teaching at the University of Strasbourg.[5] At this point, the author is about forty-three years old and has already published what still is his most famous book, the *Essay on Some Problems Concerning the Normal and the Pathological* (1943); a few years later, in 1952, the well-known collection of essays *The Knowledge of Life* will come out.

As long as Alain's student was exclusively seen as an austere science historian, his connection with Nietzsche could not but remain mysterious. However, we now know very well that Canguilhem's philosophical engagement went far beyond the theory and history of science. Only against the background of this broader spectrum of interests does it become understandable why Nietzsche

---

3 Jaspers 1957, p. 40. Jaspers quotes from *Man in the Modern Age*, the English translation of his *Geistige Situation der Zeit* (1931). According to *Philosophie*, his major work, what is "sought and required" in philosophising is "*[t]hinking*, which *transforms my self-consciousness*, because it awakens and brings me to myself in the original drives from which, acting in the existing world, I become what I am" (Jaspers 1932, Preface, VII; my translation).

4 The manuscript (shelfmark: CAPHÉS, GC, 12.1.19: "1947–48") is preserved in the *Fonds Georges Canguilhem* at CAPHÉS (*Centre d'Archives en Philosophie, Histoire et Édition des Sciences*, Paris, rue d'Ulm). For the headings of the eight sections (components), see the inventory list of the *Fonds* (Queyroux 2020). Canguilhem's legacy is accessible to scholars at CAPHÉS, but the autographs must not be published, as publication would be contrary to his will. Thus, only texts that had already appeared, howsoever, were included in the *Œuvres complètes* (cf. "Principes de l'édition," in Canguilhem 2015e, p. 51). For this reason, I will not bring any longer text excerpts and will limit myself to a few short quotations. The English translations are mine.

5 The *Faculté des Lettres*, to which Canguilhem belonged since April 1941, had temporarily (1940–1945) moved to Clermont-Ferrand during the German occupation. The second phase of his teaching at this faculty begins with its return to Strasbourg in 1945 (cf. Limoges 2015, p. 9).

was important for him. Having already dealt with this general relevance elsewhere (cf. Brusotti 2012), I will focus here on *Deviens celui que tu es*.[6]

This manuscript sheds new light on his relationship not only to Nietzsche, but also to then contemporary existential philosophy. Even if young Canguilhem was good friends with Sartre, the connection between their philosophies was rather loose. All the less relevant may seem the relation with Jaspers. Canguilhem's published writings mention him quite rarely, mostly in connection with his early *General Psychopathology*, and do not refer to his philosophy of *Existenz*. Moreover, they bear no trace of Canguilhem's reception of *Nietzsche. Einführung in das Verständnis seines Philosophierens*. Nothing seems yet to be known about his engagement with this book, even if it has not gone unnoticed that he owned the French edition (1950).

Now, our manuscript shows that Jaspers's *Nietzsche* had already engaged Canguilhem a few years before the publication of its French translation.[7] Furthermore, it shows a perhaps somewhat unexpected consonance between Canguilhem's vitalism-friendly standpoint and Jaspers's philosophy of *Existenz*. The original title of our conference, *Between Life and Existence*, could also stand for their encounter.

2. Canguilhem and Jaspers. – What did Canguilhem know of Jaspers before writing down *Deviens celui que tu es?* The *Essay* (1943), whose bibliography mentions the French translation of *General Psychopathology*, briefly discusses one single issue.[8] In a late text, he explains his interest in Jaspers as follows:

> Although, in 1943, my doctoral thesis in medicine mainly concerned problems of physiology, questions about the normal and the pathological also prompted me to refer to authors such as Karl Jaspers, Eugène Minkowski, Henri Ey (Canguilhem 2018d, p. 1234; my translation).

---

6 Whereas the present text is dedicated to Canguilhem's engagement with Jaspers, I deal more thoroughly with his own interpretation of the maxim and with *Deviens celui que tu es* as a whole in Brusotti 2021.

7 Canguilhem refers to the first edition of Jaspers's *Nietzsche* (1935), even if the second edition (1945) was already available. On the other hand, the French translation (Jaspers 1950), which is preserved in his library, appeared only years later. There is no reason to believe that Canguilhem also read *Nietzsche und das Christentum* (Jaspers 1946), Jaspers's short second book on Nietzsche. – The relevant volume of the *Gesamtausgabe* (Jaspers 2020) has recently been published.

8 For this discussion, cf. Canguilhem 1966, pp. 70, 74; the bibliography (Canguilhem 1966, p. 161) mentions Jaspers 1933 [[1]1928] (original edition: Jaspers 1923 [[1]1913]).

Alfred Schwartz, his doctoral supervisor, seems to have been acquainted with Jaspers's *General Psychopathology.*[9] However, Canguilhem must have heard of this book already in the 1920s as his friend Sartre was involved in the French edition. A few decades later, Canguilhem recalls the making of this translation:

> Curiously, it was at the *École* that two scientists, Kastler, the man of the laser, Nobel Prize winner, and Mendousse translated the *Allgemeine Psychopathologie* (1913). Sartre and Nizan corrected this translation, published in 1928. I saw them at work, in their room, decorated then with a portrait of Jules Romains. Jaspers, a physician-philosopher [médecin philosophe], brought the contribution of the psychiatric clinic to the philosophy of comprehension [philosophie de la comprehension], inaugurated by Dilthey. (Canguilhem 2018b, pp. 1129–1130; my translation)[10]

So, Canguilhem gives different and complementary reasons for his interest in Jaspers. On the one hand, it is motivated by the central question of the *Essay*, the status of the concepts "normal" and "pathological", that also concerns psychiatry, although not primarily. On the other hand, the author of this *General Psychopathology* is a physician *and philosopher* like Canguilhem himself;[11] as a clinician and, more precisely, as a psychiatrist, he makes a contribution *to philosophy*, and Canguilhem accounts his book to Diltheyian hermeneutics, whose organ is *Verstehen*. Moreover, in the *Cours de philosophie générale et de logique* (1942–1943), he considers Jaspers (alongside Weber) as one of the scholars that apply the concept of "ideal type" (Canguilhem 2015d, p. 101).

Canguilhem's first acquaintance with Jaspers may have been limited to *General Psychopathology.* Since when, then, does his engagement with Jaspers's later existential philosophy begin? This question is beyond the scope of my investigation and cannot be given a precise answer without a more detailed study of the unpublished legacy. However, Jean Cavaillès knew of Jaspers's philosophy and mentioned relevant publications by Gabriel Marcel and Jean Wahl in a bibliography that his friend and substitute would use for his teaching.[12] Therefore, it is not unlikely that Canguilhem already knew about the philosophy of *Existenz* before 1943.

---

9 Cf. the editor's comment in Canguilhem 2018a, p. 1283, note 1.

10 Cf. Bonnefoy 2015, pp. 960–961, and the editor's note, p. 961, note 1; cf. also Canguilhem 2015d, p. 101, note 1.

11 The trajectory of the two, however, is partly reversed. Jaspers went from medicine to philosophy; Canguilhem enrolled in medicine as he was already a philosophy teacher. Even after receiving his medical doctorate in 1943, he never practised as a doctor (if we leave aside his experience during the Resistance).

12 Cf. Cavaillès 2015, p. 1159, for the mention in Cavaillès' bibliography, and Canguilhem 2015b, p. 56, for Canguilhem's reference to the latter. – Besides, in 1940, Canguilhem mentions a con-

What we are interested in here, however, is only what we can gather about the early post-World War II period from the particular manuscript we will be dealing with.

3. Pindar's maxim. – Canguilhem takes the heading *Deviens celui que tu es* from "The Honey Sacrifice", the speech with which the fourth part of *Thus Spoke Zarathustra* begins. After designating himself as "the most spiteful of all fishers of human fish", Zarathustra continues:

> *That's* what I am, after all, at bottom and from the start; reeling, reeling in, raising up, raising, a raiser, a cultivator and taskmaster who not for nothing once told himself: "Become who you are!" (Nietzsche, Za-2, p. 192; cf. Nietzsche, Za IV, Das Honigopfer, KSA 4, p. 297)

Whereas *How One Becomes What One Is* [*Wie man wird, was man ist*], the well-known subtitle of *Ecce homo*, is a variation on the theme, Zarathustra's version of the formula sticks more closely to the saying from Pindar's Second *Pythian Ode* (Pyth. 2, p. 72) that accompanied Nietzsche since his youth.[13] He omits the last word of the verse – whereby Pindar's saying becomes not simply more concise, the meaning also shifts. The whole line reads: "*γένοι', οἷος ἐσσὶ μαθών*."[14] W. H. Race translates it thus: "Become such as you are, having learned what that is" (Pindar 1997, p. 244).[15] Through Pindar's Prize Song, the chariot winner, Hieron from Aitnai (Sicily), learns what kind of person he is. Is Hieron, even if he actually has already achieved a great deal, yet to become the kind of

---

tribution by Henri-L. Miéville on "Le problème de la transcendance et de la mort dans la philosophie existentielle de Karl Jaspers" (Canguilhem 2015a, p. 178).

13 For the passages in which the formula occurs, Canguilhem first generally refers to the information in the *Œuvres complètes de Frédéric Nietzsche* edited by Henri Albert. (The *Fonds Canguilhem* holds numerous works of Nietzsche in French translation [mostly from this edition]; cf. the shelf numbers CAN 1014–CAN 1016 as well as CAN 1062–CAN 1073). Afterwards, he first cites three passages: the one mentioned in the text from the fourth *Zarathustra*, aphorism 270 of the *Gay Science* and the discussion in *Ecce Homo*. Only later does he discuss other texts like *Schopenhauer as an educator*. For a more in-depth treatment of Pindar's saying and of its modern translations cf. Brusotti 2021, 185–187.

14 In *Deviens celui que tu es*, Canguilhem takes the Greek original in abbreviated form from Ernst Bertram: "*γένοι' οἷος ἐσσί*" (Bertram 1965 [[1]1918], p. 94). Canguilhem quotes from the French translation (which is kept in his library), based on the seventh edition of 1929 (Bertram 1932; CAN 1017). For Canguilhem's engagement with Bertram in *Deviens celui que tu es*, cf. Brusotti 2021, 185–189, 199–200. – Jaspers refers to Pindar as well, but without citing the Greek original (cf. Jaspers 1965, p. 155).

15 B. Williams introduces a causal relation and translates the maxim as "Become who you are *through* knowing" (Nietzsche, GS-2, p. 152 note).

person he is? Is he simply being encouraged to lend an ear to Pindar and learn from him what sort of human he is?[16] Or is he urged not only to listen to the poet who is praising him, but also, in a further step, to become the kind of person he then knows he is? If so, the already mentioned passage from Jaspers's *Man in the Modern Age* has a structure similar to Pindar's saying:

> *Existenz*-philosophy is the way of thought by which man seeks to become himself; [...] This way of thought [...] elucidates and makes actual the being of the thinker. (Jaspers 1957, p. 40)[17]

The philosopher of *Existenz* must first learn who he really is; only then, and only by means of this self-elucidation, can he become who he is. Now, precisely this two-step structure – first learning, then becoming – is rejected by Nietzsche: becoming oneself is not the result of an intellectual learning process. *Ecce Homo* tells the reader that Nietzsche has finally become himself, not that he *tried* to become himself. There was no such attempt. One should not *seek* to become oneself. Still less should one try to become oneself by elucidating oneself.

Consciously trying to find oneself would have been self-defeating. Indeed, *Ecce homo* reverses the sequence: an extraordinary person like himself learns what/who he actually is, if at all, only *after* he has become it. "Becoming what one is presupposes that one does not have the slightest inkling *what* one is. [Dass man wird, was man ist, setzt voraus, dass man nicht im Entferntesten ahnt, *was* man ist.]" Those unfortunate enough to learn it too soon would never be able to become themselves. The Socratic maxim "*nosce te ipsum* would be the recipe for decline" (Nietzsche, EH-3, Why I Am So Clever, 9; Nietzsche EH-3, 31; translation modified).

The author of *Ecce Homo* establishes only in hindsight that and how he has become what he is. Only retrospect reveals what/who he is. He was not granted this insight while he was still on his way. Does this really mean that self-becoming is merely an event that was revealed to Nietzsche himself only afterwards? Is this late report on *How one becomes what one is* not at odds with the imperative form "Become who you are!" (Za IV)? How relevant is the difference in the mode – imperative versus indicative – with the related shift from the second to the

16 According to Bremer's recent German translation, the poet simply invites the winner to learn from the Epinikion what kind of person he is: "Komm zur Kenntnis, von welcher Art du bist! [Come to know of what kind you are!]" (Pindar 2014, p. 125).

17 Obviously, the comparison should not prevent us from seeing the abyss that separates Pindar – and the concrete social context of his poetry – from Jaspers's existential engagement. The latter's reading of the saying seems more compatible with Race's than with Bremer's version. With some distinction, this also applies to Nietzsche's adaptations.

third person?[18] Is self-becoming still the object of an appeal in *Ecce Homo?* Can one still speak of a maxim to be followed ("consciously")? Is the imperative justified here? Especially this latter question is central to Canguilhem.

4. Naturalistic versus philosophical interpretations of the formula. – Taking his cue from *Ecce Homo*, the first of the five sections of *Deviens celui que tu es* deals with the "opposition between the Socratic and Nietzschean maxims" (GC, 12.1.19/3, fol. 6). Sections II and III come to the disappointing conclusion that the "'naturalistic' conception of the self" (GC, 12.1.19/3, fol. 8) in its two main variants – the biological and the psychological – is inadequate to the task of interpreting the Pindaric formula. To the "naturalistic" approach, section IV opposes philosophical views, starting with openly metaphysical theories that quickly prove to be unsatisfactory. Hence the section moves on to deal with Nietzsche and with Jaspers, before the final section (V) gives Canguilhem's own rather sceptical interpretation of Nietzsche's maxim. Whereas I sketch the whole lecture in another paper (cf. Brusotti 2021), the present will dwell on his discussion of Jaspers.

Tied to the maxim is Nietzsche's distinction between first and second nature:

> *First nature.* – The way in which we are educated nowadays means that we acquire a *second nature:* and we have it when the world calls us mature, of age, employable. A few of us are sufficiently snakes one day to throw off this skin, and to do so when beneath its covering their *first nature* has grown mature. With most people, its germ has dried up. (Nietzsche, D 455; Nietzsche, D, p. 457; translation slightly modified)

The organic metaphors deployed in this aphorism (the germ, the snake) may seem to suggest a biological process that merely needs to unfold in the course of time. However, Canguilhem questions that Nietzsche's "first nature" already contains all possible developments like a pre-formed germ. In "Become who you are!", "becoming" does not mean such a pre-determined organic maturing. Were this the case, the imperative would make no sense. As Canguilhem puts it in existential vocabulary, the aphorism of *Daybreak* is rather about "finding the authentic beneath the inauthentic" (GC, 12.1.19/3, fol. 7),[19] and "recollecting"

---

18 Besides, Pindar's "οἷος" is rendered differently in each case: in Zarathustra's maxim as "what" and in *Ecce Homo* as "who". – Some English editions translate the third person ("*Wie man wird, was man ist*") with the second, aligning the subtitle of *Ecce Homo* with Zarathustra's maxim or with aphorism 270 of *The Gay Science* ("*What does your conscience say?* – 'You should become who you are.'"). However, the difference is significant and should not simply disappear in translation.

19 On *Eigentlichkeit* in Heidegger, cf. Canguilhem 2015c, p. 141.

oneself from the "inauthentic" second nature is an existential figure of thought incompatible with a biological interpretation. The collision of biological and existential jargon is not unintentional and Canguilhem comes close to Jaspers.

Not only the biological concept of "germ", but also psychological concepts such as "aptitude", "disposition" and "*don*" ("gift" or "giftedness") all prove to be inadequate. An alternative is represented by the philosophical approach to the "first nature", the basic idea of which, starting from Plato – or at the latest from Kant – up to Jaspers, is that, at the origin, "there is not a given <fact> (germ, aptitude, gift [*don*], disposition), but a *choice.*" (GC, 12.1.19/3, fol. 8). As strange as it may sound, this "choice" is not supposed to be a psychological issue; the concept is not empirical. Accordingly, from now on, only philosophers have a say.

"Choice" is a core concept of the philosophy of *Existenz.* Before turning to Nietzsche and Jaspers, however, Canguilhem first dwells at length on Kant's and Schopenhauer's conceptions of the "intelligible character".[20] As a timeless being that somehow actualises itself in becoming, the "intelligible character" would be a hyperbolic metaphysical interpretation of the heading *How One Becomes What One Is.* However, Canguilhem points out that metaphysical approaches do not do justice to the maxim. Nietzsche's anti-Platonism involves the thesis that there is no being behind becoming. This sets him apart from Schopenhauer (and even Kant). Nietzsche stands for a non-metaphysical standpoint like the one Canguilhem himself is looking for.

5. Becoming oneself and amor fati. – With this concern, Canguilhem now brings in Jaspers. Both reject what they call a "naturalistic" reading of the maxim "Become who you are!". In the end, though, their views diverge widely. Jaspers defines philosophy as a "*via media*" that excludes atheism as well as religion. However, he holds only revealed religion as incompatible with philosophy; the ultimate horizon of his *philosophia perennis* is a "philosophical faith".[21] The metaphysical and theological proclivities involved in Jaspers's "middle way" are quite alien to

---

20 According to Aron, Sartre's "fundamental choice [choix fundamental]" is indebted to Kant's "intelligible character", a concept that was heavily discussed at the ENS (personal communication by Alfred Betschart). We do not know if Alain's student Canguilhem, for whom Kant was an important reference, had occasion to discuss the "intelligible character" with Sartre and/or Aron. – In a page close to the commentary on the Pindaric formula, Jaspers mentions in passing that "Nietzsche himself denies 'unchangeability of character' [...]" (Jaspers 1965, p. 159).

21 Jaspers sees a conflict between philosophy and religion (cf. Jaspers 1948, pp. 252–272). However, Dufrenne and Ricœur argue that his philosophy is unsatisfactory for both the atheist and the believer; the former "denounces a survival" of theological figures of thought, whereas the latter "deplores a lacuna" (Dufrenne/Ricœur 1947, p. 391; on Jaspers's "foi philosophique", cf. pp. 247–260, 388).

Canguilhem, who is interested, rather, in an alternative to the mechanistic understanding of the living. Nevertheless, he wonders whether the existential approach to "Become who you are!" might not be able to free this precept from metaphysical assumptions without interpreting it in a naturalistic sense.

Jaspers makes a principled distinction between empirical knowledge, which aims at objective truth, and philosophy, which seeks "illuminating truth". Philosophy is an existential appeal, its task is to illuminate existence, it reads ciphers of transcendence. Jaspers thus assumes a chasm between empirical claims and existential appeal and objects to Nietzsche that he constantly confuses and hybridises the two levels. Nietzsche, thus, seems inconsistent. With Jaspers, the maxim "Become who you are!" is a particularly clear and relevant case of these Nietzschean (self-)misunderstandings.

This is the point of the pages in the *Nietzsche* to which Canguilhem especially refers. The two issues on which they respectively revolve, the "*amor fati*" and "Become who you are!", are both also relevant to Jaspers's own philosophy.[22] With regard to *amor fati*, the misunderstanding that Jaspers wants to warn the reader against concerns the beloved fate: this fate is indeed necessity, but not in the sense of the causal category rejected by Nietzsche:

> The *necessity* in question is not the category of necessity that applies to causal processes subsumed under natural laws and pertaining to mechanism. It is such causal necessity that Nietzsche has in mind when he turns *against* the "deification of necessity" [...] *Fatum*, for Nietzsche, is *that* necessity which, not being a category, is neither a law of nature nor a law of purpose, neither calculable compulsion nor intention. It *includes* accident and law, chaos and purpose. (Jaspers 1965, p. 369)

Canguilhem, who writes down the keywords of this remark approvingly, defines *amor fati* as "resignation to what is coming towards me" (GC, 12.1.19/3, fol. 9). Stoic "resignation" is certainly not the attitude Nietzsche means.[23] Despite this ill-chosen word, the French philosopher interprets the maxim in terms of integral self-affirmation. His definition of *amor fati* takes up the simile of the constellation approaching the poet in *Glory and eternity*.[24] The precept "Become who you are!" "is elucidated [s'éclaire]" (GC, 12.1.19/3, fol. 9) – an unmistakable echo of

22 Dufrenne and Ricœur maintain "that the 'amor fati', taken from Nietzsche, is an important part of Jaspers's ethics." (Dufrenne/Ricœur 1947, p. 149, note 85; cf. pp. 178, 229). Cf. Jaspers 1948, pp. 482–483.

23 In another text (cf. Canguilhem 2018c, p. 230), Canguilhem closely relates Nietzsche's *amor fati* to Stoic views.

24 "I see a sign, – from farthest regions / a stellar shape sinks slowly sparkling towards me... / [...] / You come to me? –" (DD, Ruhm und Ewigkeit, KSA 6, p. 404)

Jaspers's *Existenzerhellung* – by Nietzsche's hymnic declaration of love to eternity:

> [...] eternal Yes of being,
> eternally am I your Yes:
> for I love you, oh eternity! – –[25]

At the top right of the first page of *Deviens celui que tu es*, the words "Amor fati?" (GC, 12.1.19/3, fol. 6) are pencilled in. The question mark can probably be interpreted as a consideration of whether "Amor fati" might not be a more appropriate title. Why should Canguilhem contemplate this alternative? Because unconditional self-affirmation is the valid core of Pindar's saying. For Canguilhem, the precept "Become who you are!" ultimately boils down to saying yes to one's whole being, and this self-affirmation is just *amor fati*. "This yes said to being, that is at the same time the yes said to my complete being, is summed up in the motto *amor fati*" (GC, 12.1.19/3, fol. 9). Like *amor fati*, the maxim "Become who you are!" – as well as the metaphor of the first nature – does not imply ontological or logical necessity.

6. "Creation as Freedom without Transcendence". – After dealing with the pages on *amor fati*, Canguilhem refers to an earlier paragraph, the concluding section of which is devoted to the precept "Become who you are!"; he starts his notes just with the paragraph title: "Creation as Freedom without Transcendence" (Jaspers 1965, p. 155).[26] By conceiving creation as freedom without transcendence, Nietzsche turns away from the conceptions of freedom that Jaspers's philosophy takes up.

---

25 "[...] ewiges Ja des Sein's,
ewig bin ich dein Ja:
denn ich liebe dich, o Ewigkeit! – –" (DD, Ruhm und Ewigkeit, KSA 6, p. 405)
Jaspers's paragraph on "amor fati" (Jaspers 1965, pp. 367–370) concludes with a long quotation from this *Dionysos-Dithyrambus*, from which Canguilhem takes these verses in the original German wording. Jaspers (1965, p. 370; cf. Jaspers 1981, pp. 366–367), but not Canguilhem, also quotes the verses cited in note 24.

26 The paragraph on "amor fati" belongs to "Boundaries and Sources" (Jaspers 1965, pp. 333–378), the sixth chapter of the second "book", whereas "Creation as Freedom without Transcendence [*Schaffen als Freiheit ohne Transzendenz*]" belongs to "Man" (Jaspers 1965, pp. 125–168), the first chapter of the second "book". With Jaspers, "creation without transcendence" is "self-being without God" (Jaspers 1965, p. 160; cf. also Dufrenne/Ricœur 1947, p. 151 note). – The concept of creation was already the focus of Canguilhem's own engagement with Nietzsche in the second half of the 1930s. Cf. Brusotti 2012, pp. 68–70.

> In the sense employed by the philosophy of *Existenz*, freedom, whether Christian or Kantian, exists in relation to transcendence. Freedom is the potentiality of a finite being; it is limited by transcendence, and it depends on an incomprehensible source lying at this limit (whether this be called "grace" or "being-given-to-oneself-as-a-gift"). (Jaspers 1965, pp. 155–156)[27]

Whereas the transcendence to which freedom relates may be interpreted either theologically as "grace" (Christianism) or philosophically through the concept of "gift" (the Kantian Jaspers), Nietzsche keeps aloof from any such ungraspable origin:

> The freedom that Nietzsche acknowledges and asserts amounts to being rooted in one's self and having the source of life within one's self without transcendence. Such freedom is both *negative* and *positive*. The way of freedom is *negative* insofar as it discards, breaks through, and denies what has been real and binding: [...] On the *positive* side, the fruits of freedom are of the nature of "*creation*." (Jaspers 1965, p. 156)[28]

"The creative revaluation of all values" (Jaspers 1965, p. 157), which is ultimately what is at stake, negates outmoded values and sets new ones. Jaspers now formulates the following objection:

> Because Nietzsche wishes to define, circumscribe, and interpret the creative source apart from transcendence, it constantly comes about that, in spite of his will to something more-than-life, he suddenly finds in his hands *mere nature* in the sense of what is biologically knowable, or he is left with only the corresponding mere psychological or sociological actualities. (Jaspers 1965, p. 158)[29]

Jaspers then applies this more general diagnosis to the precept "Become who you are!". It is precisely because Nietzsche interprets "the creative source without transcendence" (Jaspers, Nietzsche, p. 158) that he keeps misunderstanding Pindar's imperative.

> For example, the demand, "Become what you are!" illustrates the way in which Nietzsche's formulations lead him involuntarily from the *appeal to an existential source* to the mere as-

27 Canguilhem translates this passage from the *Nietzsche* into French – omitting the words in brackets. Perhaps "grace" is too explicitly theological to interest him. We will soon come back to "being-donated-to-oneself".

28 Canguilhem summarises this passage only in key words.

29 Canguilhem also records the content of this passage.

> sertion of *natural facts* (and hence of particular investigable existents within the world). (Jaspers 1965, p. 158)[30]

Actually, "Become who you are!" – like philosophy in general – is an existential appeal. Jaspers concedes that it is impossible to separate *Existenz* from "the physiological or causal (to let alone the psychological and sociological)" (Jaspers 1965, p. 158) dimension of human life, but, as a Kantian, he does not consider *Existenz* a possible object of empirical investigation. This also applies to "Become who you are!", a maxim that is directed exclusively to *Existenz*.[31] Nietzsche, however, claims Jaspers, confuses this existential appeal again and again with a statement involving naturalistic claims about biological and/or psychological facts.[32] Nietzsche wants to give a naturalistic foundation to the maxim, but this basis, being deterministic, deprives the modus imperative of its meaning.[33]

Canguilhem partly agrees with Jaspers that naturalistic interpretations of the formula represent a temptation which Nietzsche himself does not always resist, but which has to be rejected. However, his laconic assessment of Jaspers's achievement sounds rather reserved: in the *Nietzsche*, "becoming what you are [*deviens ce que tu es*] does not seem to need to find a rational explanation, but the explanation of its non-rational meaning" (GC, 12.1.19/3, fol. 10). According to Jaspers, who criticises "in Nietzsche a language that lets all rational thinking roll over [...]" (Jaspers 1965, p. 159), the maxim does not need a rational explanation, as Canguilhem puts it, since it is an appeal without theoretical basis, but with an existential scope – and this existential scope is what Canguilhem (not Jaspers) calls the "non-rational meaning" of the maxim.

---

30 Canguilhem also refers to this sentence, with which the final part of the paragraph "Creation as freedom without transcendence" begins; this untitled final part, typographically set off by a smaller font, is entirely dedicated to the maxim.

31 *Existenz*, the unfathomable "source of man himself" "transcends researchability [Erforschbarkeit]" (Jaspers 1965, p. 158); "this source [...] – this being that cannot be identified by objective procedures of characterology but which we ourselves most truly are – is that to which the demand 'Become what you are!' is directed." (Jaspers 1965, p. 159)

32 Jaspers quotes the following remark: "One becomes a decent man just because he *is* a decent man, i. e., because he is born with a capital of good instincts and favorable conditions. [...] Today we know better than to regard moral degeneracy as something separate from the physiological." (Jaspers 1965, p. 158; cf. Nietzsche, NL 1888, 14[113], KSA 13, p. 290)

33 "(1) The being that I 'ought' to become may simply have the significance of *just-being-that-way* [...]; *ought* then no longer has any real meaning but is an inexorable *must*. Or (2) the being that I ought to become means as well the *encompassing force of possibility* [das *Umgreifende der Möglichkeit*] which I can never know as something fixed and determinate and which no one else can know. This possibility remains open and reveals again and again what I am [...]." (Jaspers 1965, p. 159) Canguilhem does not quote this passage.

7. "Choice" and "gift"? Dufrenne, Ricœur and the "Parole of Freedom". – The basic idea of the philosophical approach to the imperative "Become who you are!" – Canguilhem had first claimed – is that, at the origin, "there is not a given <fact> ([...] gift [*don*] [...]), but a choice." After dealing with Jaspers, he comes to a different conclusion: "We said 'choice' *or* 'gift'. With Jaspers we have 'choice' *and* 'gift'" (GC, 12.1.19/3, fol. 10).[34]

How so? At first, Canguilhem counts "gift" – understood in the sense of "giftedness", "talent" – among the empirical categories; therefore, he sets up a clear alternative between this psychological concept and the existential concept of "choice". However, for Jaspers, "being given to oneself as a gift [Sichgeschenktwerden]" is an existential category not unlike "choice." Therefore, Canguilhem finally emphasises that Jaspers's philosophy insists on both existential categories, not only on "choice" but also on "gift".

With Jaspers, choice and gift belong together because "existential choice", as he calls it, *is* a gift from transcendence. In the first indent of the paragraph, the end of which addresses the maxim "Become who you are!", he claims that human freedom – and with it choice – has an unfathomable origin which philosophy can designate as a human's "being-given-to-oneself-as-a-gift" (Jaspers 1965, pp. 155–156) from transcendence.[35] However, Canguilhem owes this insight

34 Like "gift" in English and "*Gabe*" in German, the French word "*don*" is used both in the sense of "present"/"donation" and, even if not quite so often, in the sense of "talent"/"giftedness". In the French edition of the third *Untimely Meditation* used by Canguilhem, "Begabung" (Nietzsche, SE 2, KSA 1, p. 342), a word that goes back to "*Gabe*" ("gift") and means "giftedness", is (correctly) translated as "*don*". Canguilhem picks up this translation; so his text moves from using "*don*" in the sense of "talent" to using it in the sense of "present". So, by employing this French word, he connects, if only to distinguish them, Nietzsche's considerations on "Begabung" (talent, giftedness) and Jaspers's theory about "*Sichgeschenktwerden*" [literally: "being given to oneself as a gift"]. Dufrenne and Ricœur use "*don*" in their discussion of Jaspers's "Sichgeschenktwerden", a neologism that belongs to the semantic field around the substantive "*Geschenk*" (another word for "gift") and the verb "*schenken*" (to give as a present). For Dufrenne and Ricœur, the "grace-filled gift [don gracieux]" is a mode of God's "presence [presence]": "the *gift of being to freedom*" (Dufrenne/Ricœur 1947, p. 388). Dufrenne's and Ricœur's atheist, who "denounces a survival" (Dufrenne/Ricœur 1947, p. 391) in Jaspers's philosophy, will see "in the *gift of being to freedom and to the world*, the *shameful* heritage of *creationism and predestinationism*" (Dufrenne/Ricœur 1947, p. 389).

35 Canguilhem transcribes this sentence, but only partly, i.e. just without the passage in brackets where the word "*Sichgeschenktwerden*" occurs along with "grace" (see above n. 34). The *Nietzsche* ties in with the in-depth analysis in the second volume of *Philosophy*. Cf. the § "*Sichausbleiben und Sichgeschenktwerden*", in: Jaspers 1948, pp. 330–333).

in the philosophy of *Existenz* more to Dufrenne and Ricœur than to his direct reading of the *Nietzsche*.[36]

Dufrenne and Ricœur introduce Jaspers starting from the historical situation marked by the decline of Hegelian philosophy and the emergence of the two opposing "exceptions", Kierkegaard and Nietzsche, who question the confidence of philosophy in reason.[37] Jaspers claims that they must be understood through each other – even if, or just because, they thus come to constitute an untenable contradiction. But why should it be impossible to understand Nietzsche if not in conjunction with Kierkegaard? As Dufrenne and Ricœur put it: (only) in this context, the philosopher of the death of God becomes "in his own way a witness of a certain transcendence". Reading him together with Kierkegaard allows to vigorously dissociate Nietzsche "from flat atheism and vulgar materialism" (Dufrenne/Ricœur 1947, p. 24). Nietzsche and Kierkegaard stand for "the two abysms of absolute disbelief and authoritarian revelation" between which the *via media* of Jaspers's "philosophical faith advances" (Dufrenne/Ricœur 1947, p. 250).

Whereas Canguilhem does not even mention Kierkegaard, the two authors of *Karl Jaspers and the Philosophy of Existence*, who are little in tune with Nietzsche, claim that "Become who you are!" is first and foremost a Kierkegaardian theme. They obviously acknowledge that Jaspers highlights the importance of the maxim for Nietzsche; nevertheless, they surmise that the main inspiration in Jaspers's development of this issue is Kierkegaard.[38] Indeed, an analysis of Jaspers's

---

36 Dufrenne and Ricœur trace this issue back to Kierkegaard's influence. Canguilhem refers in particular to "La liberté [Freedom]" (Dufrenne/Ricœur 1947, pp. 133–152), the second chapter of the second part, and to "Les tâches de la métaphysique [The Tasks of Metaphysics]" (Dufrenne/Ricœur 1947, pp. 237–260), the first chapter of the third part. Nietzsche is also the subject of "Foi et athéisme [Faith and Atheism]" (Dufrenne/Ricœur 1947, pp. 255–260), the last paragraph of the second section of this chapter.

37 In the introduction, cf. the paragraph entitled "The shock of the exception: Kierkegaard and Nietzsche [Le choc de l'exception: Kierkegaard et Nietzsche]" (Dufrenne/Ricœur 1947, pp. 22–25). Dufrenne and Ricœur refer in particular to the first lecture in *Reason and Existenz*, "The Origin of the contemporary philosophic situation (the historical meaning of Kierkegaard and Nietzsche)" (Karl Jaspers: *Reason and Existenz: Five Lectures*, (Noonday) 1955, pp. 19–50). At that time, the association between Kierkegaard and Nietzsche was far from unusual, not only among "existentialist" thinkers. I use "existentialist" in scare quotes because of the well-known fact that "Jaspers [...] prefers to use the word *Existentialismus* to signify a view which he rejects (*Philosophie*, I, xxiii)." (Translators' note, in Jaspers 1965, p. 155)

38 "Here we recognise the Kierkegaardian theme: Become what you are, *werde, der du bist*. Jaspers noted that the same theme is found in Nietzsche, where the highest creation is a kind of fidelity: *Werde fort und fort der, der du bist* [...] (Nietzsche 346 cf. 137). Nietzsche exalts this Müssen beyond measure: *Hybris ist unsre ganz<e> Stellung zu Gott... zu uns* (N. 376), but this very hybris is overcome in the myth of the eternal recurrence." (Dufrenne/Ricœur 1947, p. 149, note views on 'choice' and 'gift' shows that Dufrenne and Ricœur are not wide of the mark. The main reason why they see in "Become who you are!" first and foremost a Kierkegaardian theme lies in Jaspers's thesis that freedom is rooted in transcendence. For Jaspers, "Become who you are!" is the "parole of freedom",[39] and without transcendence there is no freedom. Nietzsche, however, does not leave place for the concept of transcendence.

Now, for Jaspers, the claim that without transcendence there is no freedom is equivalent to the tenet that existential "choice" is a "gift" of transcendence. Dufrenne and Ricœur place a strong emphasis on the concept of "gift" when they credit him with "the double extension of the theory of choice in the sense of finitude and in the sense of gift."[40] The ultimate choice is choice of self. What Jaspers calls "existential choice [existentielle Wahl]" (Jaspers 1948, p. 449) "is the *decision* [*Entschluß*] to be myself in existence [Dasein]" (Jaspers 1948, p. 450). As Canguilhem puts it: "I come to me by choice. Choice is not a choice of anything but of myself" (GC, 12.1.19/3, fol. 10): I must choose myself in order to become myself.

Nevertheless, Jaspers points forcefully to the limits imposed on self-choice and self-becoming: "Certainly I become myself by creating myself; but if I am myself, it is because I have not created myself [...]."[41] Humans do not decide to be born, they do not decide how, where and when. This is what Heidegger calls *Ge-*

82) The two (inaccurately) quoted texts from Nietzsche are respectively Nietzsche NL 1881, 11[297] (KSA 9, p. 555) and GM III 9.

39 "Le mot d'ordre de la liberté, c'est : Deviens ce que tu es." (Dufrenne/Ricœur 1947, p. 153) This sentence introduces Chapter III, "La communication". On communication, cf. Jaspers 1948, pp. 338–396.

40 "[...] le double prolongement de la théorie du choix dans le sens de la finitude et dans le sens du don" (Dufrenne/Ricœur 1947, p. 133).

41 "[I]ch werde zwar, indem ich mich schaffe; aber bin ich ich selbst, so habe ich mich selbst nicht geschaffen" (Jaspers 1948, p. 325). In a note, Dufrenne and Ricœur comment on Jaspers's view that Nietzsche's "creation", which finds expression in the motto "Become who you are!", stands for "*freedom without transcendence*" (Jaspers 1965, p. 155). After a rather idiosyncratic translation of the passage from *Philosophie*, the two French philosophers remark: "That is why Jaspers says: I make myself, not: I create myself. Instead of Nietzsche's constant use of the expression: [...] to create is to be free without the limit of transcendence" (Dufrenne/Ricœur 1947, p. 151 note). The alleged terminological difference between Jaspers's "making" [*faire*] and Nietzsche's "creating" [*créer*] only comes about through the French translation; in the original text, Jaspers himself always uses "*Schaffen*" ["creating"]. However, he would probably agree with Dufrenne and Ricœur that the difference lies in the awareness of limits that he actually misses in Nietzsche. One could argue that Nietzsche would indeed deny that man was created, but, of course, without claiming that man created himself just in the same sense. On this point, Jaspers equivocates, and Dufrenne's and Ricœur's comment is not entirely clear either.

worfenheit: we have been thrown into the world without being asked. With Jaspers, who speaks of "situation", it sounds far more harmonious: we have been given as a gift to ourselves by transcendence. "I come to me as a gift".[42] "I'm just given as a gift to myself."[43] One can choose oneself and become oneself; however, the very freedom that allows one to choose is itself a gift from transcendence. "I'm given [*donné*] to myself. What is most mine in myself, and my very freedom, comes from elsewhere: I am 'by another', which is my transcendence" (Dufrenne/Ricœur 1947, p. 151).[44] Thus, the concept of transcendence connects the two existential categories of gift and choice: existential choice *is* a gift from transcendence.

Jaspers, who holds that the limits to finite freedom are set by transcendence, interprets the absence of transcendence in Nietzsche's philosophy as a lack of limits. Accordingly, "[a] human is only himself when he lives in relation to transcendence" (Jaspers, *Nietzsche*, p. 430; cf. Dufrenne/Ricœur 1947, 255). The maxim does not make sense if one does not acknowledge and affirm transcendence. Thus Jaspers.

Canguilhem notices that, in the philosophy of *Existenz*, the "affirmation and recognition of transcendence" (GC, 12.1.19/3, fol. 10) is the condition of meaning of "Become who you are!" The French philosopher himself, however, does not embrace Jaspers's concept of "transcendence". So he does not draw the conclusion that upholding self-becoming and denying transcendence is a contradiction.

The difference between their approach and goal is evident and more general. Among the "classical" interpreters, Jaspers emphasises Nietzsche's contradictions the most. Canguilhem, on the contrary, starts out just by introducing "Be-

42 "Je viens à moi comme un don" (Dufrenne/Ricœur 1947, p. 127). "Doch kann seine unablässige Anstrengung sich selbst *nicht* erzwingen: er *kommt zu sich wie ein Geschenk* [Yet his ceaseless effort can *not* enforce itself: he *comes to himself like a gift*]" (Jaspers 1948, p. 332).

43 "*Ich werde mir doch nur geschenkt.*" (Dufrenne/Ricœur 1947, p. 151 note.) "*Ich bin für mich verantwortlich,* [...]; und ich werde mir doch nur geschenkt, weil dieses Sichselbstwollen noch eines *Hinzukommenden* bedarf." (Jaspers 1948, "*Sichausbleiben und Sichgeschenktwerden*", pp. 330–333, here p. 333) "[...] so ist auch jenes *Hinzukommende*, durch das ich ich selbst werde, das Dunkle, zu dem ich aufblicke, wenn es sich erhellt im Werden meines Selbst." (Jaspers 1948, p. 333)

44 "Existential freedom" consists in the "existential choice"; "only in this choice I am aware of the freedom that is *original freedom* because only in it I actually know myself as myself." (Jaspers 1948, "*Freiheit als Wahl (Entschluß)* [Freedom as Choice (Decision)]", p. 449). This existential choice, "the decision to be myself in existence," "is what still gives itself to the will as the gift that I can actually be as someone who is willing: *from* which I can will, but *which* I in turn cannot will." (Jaspers 1948, p. 450)

come who you are!" as the "*seemingly* contradictory formula of a philosopher full of contradictions" (GC, 12.1.19/3, fol. 6; my emphasis), which also amounts to the suggestion that Nietzsche is only apparently "full of contradictions". Granted, rather than alluding specifically to Jaspers, Canguilhem aims at a commonplace; and the then following discussion shows that Nietzsche is not immune to positivistic self-misunderstandings; these, however, if not always utterly specious, are occasional shortcomings rather than fundamental contradictions on which his thought must "founder" in the Jaspersian sense.

8. Become who you are not! – Why should one still have to become what one already is? This is probably the contradiction at issue in Canguilhem's initial remark. But is it indeed only apparent? The "formula of a philosopher full of contradictions" would be *really* contradictory if it traced becoming back to being and/or conceived of value ontologically. So, even for Canguilhem, a twofold real problem lurks: we may be tempted to misconceive the ideal we are striving for as something we somehow already are. He points out that Nietzsche's anti-Platonism involves the claim that there is no being behind becoming. Canguilhem applies this anti-Platonic thesis to the Pindaric maxim. Unfortunately, just its wording may suggest a priority of being over becoming: the self, that we are invited to become, may seem given in advance – as a "first nature" hidden behind our "second nature".

Canguilhem questions that Nietzsche really means this. The French thinker puts forward a bold interpretative thesis. The "first nature", he claims, must not be understood as a self-identical presence, as a pre-given self, but is actually an ideal, a value. Whoever misunderstands the "first nature" as given in advance, traces both "becoming" and "value" back to the static metaphysical concept of "being" that Nietzsche firmly rejects.

This principled opposition of "value" and "being"/"fact", that Canguilhem does not owe primarily to Nietzsche himself, antedates his engagement with the philosophy of *Existenz* by far and is a reason why he takes interest in Jaspers's criticism of naturalistic interpretations and in his diagnosis that they represent a recurrent temptation for Nietzsche himself. This antithesis of value and being belongs to the chief results of the "reflexive approach", which goes back to Jules Lachelier and Jules Lagneau via Canguilhem's teacher, Alain.[45] This Kant-

---

45 Xavier Roth (2013, p. 124) shows that, in 1927, Canguilhem considered the primacy of value over being to be Kant's insight. This primacy belongs to the variant of Kantism or the peculiar "synthesis of Cartesianism and Kantism" (Roth 2013, p. 125) that Roth calls "reflexive style". This approach can be equated with the line that leads from Lachelier via Lagneau and Chartier (Alain) to Canguilhem.

ianising view underlies, say, Alain's dictum that one can only believe in the republic because it does not exist – i.e. only insofar as it is not a given reality, but an ideal that guides action (cf. Talcott 2019, p. 50). In the case of the former resistance fighter Canguilhem too, and even more clearly than in Alain's case, the opposition to the real, the existing, has less a metaphysical-religious slant than a resolute political valence; being an objection to the "passivity in the face of '*faits accomplis*'"[46] – as at the time of the German occupation – and, in general, to the inadequacies of the established social order (cf. Roth 2013).

With the principled opposition of "becoming" (and "value") to the immutable "being" of the metaphysic tradition, Canguilhem connects an apparent tautology: Pindar's maxim should actually read "*Become who you are not!* [*deviens celui que tu n'es pas!*]" (GC, 12.1.19/3, fol. 11) This tautological rephrasing is not intended to revoke the maxim, but to clarify that it is about becoming other than what one *presently* is; the possibility you may choose is something you are not yet. With a variation on Jaspers's concept, Canguilhem designates the self as a "*situation*" that is "open to being remade and resumed [ouverte sur sa refonte et sa reprise]" (GC, 12.1.19/3, fol. 10).

Here "becoming" is not organically conceived as the pre-determined unfolding of a pre-existing "essence", and the apparent tautology "*Become who you are not!*" also means that the "existential identity" at stake is not the "*logical* identity of the person" (GC, 12.1.19/3, fol. 11) in which Nietzsche does not believe any more than in ontological or logical necessity. Saying yes to life and affirming oneself does not mean that there is such a self-identical entity – be this an "ego" [*moi*] or rather an unconscious "first nature". Precisely in this respect, Jaspers's philosophy of *Existenz* – as Dufrenne and Ricœur reconstruct it – is also aimed at "overcoming logic" and at a "new logic".[47] However, Canguilhem rather compares Nietzsche with other positions. According to the then contemporary French philosopher Paul Césari, persons do satisfy the principle of identity, but only because the principle of the excluded middle does not apply to them. Canguilhem's Nietzsche is more radical: none of the three principles – identity, contradiction, excluded middle – applies to the ontology of persons.

Canguilhem refers in particular to the *Genealogy:* only a "seduction of language" makes it look "as though there were an indifferent substratum [...] which had the *freedom* to manifest strength or not. But there is no such substratum; there is no 'being' behind the deed, its effect and what becomes of it; 'the

46 Thus Canguilhem 2018e, p. 212, on Canguilhem's friend, the resistance fighter, philosopher and mathematician, Jean Cavaillès (1903–1944).

47 Cf. the paragraphs "Dépassement de la logique" (Dufrenne/Ricœur 1947, pp. 120–125) and "Vers une nouvelle logique" (Dufrenne/Ricœur 1947, pp. 125–132).

doer' is invented as an afterthought, – the doing is everything" (Nietzsche, GM I 13, KSA 5, p. 279; Nietzsche, GM 1, p. 26). Canguilhem applies this to the maxim "Become who you are!": if there is no subject, but only "doing", "becoming", one might wonder if there is anything that corresponds to the two pronouns "who" and "you". However, rather than with the straightforward denial that the "who" and the "you" exist at all, the lecture ends with the more moderate claim that, if not downright illusory, they are at least irrelevant.

> In the end, the *I* [le *moi*] is uninteresting.
> The *who* [le *celui*] is unimportant.
> Becoming is *doing*.
> Valéry (GC, 12.1.19/3, fol. 11).[48]

According to Canguilhem, the maxim "Become who you are!" has no ontological meaning, but only an existential scope. The (Socratic) idea that we can and must *try to (re)discover* ourselves is foreign to Nietzsche. If our "first nature", i.e. the self that the maxim invites us to become is not really given in advance, but is actually a value, an ideal, the maxim is simply a call to act ("becoming" is "doing") and to "re-invent" ourselves. With a quote from Paul Valéry: "There's only one thing to do: remake yourself. / It's not easy."[49]

So, "Become who you are!" is actually a norm without metaphysical or objective content. The *Essay* (1943) already dealt with a prominent case of the same confusion: positivism makes the normal and the pathological into objective categories, while in reality they are norms that refer to vital and social values. Pindar's maxim too resists such a positivist interpretation: Canguilhem considers "Become who you are!" as a norm whose supposed metaphysical or objective content dissolves upon further analysis.

Bibliography

Bertram, Ernst (1932): *Nietzsche, essai de Mythologie.* Paris.
Bertram, Ernst (1965): *Nietzsche. Versuch einer Mythologie.* 8. um einen Anhang erweiterte Ausgabe. Bonn [1918].

---

48 "Finalement le *moi* est sans intérêt.
le *celui* est sans importance.
Devenir c'est faire.
Valéry" (GC, 12.1.19/3, fol. 11).

49 "Il n'y a qu'une chose à faire : se refaire / Ce n'est pas simple." (Valéry 1947, p. 166) I go into Paul Valéry's relevance for Canguilhem in Brusotti 2021.

Bonnefoy, Claude (2015): "Entretien sur Jean-Paul Sartre" [Auszüge aus: Claude Bonnefoy: Rien ne laissait voir que Sartre deviendrait "Sartre"]. In: Georges Canguilhem: *Œuvres complètes*, vol. IV. Paris, pp. 955–963 [1961].

Brusotti, Marco (2012): "Diskontinuitäten. Nietzsche und der 'französische Stil' in der Wissenschaftsphilosophie: Bachelard und Canguilhem mit einem Ausblick auf Foucault". In: Renate Reschke/Marco Brusotti (eds.): *"Einige werden posthum geboren." Friedrich Nietzsches Wirkungen*. Berlin, Boston, pp. 51–78.

Brusotti, Marco (2021): "'Werde, der Du bist!' Selbsterkenntnis, Handeln und Selbstgestaltung bei Nietzsche in einem Ineditum von Georges Canguilhem". In: *Nietzsche-Studien* 50, 181–216.

Canguilhem, Georges (1947–48): "Deviens celui que tu es / Werde der du bist / Nietzsche", unpublished manuscript (the item belongs to GC. 12.1.19 [1947–48], Fonds Georges Canguilhem, CAPHÉS [Centre d'Archives en Philosophie, Histoire et Édition des Sciences]. Paris, rue d'Ulm.

Canguilhem, Georges (1966): *Le normal et le pathologique*. Paris [1943].

Canguilhem, Georges (2015a): [Rezension von:] "Mélanges offerts à M. Arnold Reymond". In: Georges Canguilhem: *Œuvres complètes*, vol. IV. Paris, pp. 177–178 [1944].

Canguilhem, Georges (2015b): "Certificat de philosophie générale et de logique. Indications bibliographiques". In: Georges Canguilhem: *Œuvres complètes*, vol. IV. Paris, pp. 55–63 [1942].

Canguilhem, Georges (2015c): "Commentaire au troisième chapitre de L'Évolution créatrice". In: .Georges Canguilhem: *Œuvres complètes*, vol. IV. Paris, pp. 111–170 [1943].

Canguilhem, Georges (2015d): "Cours de philosophie générale et de logique. Année 1942–1943 (extraits)". In: Georges Canguilhem: *Œuvres complètes*, vol. IV. Paris, pp. 81–109 [1942–1943].

Canguilhem, Georges (2015e): *Œuvres complètes*, vol. IV: *Résistance, philosophie biologique et histoire des sciences 1940–1965*. Paris.

Canguilhem, Georges (2018a): "Entretien de Georges Canguilhem [avec François Bing et Jean-François Braunstein]". In: Georges Canguilhem: *Œuvres complètes*, vol. V. Paris, pp. 1281–1301 [1998].

Canguilhem, Georges (2018b): "La problématique de la philosophie de l'histoire au debut des années 30". In: Georges Canguilhem: *Œuvres complètes*, vol. V. Paris, pp. 1123–1141 [1999].

Canguilhem, Georges (2018c): "L'action". In: Georges Canguilhem: *Œuvres complètes*, vol. V. Paris, pp. 215–261 [1967].

Canguilhem, Georges (2018d): "Ouverture [du colloque L'histoire de la folie 30 ans après". In: Georges Canguilhem: *Œuvres complètes*, vol. V. Paris, pp. 1232–1236 [1992].

Canguilhem, Georges (2018e): *Œuvres complètes*, vol. V: *Histoire des sciences, épistémologie, commémorations 1966–1995*. Paris.

Cavaillès, Jean (2015): "Bibliographie sommaire pour le certificat de philosophie générale et de logique". In: Georges Canguilhem: *Œuvres complètes*, vol. IV. Paris, pp. 1151–1159 [1941].

Dosse, François (2008): *Paul Ricœur: Les sens d'une vie (1913–2005)*. Édition revue et augmentée. Paris.

Dufrenne, Mikel/Ricœur, Paul (1947): *Karl Jaspers et la philosophie de l'existence*. Préface de Karl Jaspers. Paris.

Jaspers, Karl (1923): *Allgemeine Psychopathologie. Für Studierende, Ärzte und Psychologen*, 3. verm. u. verb. Aufl. Berlin [1913.]
Jaspers, Karl (1932): *Philosophie*. Vol. 1: *Philosophische Weltorientierung*. Berlin.
Jaspers, Karl (1933): *Psychopathologie générale*, traduit d'après la 3e édition allemande par A. Kastler et J. Mendousse, nouvelle édition [Bibliothèque de philosophie contemporaine]. Paris [1928].
Jaspers, Karl (1946): *Nietzsche und das Christentum*. Hameln.
Jaspers, Karl (1948): *Philosophie*. Berlin, Göttingen, Heidelberg.
Jaspers, Karl (1950): *Nietzsche. Introduction à sa philosophie (Einführung in das Verständnis seines Philosophierens)*, traduit de l'allemand. Lettre-préface de Jean Wahl. Paris.
Jaspers, Karl (1955): "The Origin of the contemporary philosophic situation (the historical meaning of Kierkegaard and Nietzsche)". In: Karl Jaspers: *Reason and Existenz: Five Lectures*, pp. 19–50.
Jaspers, Karl (1957): "Philosophical Autobiography". In: Paul Arthur Schilpp (ed.): *The Philosophy of Karl Jaspers*. New York, pp. 3–94.
Jaspers, Karl (1965): *Nietzsche: An Introduction to the Understanding of His Philosophical Activity*. Tucson.
Jaspers, Karl (1981): *Nietzsche. Einführung in das Verständnis seines Philosophierens*. Berlin, New York.
Jaspers, Karl (2020): "Nietzsche". In: Dominic Kaegi/Andreas Urs Sommer (eds.): Karl Jaspers: *Gesamtausgabe*, Abt. 1: *Werke*, Bd. 18. Basel.
Limoges, Camille (2015): "Introduction". In: Georges Canguilhem: *Œuvres complètes*, vol. IV, Paris, pp. 7–48.
Nietzsche, Friedrich (1997): *Daybreak: Thoughts on the Prejudices of Morality*. Maudemarie Clark/Brian Leiter (eds.). Cambridge et al.
Nietzsche, Friedrich (2001): *The Gay Science*. B. Williams (ed.). Cambridge et al.
Nietzsche, Friedrich (2006a): *On the Genealogy of Morality*. Keith Ansell-Pearson (ed.). Cambridge, New York et al.
Nietzsche, Friedrich (2006b): *Thus Spoke Zarathustra: A Book for All and None*. Adrian Del Caro/Robert Pippin (eds.). Cambridge New York et al.
Nietzsche, Friedrich (2007): *Ecce Homo: How To Become What You Are*. Oxford et. al.
Pindar (1997): *Olympian Odes. Pythian Odes*. Edited and translated by William H. Race. [= Loeb Classical Library 56.] Cambridge, MA, London.
Pindar (2014): *Siegeslieder: griechisch – deutsch*. Herausgegeben, übersetzt und mit einer Einführung versehen von Dieter Bremer, 2., überarbeitete Auflage. Berlin.
Queyroux, Nathalie (ed.) (2020): "Fonds Georges Canguilhem (1904–1995)". URL (26.10.2020): https://caphes.ens.fr/wp-content/uploads/2020/08/Inventaire_Canguilhem.pdf, visited: 15.06.2021.
Roth, Xavier (2013): *Georges Canguilhem et l'unité de l'expérience. Juger et agir (1926–1939)*. Paris.
Talcott, Samuel (2019): *Georges Canguilhem and the Problem of Error*. Cham.
Valéry, Paul (1947): *Mauvaises Pensées & autres*. [Paris].

Damon Boria

Reading Nietzsche and Sartre Toward an Eco-Existentialism

Abstract: Reading Nietzsche and Sartre Toward an Eco-Existentialism. Some recent readings have advanced the case against identifying Nietzsche as a central figure in the philosophical and cultural movement of existentialism. While not denying Nietzsche's influence on many existentialists, these readings offer compelling interpretations of Nietzsche that are often significantly at odds with the interpretations of Nietzsche—notably via Heidegger—that influenced the existentialists. To its credit, existentialism is an open-ended philosophy that admits many substantive disagreements, but there is still a family of concerns, methodologies, and ideas that give it philosophical and cultural distinction. Since these recent readings of Nietzsche intensify the interpretation of him as a philosophical naturalist, it may be tempting to excise Nietzsche from the existentialist conversation. However, I believe these readings of Nietzsche as a philosophical naturalist open new and fruitful directions for existentialism today. Particularly, as I aim to illustrate in this essay, an appreciation of Nietzsche's naturalism can open Sartre's existentialism—which is often accused of being anthropocentric, instrumentalist, and ecologically tone-deaf—to an ecological refashioning.

Reading Nietzsche naturalistically

According to Christopher Janaway, "Most commentators on Nietzsche would agree that he is in a broad sense a naturalist in his mature philosophy" (Janaway 2007, p. 34). Of course, "most" does not mean "all." Some scholars are skeptical of interpretations that commit Nietzsche to philosophical naturalism and some scholars are skeptical of committing Nietzsche to any kind of "-ism." Helmut Heit, for example, argues that for Nietzsche the natural sciences do not occupy a privileged epistemic position relative to philosophy and identifying him with naturalism risks depreciating his critical and perspectival project (Heit 2016, "Naturalizing Perspectives"). In this essay I do not aim to adjudicate these disputes. Rather, my aim here is to use the naturalist readings of Nietzsche to en-

https://doi.org/10.1515/9783110760163-018

gage Sartre and explore the possibility of an ecological refashioning of existentialist philosophy.[1]

One commentator who supports a naturalist reading of Nietzsche is Maudemarie Clark, who asserts that the texts that comprise Nietzsche's mature thought "exhibit a uniform and unambiguous respect for facts, the senses, and science" (Clark 1990, p. 105). Another commentator, Brian Leiter, argues that Nietzsche's philosophical methodology is Humean in the sense that it is both naturalistic and speculative (Leiter 2013, p, 577). This means that even when speculative—that is, even when not doing the empirical work required to reveal the actual naturalistic causes, philosophizing, "should be continuous with empirical inquiry in the sciences." While Janaway may not share all the details of Leiter's reading, he likewise formulates Nietzsche's methodological naturalism as follows: "[Nietzsche] is committed to a species of theorizing that explains *X* by locating *Y* and *Z* as its causes, where *Y* and *Z*'s being causes of *X* is not falsified by our best science" (Janaway 2007, p. 38). While such claims deserve to be evaluated through careful examinations of Nietzsche's texts, some representative textual evidence includes Nietzsche's labeling of those who do science (broadly conceived) as "honest workers" (Nietzsche, GM-2 III: 23) and his lament that he became a philologist rather than "a physician or something else that opens one's eyes" (Nietzsche, EH-2: "Why I Am So Clever" 2). Other representative remarks are found in *Daybreak*, where Nietzsche writes:

> Science ... compels us to abandon belief in simply causalities precisely where everything seems so easy to comprehend and we are the fools of appearance. The "simplest" things are *very complicated* – a fact at which one can never cease to marvel. (Nietzsche, D I: 6)
>
> To construct anew the laws of life and action – for this task our sciences of physiology, medicine, sociology and solitude are not yet sufficiently sure of themselves: and it is from them that the foundation-stones of new ideals (if not the new ideals themselves) must come. (Nietzsche, D III: 190)

In addition to this methodological commitment, Janaway argues Nietzsche is committed to naturalism substantively as well. He writes:

> [Nietzsche] opposes transcendent metaphysics, whether that of Plato or Christianity or Schopenhauer. He rejects notions of the immaterial soul, the absolutely free controlling will, or the self-transparent pure intellect, instead emphasizing the body, talking of the animal nature of human beings, and attempting to explain numerous phenomena by invoking drives, instincts, and affects which he locates in our physical, bodily existence. Human be-

1 Clearly, then, the plausibility of my ideas in this paper are significantly dependent on the plausibility of some naturalist readings of Nietzsche.

> ings are to be "translated back into nature," since otherwise we falsify their history, their psychology, and the nature of their values—concerning all of which we must know truths, as a means to the all-important revaluation of values. (Janaway 2007, p. 34)

We can find instances of both acceptance and rejection of many of these substantive philosophical commitments in the existentialist movement. By simply recognizing the distinction between theistic and atheistic existentialists we point to substantive disagreements over, for example, transcendent metaphysics and the immaterial soul. In other words, regardless of what position we take on some issues—such as the two noted here—we can still philosophically be an existentialist. I am not interested here in either presenting necessary and sufficient conditions for being an existentialist or adjudicating whether Nietzsche should be identified as an existentialist. My interest, rather, is to show how Sartrean existentialism can be ecologically refashioned through engagement with Nietzsche's methodological and substantive naturalism, a project that is to the benefit of existentialism's relevance in an age increasingly defined by anthropogenic, expansive, and momentous ecological crises.

Sartre, Nietzsche, and methodology

During a 1975 interview, Sartre described his work, and particularly his methodology, as "absolutely open-ended" (Sartre 1981: Schilpp, p. 24). Phenomenology and dialectics are certainly central to his methodology, but his conception of these always remained open and, additionally, they should not be viewed as comprising an exclusive, closed methodology. There is space, then, for thinking together—that is, integrating Sartre's open-ended methodology and Nietzsche's methodological naturalism. Such methodological pluralism and flexibility are virtues given the task at hand. Matthew Ally, whose book *Ecology and Existence: Bringing Sartre to the Water's Edge* precedes and influences this present paper in advancing an ecologically refashioned existentialism, quotes the limnologist G. Evelyn Hutchinson, who said of his scientific field that "many disparate things must be done at once" (Ally 2017, p. 257). Ally, quite rightly, says the same is true for existential ecology. A flexible and varied methodological approach is necessary to achieve the many disparate things that would result in an ecologically refashioned existentialism. The more pressing question is whether the proposal here is the right one—that is, whether an open-ended but integrative approach involving Sartrean phenomenology and dialectics along with Nietzsche's methodological naturalism is fit for the task. While a full-throated attempt to develop this integrative approach would take this paper too far afield, this integration is possible. Sartre already showed how to integrate phenomenology and dialectics in, most notably, *The Family Idiot.* Integrating Nietzsche's methodological naturalism is possible too, at least with the reading of Nietzsche that does not commit to mechanistic materialism, which—in Sartre's words—"produces and resolves man into an ensemble of physical laws" (Sartre 2008: ItinT, p. 37). Such mechanistic materialism is incompatible with Sartre's phenomenology of human freedom.

Another way to state this is that avoiding a methodologically-induced slip into mechanistic materialism is necessary to remain compatible with Sartre's atypical realism, a view that in 1969 Sartre said he had been trying all his life to establish a philosophical foundation for (Sartre 2008: ItinT, pp. 36–37). Sartre described his search for a philosophical foundation for "realism" as a search for "how to give man both his autonomy and his reality among real objects, avoiding idealism without lapsing into a mechanistic materialism" (Sartre 2008: ItinT, pp. 36–37). Ally labels Sartre's view "relational realism." Noting that Sartre criticized some versions of realism as "sophistry," Ally describes Sartre's version as follows: "There is no objective reality, not in the conventional, non-relational sense. There is no reality "out there" that we know "in here," because that is just the way reality *is:* reality cannot *be* other than in relation. Reality *is* this relation" (Ally 2017, p. 259). Sartre illustrated the relationality of reality in his phenomenological study of consciousness. Ally, again quite rightly, argues that we need this kind of realism if we are to face up to the ecological facts of our circumstances and imagine our future possibilities.

Take, for instance, one of Sartre's most ecological reflections: the deforestation of China's Central Plain and the ensuing flooding. Understanding it requires getting a grip on a complex and dynamic set of relations, including no less than the peasants' physiological needs, their agricultural praxis, their serial sociality (psychologically, spatially, and temporally), the forests, the rivers, and—as Sartre observes—"the whole of hydrodynamics" (Sartre 2004: CRD, p. 162). Sartre's dialectical reflection skillfully discloses these relations and the counter-finality that is their "hidden meaning." But the success of Sartre's speculative reflection is due, in part, to its continuity with empirical inquiry in the sciences, such as limnology and hydrodynamics. Likewise, the success of a broader eco-existentialism, meaning in part its ability to speak to an age of ecological crises, will require continued engagement and, to a degree compatible with Sartre's relational realism, continued epistemic continuity with empirical inquiry in the sciences.

Sartre and Nietzsche against anthropocentrism and instrumentalism

For Sartreans, a more deliberate engagement with Nietzsche's methodological naturalism should be accompanied by a more deliberate engagement with Nietzsche's substantive naturalism. One fruit of doing so will be overcoming the readings of those who are skeptical about the possibility of an ecologically refashioned (Sartrean) existentialism. Douglas Kirsner, for example, argues that Sartre:

> is echoing the view of the body and the world that has enshrined the project of Western civilization for many centuries. This is the logic of domination which views the world as there to be subdued and controlled and uses an instrumental, managerial form of rationality that regards oneself, others and the environment as objects to be quantified and manipulated. (Kirsner 1985, p. 225)

In other words, Sartre's existentialism is committed to an anthropocentrism and instrumentalism that makes it decidedly anti-ecological and anti-environmental. The popularity of this view is suggested by both Sartre's general exclusion from ecological conversations and the relative rarity of ecological discussions in Sartre scholarship. While Ally is not the only one to show that Sartrean existentialism is not committed to such views, this corrective work usually does not involve significant engagement with Nietzsche, especially the naturalistic readings of him. The benefit of engaging Nietzsche's substantive naturalism is that we can more easily see how Sartrean existentialism can reject anthropocentrism and instrumentalism without losing the fundamental sense in which it is a humanism.

Nietzsche's substantive naturalism has the effect of undermining anthropocentrism, regardless of whether it is embedded in a transcendent metaphysics. The third of the "four errors" Nietzsche identifies in *The Gay Science* is that the human being has "felt himself in a false position in relation to the animals and nature" (Nietzsche, GS: 86) In *Beyond Good and Evil* he writes:

> To translate man back into nature, to become master over the many vain and overly enthusiastic interpretations and connotations that have so far been scrawled and painted over that eternal basic text of homo natura; to see to it that the human being henceforth stands before human beings as even today, hardened in the discipline of science, he stands before the rest of nature, with intrepid Oedipus eyes and sealed Oedipus ears, deaf to the siren songs of old metaphysical bird catchers who have been piping at him all too long, "you are more, you are higher, you are of a different origin"—that may be a strange and insane task, but it is a task—who would deny that? (Nietzsche, BGE-2: 230)

Put differently, Nietzsche's substantive naturalism revokes humanity's special status. In *The Antichrist* he states the point quite starkly: "man is absolutely not the crown of creation; every creature stands beside him at the same stage of perfection" (Nietzsche, A 14). It demands we reimagine ourselves as animals largely like other conscious animals and, perhaps, significantly like non-sentient lifeforms such as plants. The greater the proximity of Sartrean existentialism to Nietzsche's substantive naturalism, the more the former can claim the anti-anthropocentrism of the latter. Nietzsche's account of drives and Sartre's account of facticity are grounds for recognizing such proximity.

Influenced by advances in human physiology, Nietzsche theorizes that our conscious experience is shaped by bodily drives. Paul Katsafanas describes Nietzsche's idea of the drive as follows:

> A drive is a disposition that induces an evaluative orientation. Drives manifest themselves by structuring the agent's perceptions, affects, and reflective thought. Moreover, drives do not simply arise in response to external stimuli: they actively seek opportunities for expression, sometimes distorting the agent's perception of the environment in order to incline the agent to act in ways that give the drives expression. (Katsafanas 2013, p. 752)

So understood, drives are not unique to human beings. This fact is not lost on Ken Gemes. In his account of Nietzsche's perspectivism, he calls attention to Section 12 of the Second Essay of the *Genealogy of Morals* wherein Nietzsche attributes perspectives to all life (Gemes 2013, pp. 563–564). After asking what it means "to say our drives have perspectives, that our drives interpret the world" (Gemes 2013, p. 564), Gemes answers using the example of a hyena. Two of the hyena's drives—for sustenance and for survival—compete for expression, creating a fight or flight scenario when in the presence of a boar. He concludes that this example "explains how animal life can be said to have drives that interpret." Gemes does not stop with conscious animals, opening the possibility for non-sentient lifeforms such as plants despite the admitted difficulty. By considering tropism—that is, plants moving or growing to reach more sunlight—he says we can "gain an idea of how Nietzsche's notion of drives and interpretation can be applied to nonsentient life forms" (Gemes 2013, pp. 565–566). Focusing on the same section of the *Genealogy*, Janaway states the relevant takeaway as follows: for Nietzsche, "at all levels of the organic world there is spontaneity, active appropriation, interpretation, and the imposition of form and meaning" (Janaway 2007, p. 38).

Nietzschean drives, especially when conceived in a way that leaves room for human freedom, are compatible with Sartre's account of facticity. Our bodies make up part of our facticity. The neediness of our bodies—our need for nourishment, for example—constitutes part of our basic *situation* from which ethics springs as a salient dimension of our experience—as Sartre observes in his lecture notes from the 1960s (Bowman and Stone 1997, p. 273)—and from which we take up one of our original free praxes—as Sartre observes early in the *Critique of Dialectical Reason* (Sartre 2004: CRD, pp. 79–83). Integrating Nietzsche's account of drives into Sartre's deforestation example, we can say that the peasants' situation is the dynamic—ecological—interplay between their facticity (which includes their bodily-rooted drive for sustenance) and their free consciousness (i. e. their capacity for transcending their facticity). This dynamic interaction is what makes our situation *meaningful*. Even though conscious animals may not be free in a Sartrean sense—I will bracket the Gemes-inspired discussion of plants—we should admit they have drives and, more broadly, a facticity.[2] In the final part of this paper I will provide two illustrations that suggest this admission supports a Sartrean case for viewing conscious nonhuman animals as creators of meaning, which is a view that should correct any latent anthropocentrism and instrumentalism in Sartrean existentialism.

Animals and beyond

Hazel Barnes raised the specter of Sartrean-inspired anthropocentrism and instrumentalism quite directly. Using the language of *Being and Nothingness* she asked: "What [are] we to say of the consciousness of nonhuman animals? ... Do all but human animals belong with the in-itself?" (Barnes 1981: Schilpp, p. 682). If Sartre denied animal consciousness and answered the latter question affirmatively, he would be a clear-cut Cartesian on the issue, likely committed to versions of anthropocentrism and instrumentalism that would prohibit, or at least severely limit, the possibility of an ecologically refashioned existentialism. Sartrean existentialism would, then, be deserving of the same criticisms that have been leveled against other Cartesian-motivated anthropocentrists and instrumentalists. We do not have to speculate about his answers to those questions. In a 1943 letter written in reply to Gabriel Marcel's critical feedback on *Being and Nothingness* (Sartre 2014: L-GM, pp. 62–63), Sartre (in a deferential tone) *excuses*, not defends, his presenting the work as "only human reality, etc." ("*seule la réalité humaine etc*"),[3] claiming it merely stemmed from his "desire to think in order," which for him meant starting with the disclosure of

2 For simplicity I use the term "animal," which I use in the sense of (to put it poorly) "nonhuman animal" or (to put it better, following Matthew Ally, who follows David Abrams) "other-and-more-than-human" (Ally 2017, p. 36).

3 I thank Rachel Fox and Nik Farrell Fox for their translation of the letter.

human reality before proceeding to issues like, as he mentions in this letter, animal consciousness and biological being more generally. He notes that he had no plans to pursue such studies, but acknowledged the possibility for them. He says the "order" of his approach is aimed at not explaining "the superior by the inferior," an expression that—like elsewhere in Sartre's writings, looks more-than-suspiciously anthropocentric. Matthew Ally, though, gives us a more charitable option for understanding the order of Sartre's approach: "his thought is most accurately described not as anthropocentric but as *anthropometric*," adding "[e]ach organism must take its first measure by means of its own toolkit" (Ally 2017, pp. 362–363). Starting the philosophical project with disclosure of human reality is "superior," then, merely in the sense that it is the reality closest to the inquirer. For Nietzsche, such an approach is hardly surprising for, as he writes, "all the products of [one's] thinking are bound to reflect the condition he is in," with "condition" referring to physiological and similar facts about the thinker (Nietzsche, D 42).

In the letter to Marcel, Sartre goes beyond the excuse and unequivocally states his view regarding animal consciousness: "I do think that animals exist as consciousness and, as such, nihilate" ("*Je pense en effet que l'animal existe comme conscience et comme tel néantise*") and, reasonably then, denies being a Cartesian on the issue (Sartre 2014: L-GM, pp. 62–63). He also notes his agreement with Marcel's example of a pet dog who has lost his owner, which is particularly interesting since Sartre, as discussed below, will say a lot more about the experience of pet dogs. Sartre's views in the 1943 letter to Marcel are not an aberration. During the 1975 interview mentioned earlier Sartre was told of some related criticism from Robert Champigny, Sartre again stated he always believed in animal consciousness even though determining the status of animal consciousness in relation to humans is "an extremely difficult problem" that he could not answer (Sartre 1981: Schilpp, p. 28). Regardless, he notes that "consciousness exists where we notice it; and there are animals that do not have it, protozoa, for example. Consciousness appears in the animal kingdom at a certain moment: in men; surely also in monkeys. But how does it appear and what is it?" (Sartre 1981: Schilpp, p. 28). Perhaps we cannot yet fully explain how animal consciousness appears or precisely what it is. But, arguably, we cannot fully explain how human consciousness appears or precisely what it is either. What we can do are phenomenological descriptions like those done by Sartre. This is sufficient for at least defending the claim that we notice consciousness in animals like, to use Sartre's example, monkeys.

Consider the following scenario: You are stranded on a small island with nothing but water, palm nuts and a few rocks. You are, in a sense, the center of the universe. The water is there for *you* to drink. The palm nuts are there for *you* to eat. The rocks are there for *you* to use as tools to crack the palm nuts. In sum, the meaning of everything on that island refers to *your* point of view and *your* intentions. But then you notice that a capuchin monkey is also on the island. While not phenomenologically guaranteed, the effect may be *relevantly similar* to the one Sartre describes when a solitary visit to a public park gets interrupted by the presence of another person. He writes, "there is now an orientation *which flees from me*" (Sartre 1984: EN, p. 342). He continues:

> ... there is a regrouping of all the objects which people my universe ... Everything is in place; everything still exists for me; but everything is traversed by an invisible flight and fixed in the direction of a new object. The appearance of the Other in the world corresponds therefore to a fixed sliding of the whole universe, to a decentralization of the world which undermines the centralization which I am simultaneously effecting. (Sartre 1984: EN, p. 343)

"*I am no longer master of the situation*," he adds (Sartre 1984: EN, p. 355). "The world disintegrates in order to be reintegrated over there as a world" (Sartre 1984: EN, p. 364). Like you, the monkey is animated by a drive for sustenance and has created a meaningful situation that rivals your own. From the monkey's perspective, the water is there for *him* to drink, the palm nuts are there for *him* to eat, and the rocks for *him* to use as tools. The monkey has its own perspective within which you are an object. Whether provoked by the consciousness of a human or a capuchin monkey, the internal hemorrhage (to use an expression from Sartre) in your center-of-the-universe orientation is relevantly analogous. Beyond illustrating that such phenomenological descriptions permit us to affirm the existence of animal consciousness with the same degree of confidence with which we affirm the existence of a human consciousness other than our own, this illustrates that Sartrean existentialism is compatible with attributing facticity, including Nietzschean drives, to at least conscious nonhuman animals and, additionally, that such beings—like us—create meaningful *worlds*. I will add, provocatively, that in the 1975 interview Sartre said: "A plant that is growing is no longer altogether in-itself. It is more complex. It is alive" (Sartre 1981: Schilpp, p. 40).

The plausibility of the preceding scenario requires neither a presupposition about *self*-consciousness in animals nor an assertion that the encounter with the-other-as-human is *equivalent* to the encounter with the-other-as-monkey. Rather, the plausibility depends merely on the phenomenologically descriptive accuracy of the encounter with some animals (e. g. monkeys) as conscious and meaningfully engaged with the world. There is precedent for such reasoning in twentieth century French philosophy. In *The Animal That Therefore I Am (More to Follow)*, Jacques Derrida writes at length about the experience of being seen naked by his cat. He writes:

> The animal is there before me, there close to me, there in front of me—I who am (following) after it. And also, therefore, since it is before me, it is behind me. It surrounds me. And from the vantage point of this being-there-before-me it can allow itself to be looked at, no doubt, but also—something that philosophy perhaps forgets, perhaps being this calculated forgetting itself—it can look at me. It has its point of view regarding me. The point of view of the absolute other, and nothing will have ever done more to make me think through this absolute alterity of the neighbor than these moments when I see myself seen naked under the gaze of a cat. (Derrida 2002, p. 380)

Derrida proceeds to criticize philosophers for their historically common blindness or denial. Citing thinkers ranging from Descartes and Kant to Lacan and Levinas, he says that even when they bothered to look at animals, they acted (in their thought, at least) as if the animals were not looking back and addressing them (Derrida 2002, pp. 383–383). He says it is "violence" or "stupidity" to think of "depriving the animal of every power of manifestation, of the desire to manifest *to me* anything at all, and even to manifest to me in some way *its* experience of *my* language, of *my* words and of *my* nudity" (Derrida 2002, p. 387). Near the beginning of this essay, he frames it as an effort to understand what he does when he says "*Je suis*." He says this effort requires moving from "the ends of man" to "'the crossing of borders' between man and animal," a move he views as approaching Nietzsche's view of the human being as "an as yet undetermined animal" and "promising animal" (Derrida 2002, p. 372). Such dual recognition of both the animality of humans and animals' varying powers of manifestation are, indeed, present in Nietzsche's thought. Consider the following passage from *Daybreak:*

> The practices demanded in polite society ... [are] to be found as social morality in a crude form everywhere, even in the depths of the animal world – and only at this depth do we see the purpose of all these amiable precautions. ... Even the sense for truth, which is really the sense for security, man has in common with the animals ... the animal understands all this just as man does, with it too self-control springs from the sense for what is real (from prudence). It likewise assesses the effect it produces upon the perceptions of other animals and from this learns to look back upon itself, to take itself 'objectively', it too has its degree of self-knowledge. (Nietzsche, D I: 26)

I will conclude with one final reflection, which is about some of the normative stakes for an ecologically refashioned existentialism. Sartre is clear that whatever is with the for-itself—freedoms—can be oppressed. That is, self-conscious human beings can be wronged. But can animals be oppressed? *The Family Idiot* contains a keen observation that promisingly points the way for a reimagining of animals in an existentialist ecology.

Sartre's surprisingly lengthy discussion of animals is a digression from the main task of the biography, which at that point is focused on Flaubert as an unloved child and how that fact shaped the quietism that would define his life and writings. After speculating about the origins of Flaubert's persistent boredom, Sartre writes:

> It seems clear that household animals are bored; they are homunculae, the dismal reflections of their masters. Culture has penetrated them, destroying nature in them without replacing it. Language is their major frustration: they have a crude understanding of its function but cannot use it; it is enough for them to be the *objects of speech*—they are spoken to, they are spoken about, they know it. This manifest verbal power which is denied to them cuts through them, settles within them as the limit of their powers, it is a disturbing privation which they forget in solitude and which deprecates *their very natures* when they are with men. (Sartre 2004: CRD, p. 137)

Sartre then recounts watching the frustration of someone's pet dog sitting amidst a group of people talking. The dog, he suggests, has its "peaceful immanence" stolen from it. Perhaps Sartre, in Nietzschean terms, would be willing to say the dog's drives have been disrupted. That which we ordinarily call "domestication" is, on this view, a form of violent instrumentalization. Through us the dog acquires a sense of itself but only as missing something, as defective in some way. Unable to acquire whatever that missing something is, the dog is trapped in limbo between the peaceful immanence of its natural life and an opaque perception of human transcendence (the restlessness of our freedom). Like the unloved child Flaubert, the dog becomes resigned and, ultimately, *bored*. Sartre writes: "Without culture the animal would not be bored—he would live, that is all," adding "Boredom with life is a consequence of the *oppression* of animals by man; it is nature grasping itself as the absurd end of a limiting process instead of realizing itself as biological spontaneity" (Sartre 2004: CRD, pp. 138–139), which is a phenomenon that has been scientifically observed in captive animals—particularly in zoos—experiencing zoochosis. For Nietzsche, culture can be a corrective to civilization (the meaning of "civilization" here is analogous to the way Sartre uses "culture" in the above passage), restoring in us our animality along with its biological spontaneities (to borrow Sartre's expression). With such a corrective, we may also have a path toward non-oppressive relations with animals such that they are able to live as the wild beings they are. We may recall here the journey of Nietzsche's Zarathustra, which sees him increasingly exist and think alongside animals—calling us to, for example, learn from ruminating cows (Nietzsche, Za-1 280).

If this brief illustration is indicative of the normative orientation of an ecologically refashioned existentialism, then we may have a philosophical movement that preserves existentialism's well-earned disclosures of human reality while simultaneously advancing, in a Nietzschean vein, a non-instrumentalist, non-anthropocentric, scientifically continuous framework that might fruitfully reorient our attention and praxes in an age of ever-expanding ecological crises. We might find ourselves with a more flexible mind for understanding our age, a greater generosity in our moral considerations, a more expansive imaginary of the world's possibilities, and a bit more hope.

Bibliography

Ally, Matthew C. (2017): *Ecology and Existence: Bringing Sartre to the Water's Edge*. New York.

Barnes, Hazel E. (1981): "Sartre as Materialist". In: Paul Arthur Schilpp (ed.): *The Philosophy of Jean-Paul Sartre*. La Salle, pp. 661–684.

Bowman, Robert V. and Elizabeth A. Stone (1997): "Making the Human in Sartre's Unpublished Dialectical Ethics". In: William McBride (ed.): *Existentialist Ethics*. New York, pp. 269–285.

Clark, Maudemarie (1990): *Nietzsche on Truth and Philosophy*. Cambridge.

Derrida, Jacques (2002): "The Animal That Therefore I Am (More to Follow)". In: *Critical Inquiry*. No. 28, pp. 369–418.

Gemes, Ken (2013): "Life's Perspectives". In: Ken Gemes and John Richardson (eds.): *The Oxford Handbook of Nietzsche*. Oxford, pp. 553–575.

Heit, Helmut (2016): "Naturalizing Perspectives: On the Epistemology of Nietzsche's Experimental Naturalizations". In: *Nietzsche Studien: Internationales Jahrbuch für die Nietzsche-Forschung*. No. 45, pp. 56–80.

Janaway, Christopher (2007): *Beyond Selflessness: Reading Nietzsche's "Genealogy"*. Oxford.

Katsafanas, Paul (2013): "Nietzsche's Philosophical Psychology". In: Ken Gemes and John Richardson (eds.): *The Oxford Handbook of Nietzsche*. Oxford, pp. 727–755.

Kirsner, Douglas (1985): "Sartre and the Collective Neurosis of Our Time". In: *Yale French Studies*. No. 68, pp. 206–225.

Leiter, Brian (2013): "Nietzsche's Naturalism Reconsidered". In: Ken Gemes and John Richardson (eds.): *The Oxford Handbook of Nietzsche*. Oxford, pp. 576–598.

Sartre, Jean-Paul (1981): "An Interview with Jean-Paul Sartre". In: Paul Arthur Schilpp (ed.): *The Philosophy of Jean-Paul Sartre*. La Salle, pp. 2–51.

Sartre, Jean-Paul (1984): *Being and Nothingness: A Phenomenological Essay on Ontology*. New York.

Sartre, Jean-Paul (2004): *Critique of Dialectical Reason, Volume 1: Theory of Practical Ensembles*. New York.

Sartre, Jean-Paul (2008): "The Itinerary of a Thought." In: *Between Existentialism and Marxism*. New York.

Sartre, Jean-Paul (2014): "Lettre de Jean-Paul Sartre à Gabriel Marcel". Bibliothèque nationale de France. Paris, pp. 62–63

Christian Niemeyer

Sartres Flaubert im Vergleich zu jenem Nietzsches. Oder: Über einen Existenzialisten *avant la lettre* unter Bezug auf Nietzsches Madame Bovary und seinem auf sie bezüglichen „Experimentalismus“

Abstract: Sartre's Flaubert in Comparison to Nietzsche's. Or: About an Existentialist *avant la lettre* with Reference to Nietzsche's Madame Bovary and his „Experimentalism“ in this regard. Sartre's *Flaubert* is vastly more voluminous than Nietzsche's. Sartre wrote more about Flaubert than about any other literary figure. Nietzsche, on the other hand, hardly read anything by Flaubert in the original and generally relied on the impressions of others, on Guy de Maupassant's obituary, for example, or on entries in the *Journal des Goncourts*. This slim input was enough for Nietzsche's genius to generate a number of important, partly encoded insights. Decoded, they support the proposition put forward here that in his relationship with *his* Madame Bovary, Lou von Salomé, Nietzsche played the role of a Rodolphe in framing Lou in a kind of visionary „dangerous liaison“ which could just as well have come from Flaubert. Nietzsche's last word on the matter, in 1888, was an anxious denial of this vision. We are left with the fact that Nietzsche may be read as an existentialist *avant la lettre* in the „experimentalism“ vein revealing connections that Sartre did not contemplate, possibly due to his insufficient knowledge of Nietzsche.

Vorab: Mein Zugang ist der eines Nicht-Literaturwissenschaftlers und, wie das Thema Nietzsche, so wie ich es bisher traktiert habe, ausweist, Nicht-Philosophen. Positiv geredet: Mein Zugang ist der eines Psychologen mit stark biographischem Zugang. Aus Zeitgründen möchte ich auf weitere Sie womöglich schockierende Vorbemerkungen verzichten, außer vielleicht: Sartres Kurzdefinition des Existentialismus – dass „der Mensch *nicht ist*, sondern *sich schafft*“, ergänzt um die Setzung, dass es „weder einen Wert noch eine Moral gibt, die *a priori* gegeben sind“ (Sartre 2000a: PropE, S. 117) – könnte von Nietzsche stammen. Punkt. Das Komma dazu folgt. Sowie, als dritte Vorbemerkung: Ich hoffe, Sie gestatten mir, nicht nur aus Unterhaltungsgründen, sondern auch getreu der Variante, dass „der Referent *nicht ist*, sondern *sich schafft*“, hin und wieder ein

https://doi.org/10.1515/9783110760163-019

Kaninchen aus dem Hut zu zaubern, etwa, im II. Teil und mit dem Untertitel bereits angedeutet, ein solches namens Lou. Damit ungesäumt zur Sache. Ich beginne mit „Nietzsches Flaubert im Vergleich zu jenem Sartres" (I.), setze fort mit „Flauberts *Madame Bovary* im Vergleich zu Nietzsches" (II.) und bereite damit ein Fazit vor, in welchem ich meine Lesart von Nietzsches Existentialismus als „Experimentalismus" zu begründen suche.

Nietzsches Flaubert im Vergleich zu Sartres

Über Nietzsches Flaubert zu handeln ist keine große Sache – wenig erstaunlich, wenn man bedenkt, dass Nietzsche gerade einmal zwölf Lenze zählte, als Flauberts Hauptwerk *Madame Bovary* erschien. Und dass dasjenige, was Nietzsche via Guy de Maupassant (1884) über ihn erfuhr (vgl. Niemeyer 2020, S. 400), normalerweise kaum ausgereicht hätte für ein ernstzunehmendes Urteil, selbst unter Hinzunahme dessen, was Nietzsche dem *Journal des Goncourt* entnahm. Es kommt hinzu, dass Flaubert damals in Deutschland kaum bekannt war, abgesehen von seinem 1882 auf Deutsch erschienenen Roman *Salammbô* (1862). Kurz: Nietzsche war im Fall Flaubert auf Wissen vom Hörensagen angewiesen, nebst der Brille Dritter, etwa, neben jener Maupassants, auch derjenigen Paul Bourgets (1926) oder Ferdinand Brunetières (1887). Von Bourget entlehnt ist beispielsweise Nietzsches flammende Rede aus *Nietzsche contra Wagner* gegen Flauberts „impersonnalité" („L'homme n'est rien, l'oeuvre est tout"; NW, *Wir Antipoden*, KSA 6, S. 426), deren strategische Bedeutung auch darin liegen könnte, dass er von seinem gleichgerichteten Grundsatz aus *Ecce homo* und den Vorteilen, die er aus ihm zog, ablenken wollte. Weiterführender ist da schon Nietzsches Urteil aus *Jenseits von Gut und Böse*, Flaubert habe, wie andere „Psychologen Frankreichs", sein „bitteres und vielfältiges Vergnügen an der bêtise bourgeoise" kultviert, als „braver Bürger von Rouen", der „zuletzt nichts Anderes mehr [hörte und schmeckte]: es war seine Art von Selbstquälerei und feiner Grausamkeit." (JGB 218, KSA 5, S. 153) Denn die kunstvoll eingebaute, auf ein dann doch nicht offenbartes Geheimwissen abstellende Verzögerung („gleichsam als wenn"), schreit vielleicht nicht, ruft aber doch immerhin nach der Frage, was Nietzsche über das dunkle Geheimnis dieses „braven Bürgers" wusste – oder soll man besser variieren: was er wissen *wollte?* Eine Frage vom Typ Kaninchen, auf die resp. auf das ich gleich zurückkommen werde.

Vorerst aber zu unserer zweiten Auskunftsquelle, Sartre resp. dessen sich auf ca. 3000 Druckseiten erstreckendes Spät- und Monumentalwerk *Der Idiot der Familie. Gustave Flaubert 1821–1857* (1971/72), welches ihn, Sartre, was Flaubert-Wissen angeht, im Vergleich zum David namens Nietzsche als Goliath kenntlich macht. Gewiss: Sartres *Idiot* gilt Sarah Bakewell als „eines der unlesbarsten Bücher überhaupt“ (Bakewell 2016, S. 251), ist jedenfalls aktuell ein wenig aus der Mode geraten, im Vergleich zu der Zeit vor gut vierzig Jahren, als es verbreitet als „Meisterwerk“ (Frank 1980, S. 84; Oehler 1980, S. 149) rubriziert wurde und man sich Sartres Impulssetzung im Bereich der literaturwissenschaftlichen Methodologie ungeachtet des auch gegen ihn gerichteten Verdachts des „Biographismus“ (Grimm 1980, S. 109), des „exzessiven“ sogar, noch mit Verve annahm, neugierig geworden allein schon wegen der zehn, wenn nicht gar fünfzehn Lebensjahre, die Sartre in dieses Projekt über einen aus marxistischer Sicht – etwa jener von Georg Lukács (vgl. Oehler 1980, S. 151) – längst schon, allein durch seine Klassenzugehörigkeit, hinreichend Gerichteten wie Flaubert investiert hatte. Aber: Dass Sartre derlei (leicht antizipierbare) Einwände unbeachtet ließ oder durch den Hinweis meinte erledigen zu können, es habe halt seine Zeit gebraucht, ehe sich seine „ursprüngliche Antipathie [gegenüber Flaubert] in *Empathie* verwandelt“ habe (Sartre 1977: IF, Bd. I, S. 8), musste, wie es seinen Verteidigern schien, akzeptiert werden – womit die Sache vorerst erledigt war.

Temps perdu: In neuerer Zeit wird Sartres *Idioten* zumal in der (deutschsprachigen) Flaubertforschung bevorzugt mittels Nekrologen gedacht. Führend auf diesem Feld seit 2009: Barbara Vinken mit ihrem die Intention von Sartres Flaubert-Studie als auch deren Verhängnis auf den giftigen Punkt bringenden Satz: „Sartre verbrachte zehn Jahre damit, Flaubert als typischen Irrweg, als eine neurotische Fehlentwicklung der Moderne aus dem Weg zu räumen.“ (Vinken 2009, S. 13) Knapp dahinter ist knapp davor, genauer: ist zwei Jahre früher, also 2007, als Wolfgang Matz in Sachen von Sartres *Idioten* zwar die späte Würdigung nicht vergaß (vgl. Matz 2007, S. 393–394), ansonsten aber mit ihm nichts Rechtes anzufangen wusste – ebenso wie Vinken, die übrigens analog mit Matz verfuhr (dem sie noch nicht einmal in ihrem Literaturverzeichnis Unterkunft gewährte). Irgendwie nicht ganz fein, könnte man hier einwenden – aber irgendwie auch nicht ganz ohne Recht. Denn was Matz geboten hatte, war kaum mehr als Sartre II: Flaubert, dieser durch schwere Kindheit und Zurücksetzung gehandicapte ‚Idiot‘, habe womöglich eine Epilepsie simuliert, um gegen den Willen des Vaters sich schreibend eine imaginäre Welt erfinden zu dürfen. Sowie, und das macht die Sache nicht besser: Matz’ Flaubert war in manchen Passagen kaum mehr als einer minus Sartres, so dass man via Matz am Ende als Leser nicht wirklich weiß, was Flaubert zwischen September 1849 und Mai 1851 in Nordafrika „unter“ (mit Zarathustra resp. Nietzsche gesprochen) „Töchtern der Wüste“ trieb. Erstaunliches, wie wir noch von Julian Barnes erfahren werden. Immerhin, in einem Punkt war Matz genauer als Sartre: Die wildeste Sexszene der unglücklich verheirateten und sich am Ende mit Gift richtenden Arztfrau Emma Bovary mit ihrem Lover Léon während einer stundenlangen Kutschfahrt durch Paris fehlte zu Flauberts Empörung beim Vorabdruck in der *Revue de Paris* (vgl. Matz 2007, S. 122–123) – was die Zensur gleichwohl nicht hinderte, den Roman gerichtlich anzufechten, erfolglos, wie man von Matz ausführlichst berichtet bekommt.

Nur: Ist dies wirklich wichtig? Oder, um nun noch einmal einen Blick auf Barbara Vinken zu werfen: Ist es wichtiger als die zahlreichen subtilen (sprachanalytischen) Beobachtungen dieser Matz-Ignorantin wie Sartre-Kritikerin? Deren lustigste (von den Folgen her, allerdings wohl nicht für Sarah Bakewell) Beobachtung den Umstand betrifft, dass die Aprikose im Französischen eine Metapher ist für das „weibliche Geschlecht" (wie im Deutschen die Pflaume) und der Womanizer Rodolphe aus Flauberts *Madame Bovary*, bezeichnenderweise unter einem Hirschgeweih sitzend, ‚seiner' Madame Bovary einen Abschiedsbrief schickt, den er unter Weinblättern und Aprikosen versteckt. Heißt: Dem Aprikosencocktail, von Bakewell in ihrem wunderbar zu lesenden Cocktail *Das Café der Existenzialisten* (2016) als Existentialistenabzeichen der 1950er Jahre ausgewiesen, kommt womöglich doch eine tiefere Bedeutung zu als gedacht. Oder, so könnten Sie sich und mich nun fragen, steht diese Überlegung nur für eines jener einleitend beschworenen Kaninchen?

Ich weiß es, ehrlich gesagt, selber nicht – und fahre lieber fort: Was man – um nach diesem kurzen Bericht zum Stand der Forschung zum Wichtigsten zu kommen und wieder etwas ernster zu werden – bei Wolfgang Matz gleichfalls nicht findet, auch nicht bei Barbara Vinken oder Wolf Lepenies, Dieter Thomä et tutti quanti bis zurück zu Viktor Klemperer, ist Sartre à la Sade, soll heißen: was ihnen mangelt, den Vorgenannten, ist die Abgebrühtheit des an Stoffen wie diesem Gestählten. Denn man muss hier bedenken: Gut zwanzig Jahre, nachdem Sartre mit dem Zeitaufwand von immerhin zwei Lebensjahren als eine Art Vorstudie zu seinem *Idioten* in Gestalt von Jean Genet einen Schriftsteller umfänglich rehabilitiert hatte, der „die für die Literatur bisher tabuisierte Welt der Homosexuellen, Strichjungen und Zuchthäusler besang" (Sartre 1982: SG, U2), damit erneut dem Vorwurf Auftrieb gebend, Existentialisten würden immer nur „überall das Schäbige, Trübe und Klebrige" (Sartre 2000c: EH, S. 145) sehen – ein Vorwurf, dem übrigens schon Nietzsche an den Beispielen Maupassant und Émile Zola die Spitze gebrochen hatte (vgl. Niemeyer 2020, S. 425–426), ohne dass Sartre dies sonderlich zur Kenntnis genommen hätte –, offenbarte Sartre den Mut, erneut jenem ‚Schäbigen' auf die Spur zu kommen, und zwar bei Flaubert, aber auch im Herzen des Bildungsbürgertums allgemein, nämlich im Kreis um das legendäre *Journal des Goncourt*.

Um die Diskrepanz ermessen zu können, die sich von Sartre ausgehend in der Bewertung desselben auftut: Wolf Lepenies las Nietzsches Freude über das *Journal des Goncourt* – „Exasperirter Pessimismus, Cynismus, Nihilismus, mit viel Ausgelassenheit und gutem Humor abwechselnd; ich selber gehörte gar nicht übel hinein" (Nietzsche, Bf. an Heinrich Köselitz, 10.11.1887, KGB III/5, Bf. 948), – als Indiz für seinen „Neid auf diese Abendgesellschaften" (Lepenies 1997, S. 98), ohne deren dunkle Seite in Betracht zu ziehen. Ganz ähnlich Wolfgang Matz: Der „alte Freigeist" (= Nietzsche) blickte „neidvoll [...] auf den ausgelassenen Zynismus am Pariser Restauranttisch" (Matz 2007, S. 378), lesen wir hier zehn Jahre nach Lepenies zum nämlichen Sachverhalt.

Neid? Oh heilige Unschuld! Hat denn niemand bisher Zeit gehabt, sich ein wenig zu vertiefen in Nietzsches (Lese-)Not, der zuzuschreiben ist, dass er womöglich im November 1887 in Nizza, als er das soeben erschienene *Journal des Goncourt* (der Jahre 1862–65) als „interessanteste Novität" (Nietzsche, Bf. an Heinrich Köselitz, 10.11.1887, KGB III/5, Bf. 948) feierte, nur das ihm Genehme gelesen hat und, wie die genannten Sekundärliteraten, über das andere, Schmutzige, vornehm schwieg? Übrigens anders als Sartre. Der in seinem *Idioten* ausführlich aus den Gesprächen dieser trefflichen ‚Abendgesellschaft' referiert, ausgehend von dem – und dies kann man ihm gerne anhängen: moralisierenden, also unnietzscheanischen – Ausruf: „Falsch, völlig falsch die Beziehungen dieser Intellektuellen zur ‚Halbwelt'" (Sartre 1977: IF, S. 536). Letzterer zugehörig, nach Sartre: Die Schauspielerin Suzanne Lagier (1833–1893), die, so Sartre, „bei Flaubert und vor den Goncourts [erklärt], das Theater sei ‚der Absinth des Bordells'" und die Flaubert „nicht so sehr durch die Sinne [gefiel], (sie gehörte ihm wie jedermann und wahrscheinlich für nichts) als durch die Vulgarität ihrer Konversion." (Sartre 1977: IF, S. 536) Mehr als dies, und hier beendet Sartre leider seinen Bericht. Hätte er dies nicht getan, wäre ihm womöglich Mario Praz in den Sinn gekommen. Dieser hatte unter der Headline *Die schwarze Romantik* (1930) Flauberts Vorlieben für das Sado-Masochistische skandalisiert. Und er hatte auch einen skeptischen Blick geworfen auf Flauberts Günstling Maupassant, im Einzelnen: auf dessen im *Journal de Goncourt* breit ausgemalte Potenz-Protz-Geschichten inklusive von Andeutungen auf Marquis-de-Sade ähnliche Rituale mit aus England importierten minderjährigen Sexsklaven männlichen Geschlechts in einem Kreis um den englischen Lyriker Algernon Swinburne (vgl. Paz 1960), kurz: auf Fälle von Päderastie, Sodomie, Flagellantentum und sexuellem Missbrauch, um die herum Richard von Krafft-Ebing locker seinen Skandalbericht *Psychopathia sexualis* hätte füllen können (und es teilweise ja auch tat; vgl. Krafft-Ebing 1997, S. 111). Kurz: Der 1870 in allen Details im *Journal* von seinem Bruder geschilderte Syphilistod des Jules de Goncourt wird in der Linie von Treiben dieser Art fast zu einer folgerichtigen Angelegenheit. Eingerechnet den Verfall Edmond de Goncourts als Literat. Denn es lässt sich kaum in Abrede stellen, dass Edmond in der Folge kaum noch ein anderes Thema kannte als dieses, bis hin zum Roman *Juliette Faustin* (1881/82). Der kaum für mehr steht als für den bis an die Grenze des Lächerlichen gehenden Versuch, Jules' Tod als durch alles Mögliche bedingt zu erklären – nur nicht durch Syphilis. Und ohne dass Nietzsche, wie andernorts gezeigt (vgl. Niemeyer 2019a), seinen sonstigen Scharfsinn einsetzte, um diesem (Selbst-) Betrug auf die Spur zu kommen.

Was zugleich meint: Zum Lachen ist die Sache sicherlich nicht – was für einen Engländer indes nichts heißen muss. Auftritt Julian Barnes, der ein paar Jahre nach dem Höhepunkt der deutschsprachigen Debatte um Sartres *Idioten* in seinem Roman *Flauberts Papagei* (1984) zum Stichwort ‚Prostituierte' notiert:

> Im neunzehnten Jahrhundert notwendig zur Erwerbung von Syphilis, ohne die kein Anspruch auf Genialität erhoben werden konnte. Träger der Roten Tapferkeitsmedaille waren u. a. Flaubert, Daudet, Maupassant, Jules de Goncourt, Baudelaire. Gab es irgendwelche Schriftsteller, die nicht damit behaftet waren? Wenn ja, waren diese vermutlich homosexuell. (Barnes 1984, S. 221)

Noch einmal, verlangsamt: Flaubert, Daudet, Maupassant, Jules de Goncourt, Baudelaire waren Syphilitiker, des Gleichen, wie ich mir hier erlaube zu ergänzen: Nietzsche – gesetzt jedenfalls, was hier gesetzt resp. ad Barnes extrapoliert werden kann: dass er entgegen der hiergegen sprechenden, aber nicht wirklich überzeugenden Argumente etwa von Günther Schulte (zur Kritik: Niemeyer 2020, S. 146) oder von Joachim Köhler (zur Kritik: Niemeyer 2020, S. 148–149) *nicht* homosexuell war und also – dies geht gegen Richard Wagner, der nur auf Haltlosigkeit hin kalkuliert und Syphilis, wohl aus Tugendhaftigkeit, gar nicht auf dem Schirm hat – aus anderen Gründen onanierte, etwa denen des, etwas lax geredet, „Objektschutzes" (was sich bei venerischen Erkrankungen aller Art anempfiehlt, dies zumal in Zeiten unsicherer Kondome). Und nochmals, zum Mitschreiben und mit der Frage an Sie im Publikum: „Sehen Sie es auch, das Kaninchen, das soeben durchs Bild läuft und auf dem geschrieben steht: ‚Nietzsche hatte Syphilis', meint Herr Niemeyer in seinem neuen Buch gegen Richard Schain & Co. Bloß nicht nachfragen, sonst hört er nie auf!'"

So weit, so klar, nur: Ist dieser Nachweis, gesetzt, er wäre gelungen, wirklich wichtig? Offenbar schon, wenn man Nietzsche auch seiner dunklen Seite nach verstehen will, in dem etwa, was ihn letztlich, in seinem ‚Spätwerk', insbesondere in seinen Aufzeichnungen zu *Der Wille zur Macht*, aber auch in mancherlei Passagen aus *Der Antichrist*, *Götzen-Dämmerung* sowie *Ecce homo*, in einen Euthanasie-nahen Denker verwandelte. Der Nietzsche also, den, beispielsweise, Barbara Vinken ausschließlich zu kennen scheint und ablegt unter „verhängnisvolles Gerede" (Vinken 2009, S. 21), das Flaubert peinlich gewesen wäre. Einverstanden, was Flaubert angeht. Aber kann dies ein Grund sein, „peinliches Gerede" Flauberts, wie es Sartres *Idiot* in Erinnerung brachte, sowie darauf bezügliche Einlassungen, etwa jene von Julian Barnes, unter den Tisch fallen zu lassen, wie bei

Vinken und in der Flaubertforschung des neueren deutschsprachigen Mainstreams beobachtbar?

Im Übrigen: Nicht jeder Syphilitiker tickt wie der andere. Und: Nicht jeder erkrankt gleich schwer wie der Andere – und mischt sein Leid beharrlich in fast jedes seiner Themen hinein. Schließlich: Um dies gleichwohl zu tun, muss man nicht notwendig Syphilitiker sein. Mitunter reicht dafür auch, wie Michel Onfray (2013, S. 92) für Albert Camus im Vergleich zu Nietzsche als wahrscheinlich erklärte, eine einfache Tuberkulose. Eines aber dürfte klar sein: Gar nicht darüber reden geht heutzutage, 150 Jahre nach Erfindung der Moderne in der Linie von Zarathustras Warnung: „Alle verschwiegenen Wahrheiten werden giftig!", nicht – was, um nochmals auf Vinken zu kommen, des Weiteren tadelnswert macht, dass im von ihr und Cornelia Wild edierten *Flaubert-Wörterbuch* (Vinken/Wild 2010) nicht nur Barnes ‚Prostituierten'-Lemma fehlt, sondern auch das auf Flauberts Reisetagebücher vom Oktober 1849 bis Juni 1851 sowie den dazugehörenden Briefwechsel bezügliche Stichwort ‚Kuchuk-Hanem' aus Barnes' Flauberts *Papagei.* Ich zitiere und bitte Sie, sich zur Not die Ohren zuzuhalten:

> Ein Lackmustest. Gustave musste sich entscheiden zwischen der ägyptischen Kurtisane und der Pariser Poetin – Wanzen, Sandelöl, rasierte Pudenda, Klitoridektomie und Syphilis gegen Reinlichkeit, lyrische Poesie, relative sexuelle Treue und Frauenrechte. Ihm schienen Pro und Contra bestens austariert. (Barnes 1984, S. 221)

Ja, Sie, die Ungehorsamen und Überneugierigen, haben richtig gehört und gefolgert: Barnes, der Mann, will damit sagen, fast im Stil der österreichischen Kabarettistin Lisa Eckart, Flaubert hätte, jedenfalls als Endzwanziger, die Syphilisation ausgespielt gegen die Zivilisation, insonderheit gegen die „Frauenrechte". Dies war damals, nach 1984, offenkundig zu frech, wahlweise: zu misogyn für das *Flaubert-Wörterbuch.* Ob man derlei Romanistik, gleichsam eine jenseits von Sartre und Barnes, indes unter Fortschritt abbuchen sollte, scheint mir doch sehr fraglich.

Kommen wir auf den werkanalytisch entscheidenden Punkt: Flaubert war in seiner Jugend seinerseits ein mit Maupassant in Vergleich zu setzender, aber, was die Schwere seiner venerischen Erkrankung angeht, von Maupassant deutlich unterschiedener bedenkenloser Womanizer von Rodolphe-Aprikosen-Format gewesen, und er richtete in *Madame Bovary* in den auf diesen bezüglichen kritischen Passagen also auch über sich und jene. Sartre, dem Syphilisthema gegenüber blind sich stellend, erkannte zahlreiche der hier sich stellenden Zusammenhänge, dies vor allem durch subtile Rekonstruktion von Flauberts posthum (1910) erschienener Erzählung *November* (1840 – 42) (vgl. Sartre 1977: IF, Bd. II, S. 1104 – 1149). Flaubert schilderte hier auf vergleichsweise zarte Weise sein sexuelles Erwachen, um den Höhepunkt seiner Erzählung zu erreichen mit der Schilderung einer als Traum erlebten Begegnung mit einer etwas in die Jahre gekommenen, allein von Wollust dominierten Schönheit namens Marie mit recht dunklem Geheimnis. Es enthüllt sich im weiteren Textverlauf als weitgehend missglückte Suche nach sexueller Erfüllung – ein *Madame-Bovary*-Motiv, nur dass es hier noch deftigere Farben gewinnt. Marie nämlich trieb es, so der blutjunge Flaubert, mit so gut wie allen, auch mit „vom Weib erschöpften Paralytikern, die auf dem Bettlaken zu sterben fürchteten" (Flaubert 1996b, S. 87) – vor Überanstrengung, wohlgemerkt. Syphilis, zumal in der später – etwa von Arthur Canon Doyle (1894) – vielbeschworenen ‚dritten Generation', sei kein Problem mehr, sollte dies wohl heißen in der Fantasie eines Zwanzigjährigen, der dieses Trostes bedurfte, um wenigstens nicht sehenden Auges in sein Triebschicksal namens Syphilis zu laufen, wie so viele andere seines Ranges und seiner Zeit (vgl. Niemeyer 2020, S. 410). Nichts davon gelangt allerdings bei Sartre zu wirklicher Klarheit, auch nicht, dass Flaubert sich in seinem Roman *Madame Bovary* in Gestalt des skrupellosen Verführers Rodolphe als einen über den gehörnten Gatten Charles – für Flauberts Vater stehend – triumphierenden Liebhaber feiert, unter dessen glutvollen Küssen der Groll Emmas auf Charles „wie Schnee dahin[schmolz]" (Flaubert 1996a, S. 223). Auffällig dabei und resümierend gesprochen: Es scheint fast so, als habe Flaubert zwar die neue (Liebes-)Praxis bedacht, aber er und die Flaubertforschung insgesamt die Theorie derselben vergessen, die, wenn sie im Rahmen dieser Tagung von Interesse sein soll, nur eine existentialistische sein kann. Versuchen wir, einer solchen Theorie näherzukommen, unter Beiziehung von Nietzsches Lou von Salomé – ein Kaninchen, wie einleitend bemerkt, und sei es nur, weil die Nietzscheforschung die Bedeutung dieser als Lou Andreas-Salomé bekannt gewordenen Literatin, insonderheit jene für Nietzsche, so gut wie komplett ignorierte, selbstredend auch Sartre.

Flauberts *Madame Bovary* im Vergleich zu Nietzsches

Vorab und nur der Deutlichkeit halber: Meine These lautet nicht, Nietzsche habe einen bisher noch unbekannten Roman dieses Titels *geschrieben*. Sie lautet vielmehr, er habe einen Roman dieser Art *erlebt*. Der ihn, des tragischen Ausgangs wegen – Lou von Salomé erwiderte seine Gefühle nicht, seine Mutter wie Schwester taten alles dafür, diese Beziehung zu blockieren – existentiell erschütterte und sein ferneres Werk, zumeist negativ, beeinflusste (vgl. Niemeyer 2017, S. 331–340). So sehr übrigens, dass man beinahe sagen könnte: „Friedrich

Nietzsche, das ist der unglücklich in Lou von Salomé verliebt gewesene und wegen der dabei von seinen Nächsten ins Spiel gebrachten Intrigen zum Menschenfeind mutierende vormalige Basler Professor und Wagnerianer", korrespondierend zu dem Satz Sartres: „Gustave Flaubert, das ist der Autor von *Madame Bovary*" (Sartre 1977: IF, Bd. I, S. 7). So zitierte Sartre gleich zu Beginn von *Der Idiot der Familie* den damaligen Mainstream – um sich auf den nächsten gut dreitausend Seiten in derart vielen Details der Vorgeschichte dieses 1857 erschienen Meisterwerks zu verlieren, dass am Ende die Zeit knapp wurde und die Biologie mitredete: Sartre erblindete allmählich und konnte den versprochenen 5. Band, der diesem Roman zugedacht war, nicht mehr schreiben. (vgl. Sartre 1977: IF, Bd. IV, S. 683)

Der Vorteil, wenn man so will: Ich kann vergleichsweise ungestört zu *meiner* These kommen, ausgehend vom Textbuch Flauberts. In dessen Zentrum: Die junge, betörende Arztfrau Emma Bovary, die ihrem ihr anvertrauten Gemahl Charles in der Hochzeitsnacht den Himmel auf Erden zu bereiten vermag – ihn damit allerdings in ein Schoßhündchen verwandelnd mitsamt weiterer unvermeidlicher Domestizierungseffekte: Charles wird kugelrund vor Glück, ihm wachsen unvermeidlich Scheuklappen, die ihm helfen, seine Ehe, die sich mehr und mehr als Glück über ein gemeinsames Nachtmahl als Abschluss eines anstrengenden Berufsalltags des Arztes mit Vorlieben für Hausbesuche erweist, bei entsprechend zurückgehender Kopulationsbereitschaft, als ungetrübt zu fingieren. Kurz: Im Schatten dieses so motivierten Nicht-wissen-wollens gedeiht das umso wachere Schwiegermuttermonster sowie die allzu bekannte Figur des Hausfreundes, zunächst in Gestalt eines klassischen Womanizers resp. *latin lovers* namens Rodolphe, später, nach der Weigerung desselben, sich mit Emma nach Art eines Pärchens vom Typ „Bonny & Clyde und ihre immerwährenden Hippie-Liebe" auf Weltreise zu begeben, mit Léon ihr Glück versucht, also mit dem uns bereits bekannten Lover aus der Kutsche. Tatsächlich aber ist Léon kaum mehr als dies, also ein jugendlich-schwärmerischer, gutaussehender Notariatsgehilfe, dem im entscheidenden Moment der Mut zum letzten Schritt fehlt und der, wie je auf ihre Art auch Rodolphe und Charles, sich am Ende die Mit-Verantwortung für Emmas schrecklichen Suizid aus Verzweiflung und Verbitterung zurechnen lassen muss. Kurz und unter Einbezug der am Bankerott Emmas beteiligten geldgierigen Männer, darunter des unvermeidbaren Typs des ältlichen Lüstlings: Neue Männer braucht das Land, wenn es sich als tauglich erweisen will für die neue Frau und das von ihr reklamierte und bisher allein Männern vorbehaltene Recht auf, so Barbara Vinken, ein „ausgeprägtes Sexualleben" als (bisher) „Ausweis von Männlichkeit" (Vinken 2009, S. 139) – so könnte man das Credo dieses Romans auf den Punkt bringen. Dass dies mit dem alten Bürgertum nicht gut gehen wird, genauso wenig wie mit dem Klerus (Stichwort: Augustinus), ist dabei klar und erklärt das weitergehende Credo: Den ‚neuen Menschen' braucht das Land, jenseits der herkömmlichen Setzungen von Gut und Böse und damit jenseits von Christen- wie Bürgertum – ein Programmpunkt Nietzsches, wie leicht erkennbar, jedenfalls wenn man, wie Sartre, dessen Übermenschenkonstrukt nicht missversteht.

Wichtig im Blick auf die angedeutete, nun via Nietzsche nachzuliefernde Theorie zu dieser Praxis und dabei spaßeshalber gesetzt, Flauberts Madame Bovary entstamme dem Land seines Freundes Turgenjew. Dann nämlich passt zumindest der Auftakt von Nietzsches Lou-Affäre: „Grüssen Sie diese Russin von mir wenn dies irgend einen Sinn hat: ich bin nach dieser Gattung von Seelen lüstern. Ja ich gehe nächstens auf Raub darnach aus" (Nietzsche, Bf. an Paul Rée, 21.03.1882, KGB III/1, Bf. 215), schreibt Nietzsche gleich zu Beginn dieser Affäre seinem damals besten Freund, der ihm von der einundzwanzigjährigen Petersburger Generalstochter Lou von Salomé vorgeschwärmt hatte, die er soeben in Rom kennengelernt habe und die ganz begierig sei, ihn kennenzulernen. Dieses Porträt klingt ein wenig nach Emma Bovary und Nietzsche in seinen Briefen an Rée wie Rodolphe, dem man problemlos auch Sätze zutraut wie: „Geist? Was ist mir Geist! Was ist mir Erkenntniß! Ich schätze nichts als A n t r i e b e – und ich möchte darauf schwören, daß wir darin etwas Gemeinsames haben" (Nietzsche, Bf. an Lou von Salomé, verm. 24.11.1882, KGB III/1, Bf. 335) – so, man mag es glauben oder nicht, unser von seiner Schwester in ein Asketen-Monster verwandelter Pastorensohn aus Röcken b. Leipzig, in einem seiner kühnsten Briefe an eine junge Studierende der Universität Zürich, die zu jener Zeit dem Grundsatz huldigt:

> Ich kann weder Vorbildern nachleben, noch werde ich jemals ein Vorbild darstellen können für wen es auch sei, hingegen mein eignes Leben nach mir selber bilden, das werde ich ganz gewiß [...]. Damit habe ich ja kein Prinzip zu vertreten, sondern etwas viel Wundervolleres, – etwas, das in Einem selber steckt und ganz heiß von lauter Leben ist und jauchzt und heraus will. (Andreas-Salomé 1984, S. 78)

Besser, so will mir scheinen, ist über Existentialismus nie gesprochen worden, auch nicht von Emma Bovary.

Wie das Ganze weiterging, weiß man: Nietzsche, gehemmt durch sein dunkles Geheimnis, *spielte* nur Rodolphe, war es aber nicht, in ihm steckte noch viel zu viel Charles und insoweit die Hoffnung auf Spießerglück, also auf ein „Weib", das, wie es Zarathustra kurze Zeit später, Nietzsches Resignation aufnehmend, nennen wird, schließlich gedacht sei „zur Erholung des Kriegers" (Nietzsche, Za, Von alten und jungen Weiblein, KSA 4, S. 85). In der Umkehrung gesprochen: Lou war viel zu wenig Emma, um eines Spießers wie jenem Charles zu verfallen oder auch nur dem Versprechen eines frühpensionierten und kränkelnden Baslers Professors, sie müsse ihn, diesen Frosch, nur küssen, schon werde eine Art Rodolphe daraus. Mehr als dies: Lou sich aus Liebes- wie Lebensenttäuschung mit Arsen in einem gar schrecklichen Todeskampf den irdischen Abläufen entziehen zu sehen, ist eine geradezu absurde Vorstellung bei einer, so Viktor Tausk, *femme fatale* wie ihr, die eher schon daran gewohnt war, dass um sie herum alles, etwa ihr pro forma geheirateter impotenter Mann, erstarb oder gar, aus ungestillter Liebessehnsucht, sich, wie Rainer Maria Rilke, zu mancherlei erniedrigen ließ, wenn der nicht erhörte Lover nicht gar still und leise aus dem Leben schied, was für Paul Rée, Nietzsches Wegbahner bei Lou und späterer Rivale auf Leben und Tod, wohl als gewiss gelten darf (vgl. Niemeyer 2017, S. 122–123).

Insoweit: Nein, Lou war nicht Emma, Nietzsche seinerseits viel zu sehr Charles und viel zu wenig Léon oder gar Rodolphe, als dass wir mehr Funken als die genannten aus diesem Vergleich schlagen können – abgesehen von dem nun ein wenig genauer in Betracht zu ziehenden Aspekt: Nietzsche wie Flaubert, Letzterer als Zwanzigjähriger, Ersterer als ‚Spätling' mit 37 Jahren, sich endlich – was für ein Selbstbetrug für einen Syphilitiker wie ihn! – als genesen wähnend, waren unisono der Meinung, dass Neues und zumal perfektes Lebens- wie Liebesglück nur aus überschießender Kraft und Lebens- wie Liebesfreude entspringen könne, dem also, was den zwanzigjährigen Flaubert davon träumen ließ, er werde dereinst „alle Horizonte durchspähen", „o daß ich unterginge beim Umsegeln des Kaps, in Kalkutta an der Cholera stürbe oder in Konstantinopel an der Pest!" (Flaubert 1996b, S. 110) Sicherlich: Dies war allererst eine Vorstellung, aus der Flaubert Honig zu saugen wusste angesichts der damals zumal dem Freier allüberall drohenden Gefahr, sei es in Paris, sei es in Nordafrika resp. Südeuropa. Aber es war auch eine Vorstellung, die man als Vorwegnahme von Nietzsches Imperativ „gefährlich leben!" würdigen sollte, was gegen Wolfgang Matz geht, der jene Passagen extensiv zitiert (vgl. Matz 2007, S. 41–42), aber ihrer Parallelen zu Nietzsche nicht inne wird. Dies sei abschließend noch mit wenigen Strichen erläutert.

Anknüpfen könnte man dabei an Überlegungen Nietzsches im unmittelbaren Vorfeld der Lou-Affäre, darunter den Plan: „Ich wuerde gern eine Colonie nach den Hochlanden Mexikos fuehren: oder mit Rée in die Palmen-Oase Biskra reisen – noch lieber kaeme mir ein Krieg", ragt ein resignativer Befund heraus: „Die Gesundheit sagt zu allem Nein" (Nietzsche, Bf. an Heinrich Köselitz, 04.03.1882, KGB III/1, Bf. 205) – aber, so ließe sich ergänzen: die Theorie, sie sagt eigentlich Ja. Denn Nietzsche steckte zu dieser Zeit mitten in der Arbeit an *Die fröhliche Wissenschaft* und huldigte insoweit bereits der hier begründeten Losung: „gefährlich leben!" (Nietzsche, FW 283, KSA 3, S. 526) In der Linie dieser Selbstüberwindungsrhetorik bewegt sich auch Nietzsches Wort von der „grossen Gesundheit", die der „Besieger Gottes und des Nichts" aufzuweisen hätte:

„Geister, durch Kriege und Siege gekräftigt, denen die Eroberung, das Abenteuer, die Gefahr, der Schmerz sogar zum Bedürfniss geworden ist" (Nietzsche, GM II 24, KSA 5, S. 336). Gewiss: Dies war über weite Teile Mutmacher-Rhetorik am Rande der absehbaren – um an den Syphilitiker Heinrich Heine zu erinnern – ‚Matratzengruft'. Aber es ist gleichwohl dem Theoriekern nach ernst zu nehmen als Indiz für den Nietzsche der Post-Wagner-Ära, also für den, mit Karl Schlechta, aber auch mit Mazzino Montinari geredet, zu sich selbst gekommenen Nietzsche. Er weiß um sich als „Erstling und Frühgeburt des kommenden Jahrhunderts" (Nietzsche, FW 343, KSA 3, S. 574), „welches wir" – so ein berühmter Passus aus Nietzsches unter dem Titel *L'Ombra di Venezia* überlieferten Nachlassaufzeichnungen vom Frühjahr 1880 – „das bunte nennen wollen und das viele Experimente des Lebens machen soll" (Nietzsche, NL 1880, 3[1], KSA 9, S. 48). Zwei Jahre später, in *Die fröhliche Wissenschaft*, erklärt Nietzsche, dass es „vieler vorbereitender tapferer Menschen [bedarf], welche doch nicht aus dem Nichts entspringen können – und ebensowenig aus dem Sand und Schleim der jetzigen Civilisation und Grosstadt-Bildung: [...] gefährdetere Menschen, fruchtbarere Menschen, glücklichere Menschen!" (Nietzsche, FW 283, KSA 3, S. 526) Man ist hier versucht zu ergänzen: wie schließlich auch Emma Bovary als ‚vorbereitender tapferer Mensch' nicht aus dem Nichts entstehen konnte und also, wie ihr Selbstmord drastisch anzeigt, scheitern musste. Von weiteren, in diese Richtung weisenden Textstücken sei hier abgesehen – nicht aber von jenen Programmsätzen, die Nietzsche im Frühjahr 1880 zu Papier brachte:

> [E]rstens die Enthaltung in Bezug auf die letzten Entscheidungen [...]; zweitens die Voreingenommenheit gegen alle Sitten und alles nach Art der Sitte Bindende; drittens eine größere Ehrlichkeit im Sichtbar-werden-lassen sogenannter böser Eigenschaften. (Nietzsche, NL 1880, 3[6], KSA 9, S. 48 – 49)

Nochmals und mit mehr Betonung geredet: Diese drei Programmpunkte, nicht wortwörtlich, aber dem Sinn nach, hätten auch auf der Agenda Flauberts stehen können, als er in den 1850er Jahren in der Gestalt der Emma Bovary einen neuen Frauentyp schuf, für den eben diese Punkte kennzeichnend sind nebst des vierten, von Nietzsche zwei Seiten später nachgereichten, der wie folgt lautet:

> In den Leidenschaften des Menschen erwacht das Thier wieder; die Menschen kennen nichts Interessanteres, als diesen Rückgang ins Reich des Unberechenbaren. Es ist, als ob sie sich an der Vernunft allzusehr lan<g>weilten. (Nietzsche, NL 1880, 3[21], KSA 9, S. 50)

So – um nun die Wende hin zu Flauberts Roman *Madame Bovary* (1857) vollständig zu vollziehen –, wie sich Emma ob der Vernunft ihres Gatten Charles allzu sehr langweilte im Kostüm ihrer Ehe.

Wie das Ganze endete, wissen wir, zunächst, was Nietzsche angeht, von seinem schrecklichen Scherz über *seine* Madame Bovary, ein Scherzwort des Liebesenttäuschten (vgl. Niemeyer 2017, S. 132–133) aus dem Jahr 1883: „Diese dürre schmutzige übelriechende Äffin mit ihren hängenden Brüsten – ein Verhängniß!" (Nietzsche, Bf. [Entwurf] an Georg Rée, Mitte Juli 1883, KGB III/1, Bf. 435) Bitte die Reihenfolge beachten bei meiner Antwort auf diesen unlängst im Film *Lou Salomé* (2016) von Cordula Kablitz-Post in Erinnerung gebrachten und aus frauenbewegter Sicht vielgetadelten Satz: Nein, war sie nicht, nämlich eine ‚dürre schmutzige übelriechende Äffin'; nein, hatte sie nicht, nämlich ‚hängende Brüste'; sowie: Ja, war sie sehr wohl, nämlich Nietzsches Verhängnis. Der Rest ist Psychologie, als Abwertung des ohnehin nicht erreichbaren Liebesobjekts; oder, auch dies zu beobachten: als Teil der Freude darüber, dass es auch mit der *Madame Bovary* aus der Fiktion, mit Flauberts Traumfrau also, sich kaum besser verhielt, jedenfalls dem allerletzten Scherz Nietzsche in dieser Angelegenheit zufolge, vorgetragen in *Der Fall Wagner:* Flaubert hätte es freigestanden, *Madame Bovary* „mythologisiert, Wagnern als Textbuch anzubieten" (Nietzsche, WA, Turiner Brief vom Mai 1888, KSA 6, S. 34), lesen wir hier aus der Feder eines Spötters, der längst der Hoffnung auf irdisches Glück von Lou-von-Salomé-Format entsagt hat und dem es offenbar ein Trost besonderer Art ist, dass auch Flauberts Vision, sichtbar zu machen, sobald man sie in die Wagnersche Welt übersetzt, eine zum Scheitern verurteilte wäre. Nietzsche, weniger analytisch korrekt denn von der Hoffnung umgetrieben:

> Würden Sie es glauben, dass die Wagnerschen Heroinen sammt und sonders, sobald man nur erst den heroischen Balg abgestreift hat, zum Verwechseln Madame Bovary ähnlich sehn! [...] Immer fünf Schritte weit vom Hospital! Lauter ganz moderne, lauter ganz grossstädtische Probleme! [...] Haben Sie bemerkt [...], dass die Wagnerischen Heldinnen keine Kinder bekommen? – Sie können's nicht... (Nietzsche, WA, Turiner Brief vom Mai 1888, KSA 6, S. 34)

Muss man hier wirklich noch ergänzen: So wie es auch mir, Nietzsche, mit *meiner* Madame Bovary gegangen wäre, mir, der doch nur so wenig von Rodolphe hat und der so sehr des Stubenglücks eines Frühpensionärs bedarf?

Zusammenfassend geredet: Es gibt gute Gründe, Lou von Salomé in Analogie zu Flauberts *Madame Bovary* allererst als von Nietzsche gefeiertes, Fleisch gewordenes Symbol der Aufklärung à la Voltaire zu lesen, gerichtet gegen den Klerus und die „Sittlichkeit der Sitte" (Nietzsche), insonderheit gegen das von Barbara Vinken (2009, S. 135–140) ins Zentrum gerückte Ehebruchverbots à la Augustinus. (dem sie, die Salomé, im Verlauf ihrer Pseudoehe systematisch zuwiderhandelte.) Insoweit und zunächst unbedacht um die Frage, ob damit den 1857er Anklägern Auftrieb gegeben wird oder nicht: Ja, Flauberts Roman – erinnert sei an die freizügige Kutschfahrtszene – feiert die sexuelle Libertinage und huldigt dem Frauentyp, der ihrer bedarf, sei es in, sei es außerhalb der Ehe. Zarathustras Wort: „Und besser noch Ehebrechen als Ehe-biegen, Ehelügen! – So sprach mir ein Weib: ‚wohl brach ich die Ehe, aber zuerst brach die Ehe – mich!'" (Nietzsche, Za III, Von alten und neuen Tafeln 24, KSA 4. S. 246.) könnte insoweit auch Flauberts sein, der angeführte Nachsatz des ‚Weibes' wirkt wie abgeschrieben aus Emma Bovarys Textbuch, das auch vorsieht, einem vom Ballast der Mütterlichkeit befreiten Frauenbild Raum zu schaffen. So entledigt sich Madame Bovary ihres Töchterchens mit einer Beharrlichkeit, die einem konventionellen Sozialpädagogen schockieren muss. In der Umkehrung gesprochen: Flaubert verficht eine fast vollständig auf die Erotik konzentrierte Männerfantasie, wie es so einige gab in der Folge, etwa von Ola Hansson in *Sensitiva amorosa* (1889) oder von Arthur Schnitzler in Gestalt der sich ans Fensterkreuz nagelnden liebestollen Friederike aus *Doktor Gräsler, Badearzt* (1913) – übrigens im letztgenannten Fall gezielt gerichtet gegen die Empörung des Bürgertums, die Schnitzler, der „österreichische Maupassant" (Torberg), gezielt wachrief, im Gegenzug zum 1933 als NS-Sexualaufklärer reüssierenden schwedischen Arzt und Sexualaufklärer Seved Ribbing (1845 – 1921), der noch 1890 Leute wie Hansson oder August Strindberg ins Gefängnis zu werfen oder in die Irrenanstalt zu überstellen für angezeigt hielt. (vgl. Niemeyer 2020a) So gesehen steht der 1857er Prozess gegen Flaubert fast für eine Harmlosigkeit, anders als der 1888er Prozess Nietzsches gegen diesen. Er nämlich spricht eher dafür, dass da endlich jemand seine Ruhe haben will und sich zurücksehnt nach dem Hausfrauenideal.

Fazit

Nietzsche hat zwischen 1880 und 1882 in offenkundiger Unkenntnis von Flauberts *Madame Bovary* die Metatheorie der hier im Zentrum stehenden Ehebruchsgeschichte skizziert, mittels, wie gesehen, von vier Programmsätzen, die einem ein kommendes Zeitalter begründenden Experimentalismus zugerechnet werden können, wobei gerne eingeräumt sei, zumal angesichts der Zurechnung dieses Textes zum existentialistischen Kanon durch Walter Kaufmann (1957, S. 127): Existenzialismus und Experimentalismus gehen vermutlich weitgehend überein. Gründe indes, Nietzsche aus diesem Kanon auszuschließen, sehe ich nicht, auch nicht via Sartre, dessen Vertrauen auf Descartes *cogito* (Sartre 2000b: LibCar, S. 165) – um nur dies noch zu nennen – nach Nietzsches *vivo, ergo cogito* sowie seinem später nachgereichten: „Es denkt!" (vgl. Niemeyer 2019, S. 400 – 401) altertümlich wirkt, sprich: anti-nietzscheanisch als auch anti-existentialistisch.

Insoweit können wir nun, den einleitend erwähnten Einwand Barbara Vinkens gegen Sartre aufnehmend, folgern: derlei Einsicht in die diesbezügliche Vorläuferschaft Nietzsches gibt es womöglich allein deswegen nicht in Sartres *Idioten*, weil, seiner Setzung zufolge, Flaubert nicht am Beginn der Moderne stehen *durfte und ihn deswegen auch Nietzsches Flaubert nicht interessierte*. Derlei indes, also persönliche Ranküne, taugt nicht als Argument, besser vielleicht: taugt nicht im Vergleich zu den im Vorhergehenden vorgetragenen Argumenten. Vor allem jenen, die auf den von Sartre systematisch Ignorierten zurückgehen: eben Nietzsche. Der allerdings seinem selbstgesetzten Ideal 1883 der Praxis nach und 1888 auch der Theorie zufolge entsagte, Ersteres aus Resignation ob des Abgelehntwordenseins, Letzteres aus Resignation eines unheilbar Erkrankten sowie Liebesenttäuschten.

Bibliographie

Andreas-Salomé, Lou (1984): *Lebensrückblick*. 5. Aufl. Frankfurt a. M.

Bakewell, Sarah (2016): *Das Café der Existentialisten. Freiheit, Sein und Aprikosencocktails*. München.

Barnes, Julian (1984): *Flauberts Papagei*. Roman. Köln.

Bourget, Paul (1926): *Essais de Psychologie Contemporaine*. Paris.

Brunetière, Ferdinand (1883): *Le Roman Naturaliste*. Paris.

Flaubert, Gustave (1996a): *Madame Bovary*. Frankfurt a. M., Leipzig (1856/1857).

Flaubert, Gustave (1996b): *November*. Erzählung. Frankfurt a. M., Leipzig (1910).

Frank, Manfred (1980): „Das Individuum in der Rolle des Idioten. Die hermeneutische Konzeption des *Flaubert*“. In: Traugott König (Hrsg.): *Sartres Flaubert lesen. Essays zu „Der Idiot der Familie“*. Reinbek b. Hamburg, S. 84–108.

Grimm, Reinhold R. (1980): „*Der Idiot der Familie* als Herausforderung der Literaturgeschichtsschreibung“. In: Traugott König (Hrsg.): *Sartres Flaubert lesen. Essays zu „Der Idiot der Familie“*. Reinbek b. Hamburg, S. 109–148.

Kaufmann, Walter (1957): *Existentialism from Dostoevsky to Sartre*. London.

Krafft-Ebing, Richard v. (1997): *Psychopathia sexualis. Mit besonderer Berücksichtigung der konträren Sexualempfindung*. Hrsg. von A. Fuchs. München (14. A. 1912).

Lepenies, Wolf (1997): *Sainte-Beuve. Auf der Schwelle zur Moderne*. München.

Matz, Wolfgang (2007): *1857. Flaubert, Baudelaire, Stifter*. Frankfurt a. M.

Maupassant, Guy de (1884): „Gustave Flaubert“. In: *Lettres de Gustave Flaubert à George Sand*. Paris, S. I–LXXXI.

Niemeyer, Christian (2017): *Nietzsche auf der Couch. Psychologische Lektüren und Relektüren*. Weinheim, Basel.

Niemeyer, Christian (2019): *Sozialpädagogik als Sexualpädagogik. Beiträge zu einer notwendigen Neuorientierung des Faches als Lehrbuch*. Weinheim, Basel.

Niemeyer, Christian (2019a): „Edmond de Goncourts *Juliette Faustin* (1881/82), gelesen, aus Nietzsche-Perspektive, als Syphilis- und Schlüsselroman“. In: *Comparatio* 11, S. 331–342.

Niemeyer, Christian (2020): *Nietzsches Syphilis – und die der Anderen. Eine Spurensuche.* Freiburg i. Br.
Niemeyer, Christian (2020a): „Die Bedeutung der Syphilis im schriftstellerischen Werk des Arztes Arthur Schnitzler – eine Spurensuche". In: *Jahrbuch Literatur und Medizin* 12, S. 57–84.
Oehler, Dolf (1980): „Zum gesellschaftlichen Standort der Neurose-Kunst. Sartres *Idiot der Familie* und Flauberts *Éducation sentimentale:* Versuch einer vergleichenden Lektüre". In: Traugott König (Hrsg.): *Sartres Flaubert lesen. Essays zu „Der Idiot der Familie".* Reinbek b. Hamburg, S. 149–192.
Onfray, Michel (2013): *Im Namen der Freiheit. Leben und Philosophie des Albert Camus.* München.
Praz, Mario (1960): *Liebe, Tod und Teufel. Die schwarze Romantik (1930).* München.
Sartre, Jean-Paul (1977): *Der Idiot der Familie. Gustave Flaubert 1821–1857. Bde. I–IV.* Reinbek b. Hamburg.
Sartre, Jean-Paul (1982): *Saint Genet, Komödiant und Märtyrer.* Reinbek b. Hamburg.
Sartre, Jean-Paul (2000a): „Zum Existentialismus – Eine Klarstellung". In: Jean-Paul Sartre: *Der Existenzialismus ist ein Humanismus und andere philosophische Essays 1943–1948.* Reinbek b. Hamburg, S. 113–121.
Sartre, Jean-Paul (2000b): „Die cartesianische Freiheit". In: Jean-Paul Sartre: *Der Existenzialismus ist ein Humanismus und andere philosophische Essays 1943–1948.* Reinbek b. Hamburg, S. 122–144.
Sartre, Jean-Paul (2000c): „Der Existentialismus ist ein Humanismus". In: Jean-Paul Sartre: *Der Existenzialismus ist ein Humanismus und andere philosophische Essays 1943–1948.* Reinbek b. Hamburg 2000, S. 145–192.
Vinken, Barbara (2009): *Flaubert. Durchkreuzte Moderne.* Frankfurt a. M.
Vinken, Barbara/Wild, Cornelia (Hrsg.) (2010): *Arsen bis Zucker. Flaubert-Wörterbuch.* Berlin.

Personenregister

Adler, Alfred 77
Adorno, Theodor W. 70
Agamben, Giorgio 157
Alain (Chartier, Émile) 250, 256, 265 f.
Ally, Matthew 10, 14, 26 f., 29 f., 273–275, 277 f.
Andler, Charles 8, 37, 82
Anouilh, Jean 228
Arendt, Hannah 71, 75
Aristoteles 125
Aron, Raymond 8, 256
Augustinus 55

Badiou, Alain 23
Baeumler, Alfred 2, 12
Bauer, Bruno 73
Beauvoir, Simone de 1, 9, 35, 37, 39, 41–44, 47, 73 f., 76 f., 79 f., 82, 91, 151, 156, 161–167, 169 f., 172–176, 178, 180 f., 186, 193 f., 196
Bianquis, Geneviève 8
Binswanger, Ludwig 77
Bolle, Jean 8
Brunschvicg, Léon 8
Burckhardt, Jacob 181
Burnier, Antoine 7

Camus, Albert 1, 53–64, 80, 85, 89–102, 105, 107–120, 126, 133–135, 141 f., 144, 146, 150–154, 156, 158, 184, 186, 219–226, 228–231, 289
Canguilhem, Georges 1, 249–253, 255–267
Cassirer, Ernst 228 f.
Chestov, Léon 60, 109, 119, 191
Cioran, Emil 1, 123 f., 126 f., 129 f., 132–136
Cohen-Solal, Annie 8
Colli, Giorgio 55, 128
Contat, Michel 8, 37

Daigle, Christine 8 f., 72, 161, 171, 192, 233
Darwin, Charles 29, 76
Deleuze, Gilles 12, 18, 21, 24 f., 27, 58, 127
Derrida, Jacques 12 f., 15, 199, 279 f.
Descartes, René 12, 16, 40, 58, 61 f., 96, 280, 296
Dilthey, Wilhelm 252
Dostoevskij, Fëdor 72, 80, 84
Dufrenne, Mikel 249 f., 256–258, 261–264, 266

Eisner, Kurt 81

Fanon, Frantz 156 f., 186
Feuerbach, Ludwig 73, 82
Fichte, Johann Gottlieb 71
Fondane, Benjamin 60
Förster-Nietzsche, Elisabeth 12, 128
Foucault, Michel 12, 73, 75, 84, 145, 206
Freud, Sigmund 75–77, 151
Frisch, Max 70

Gadamer, Hans-Georg 244
Genet, Jean 8, 45, 70, 76 f., 79, 234, 286
Gerassi, John 8, 35, 37, 43, 193
Geuss, Raymond 209 f.
Gide, André 57, 228
Grenier, Jean 54, 56 f.
Guattari, Félix 16, 21, 25, 27

Halévy, Daniel 8, 37
Hegel, Georg Wilhelm Friedrich 12, 24, 136, 143, 151, 153, 235
Heidegger, Martin 12, 15, 27, 60, 71, 76, 85, 107, 124, 127 f., 132–135, 144, 163, 171, 191, 194, 200, 204, 234, 255, 263, 271
Heraklit 13, 21, 229 f., 243
Herr, Lucien 82
Hobbes, Thomas 131
Honecker, Erich 69–71, 86
Howells, Christina 13
Husserl, Edmund 16, 18, 37, 119, 178, 200, 204

Jacobi, Friedrich Heinrich 71 f.

https://doi.org/10.1515/9783110760163-020

Jaspers, Karl 2, 10 f., 60, 71, 85, 105–107, 147, 191, 234, 249–266
Jaurès, Jean 82

Kant, Immanuel 12, 16, 18, 46, 74 f., 78, 82, 226, 256, 265, 280
Kierkegaard, Søren 2, 60, 69, 74, 82, 105–114, 117–121, 123, 141 f., 144–147, 149, 151, 156–158, 191 f., 200, 220, 262
Kofman, Sarah 12
Köselitz, Heinrich 128, 287, 293

Leibniz, Gottfried Wilhelm 127, 182, 187
Lichtenberger, Henri 54
Louette, Jean-François 43, 81, 233
Löwith, Karl 2, 106 f., 147
Lukács, György 70 f., 285

Marcel, Gabriel 186, 252, 277 f.
Marx, Karl 46, 75, 145, 151, 153, 180
Menthon, François de 70
Merleau-Ponty, Maurice 185
Montinari, Mazzino 55, 128, 294

Nishitani, Keiji 27
Nizan, Paul 8, 37, 81, 252
Noudelmann, François 8 f., 233

Oehler, Richard 12
Onfray, Michel 89 f., 289
Overbeck, Franz 175 f.

Parmenides 77, 244
Pascal, Blaise 55, 60, 85, 220
Platon 12, 75, 200, 256, 272
Protagoras 75

Rauschning, Hermann 71
Ricœur, Paul 249 f., 256–258, 261–264, 266
Rosa, Hartmut 144–146, 149, 157
Rosenberg, Alfred 101, 230
Rybalka, Michel 8, 37, 192

Salomé, Lou von 283, 290–293, 295
Sartre, Jean-Paul 1, 7–21, 23–31, 35–47, 60, 69–86, 91 f., 126, 141 f., 144–146, 149–154, 156 f., 162 f., 165, 175–181, 184–187, 191–197, 199–201, 203–208, 210–213, 220, 222, 228, 233–235, 239–245, 251 f., 256, 271–281, 283–292, 296 f.
Schestow, Leo *Siehe* Chestov, Léon
Schiller, Friedrich 20, 182, 187
Schnädelbach, Herbert 2
Schopenhauer, Arthur 55, 62, 76 f., 143, 146, 178, 221, 237, 243, 253, 256, 272
Sextus Empiricus 135
Shestov, Lev *Siehe* Chestov, Léon
Sloterdijk, Peter 144, 155–157
Spencer, Herbert 76
Staiger, Emil 69–71, 86
Stirner, Max 114 f., 145
Strauß, David Friedrich 73

Turgenev, Ivan 72

Ulbricht,Walter 69

Wagner, Richard 54 f., 74, 180, 227, 238, 284, 288, 294 f.
Weber, Max 182, 187, 252
Weiss, Peter 70